# GEOGRAPHY AND ENVIRONMENT ISSUES AND CHALLENGES

"We are species on the earth. Have we right to squander its resources, to pollute water and air, to extinguish animal and plant life, to upset the delicate balance in nature and mar its beauty? Cannot an organisation such as U.N. direct our thinking in terms of the larger well-being".

—*Indira Gandhi*

---

**About the Editors**

**Prof. H.H. Singh,** Department of Geography, Banaras Hindu University, Varanasi.

**Dr. Prithvish Nag,** National Atlas and Thematic Mapping Organisation, Kolkata.

**Dr. V.K. Kumra,** Department of Geography, Banaras Hindu University, Varanasi.

**Dr. Jagdish Singh,** Department of Geography, Banaras Hindu University, Varanasi.

**Dr. Haushila Prasad,** Department of Geography, Banaras Hindu University, Varanasi.

**Dr. S.K. Singh,** Department of Geography, Kumaun University, Almora Campus.

# GEOGRAPHY AND ENVIRONMENT

## *Issues and Challenges*

Prof. S. L. Kayastha Felicitation Volume

*EDITORS*

H. H. SINGH
V. K. KUMRA
HAUSHILA PRASAD
P. NAG
JAGADISH SINGH
S. K. SINGH

**CONCEPT PUBLISHING COMPANY PVT. LTD.**
**NEW DELHI-110059**

**ISBN-13: 978-81-7022-049-7**

Edition 2017

*Published and Printed by*

Since 1974

**Concept Publishing Company Pvt. Ltd.**
A/15-16, Commercial Block, Mohan Garden
New Delhi-110059 (India)
*T*: 25351460, 25351794, *F*: 091-11-25357109
E: publishing@conceptpub.com, W: www.conceptpub.com
**Editorial Office:** H-13, Bali Nagar, New Delhi-110 015, India.

**Cataloging in Publication Data**--Courtesy: D.K. Agencies (P) Ltd. <docinfo@dkagencies.com>

**Geography and environment : issues and challenges** : Prof. S.L. Kayastha felicitation volume/ editors, H.H. Singh [and five others].
pages cm
Includes bibliographical references and index.
ISBN 9788170220497

1. Human ecology--India. 2. Environmental protection--India. 3. Environmental geography--India. I. Kayastha, S. L. (Shanti Lal), 1924- honouree. II. Singh, Harihar, 1932- editor.

DDC 304.20954 23

GF661.G46 2016

# PROFESSOR S.L. KAYASTHA
## A Profile

PERHAPS there cannot be a better way to felicitate a teacher than this when his students of different periods, stretching over four decades, decide to bring out a felicitation volume in his honour. Efforts of the students, colleagues and admirers of Professor Shanti Lal Kayastha, distributed all over the country to felicitate him is surely a sign of love, respect and regard he commands. He is known as a soft spoken person, an acclaimed scholar, a keen researcher, a dedicated teacher, and a talented personality of versality and charm. He is a Senior Geographer of the country. Many of his students occupy positions of distinction in India and elsewhere.

Professor Kayastha was born on 30th March 1924, and educated in the famous Government College, Lahore, and later in Aligarh Muslim University. He is the first Ph. D. from the Department of Geography, Banaras Hindu University. Later, he became the head of this department. For his contribution in geography, the National Geographical Society of India conferred him the Sardar Vallabh Bhai Patel Medal in 1971.

He has to his credit several pioneering endevours in geography. He was the first in the country to initiate study on river basin concept as a viable unit for integrated natural resource development and overall planning. His comprehensive study, *The Himalayan Beas Basin : A Study in Habitat, Economy and Society*, is a landmark in this direction. He introduced 'resources', 'population' and 'Indian Ocean' as important fields of study in university syllabi, and contributed research studies on both in numerous journals. His interest in integrated regional development and planning found fulfilment in the coordinated team study on *South Mirzapur : An Integrated Area Development Project* entrusted by U.G.C. as a major project of the Department of Geography, Banaras Hindu University, of which he was the Principal Investigator.

Professor Kayastha has the credit of initiating 'Environmental

Studies' in geography in India. He guided several researchers relating to concepts, perceptions and problems of urban pollution, flood plain and human adaptation, rural water supply, the Himalaya, and studies related to impact of development on environment. The ecological-environmental mapping was initiated by him in the University Grants Commission (U.G.C.) as pilot study in several universities. He was the Chairman of the Geography Panel of Experts of the U.G.C. Most significant contribution of Professor Kayastha is his pioneering attempt for policy-oriented studies in geography and environment. His studies on national forest, water and environment policies have been considered as one of the most significant contributions in this field.

His Presidential Address to the Indian Council of Geographers in Mysore in 1982 on 'Perspectives on Environment and Development' gave a clarion call to geographers to involve themselves in environmental studies and research in the interest of the subject and for the well-being of man and his planet, Earth. His main interest remains on the problems and perception of environment. Incidently, he has been a full member of the I.G.U. Commission on Environmental Problems. During various seminars and symposia in India and abroad, his contribution and delibrations received high recognition and were included in the proceedings and publications. Professor Kayastha went abroad several times for various reasons of academic nature. Some of the countries are Romania, Mexico, U.S.A., England, Japan, Brazil, Spain, Nepal, Pakistan, U.S.S.R. and France. His contribution during the 24th International Geographical Congress, Japan was remarkable. During the activities of this Congress, he presented 8 papers, took part in United Nations University—IGU Panel discussion, honoured as an invited speaker at the Centenary Celebrations of the Hosie University, and chaired scientific sessions. For this occasion, he brought out a special issue of *NGJI* on Perception and Environmental Problems. He has been the Secretary of the National Committee for Geography of the country.

His name rings familiar in India and abroad due to his long association with the *National Geographical Journal of India* as its Honorary Editor. His services in the dissemenation of geographical knowledge, as an editor for about four decades, deserve all honour. He undertook this challenging task at a time when research work found little scope for seeing the light of the day. Many a young geographer remember with gratitude seeing their first article in print in *NGJI* The Journal, in his hands, was nurtured with love of labour and knowledge, and now finds international reputation as a place of publication of original research. The long selfless services rendered by Professor Kayastha, in various ways, earn him an ever-abiding place in the development of geography and scholarship.

Professor Kayastha has consistently risked his popularity by questioning the dominant paradigms rather than falling into line to get the most quick mileage possible out of what was most vogue at the moment. His independence of thought is a contribution of great value for which the geographical community owes Professor Kayastha recognition and gratitude. He has been a leader in invoking new approaches to inquire into unsolved problem areas. His work has stimulated both, those closely associated with him and those who have become aware of his work, to seek deeply and more widely.

Professor Kayastha is now the Chairman of Advisory Board of the National Atlas & Thematic Mapping Organisation.

# PREFACE

> **"Geographical environment is the cradle in which man's genius awaits the promptings of motives which give him mastery over his fate"**
>
> ***—Willson D. Wallis.***

**GEOGRAPHERS** have a special inclination for environment, the visible, part of nature. Environmental includes atmosphere, hydrosphere, lithosphere and bioshere. Geographers have been studying environment for a long time, but they are considered to be late-comers in this field. The fact that geography is an environment-related discipline needs to be emphasised. Man-environment relationship has been one of the basic concepts in geography. The different schools of thought in geography are based on the varied perception of the degree of man's control over the environment.

Recently, the study of the environment has come close to the basic human problems and the quality of life. As a result, there has been a flood of researches on the environment. Environmental science has developed into a multi-discipline, surpassing-like researches, namely the study of ecology or bioshpere. The study of environment indicates the multifarious nature of this science. Apart from mere description of the physical elements, the study in relation to man can be put into two complementary phases of the 'environmental coin' : one is the influence of the environment upon man ; the other is man's influence on the environment.

Environmental scientists have coined slogans for the so-called continuously deteriorating environment. Environmental crises, stress, protection, management, monitoring and the like have become catchwords or rather watchwords in the scientific literature. It has been a trend to highlight the gloomy picture of the environment by describing about the excessive fallouts in the troposphere ; human valcano ; pollution domes and heat islands over big cities ; acid rains ; energy transformation in ecosystem, hence, in food chain ; ecological successions ; genetic damages, disposal of liquid, solid and radioactive wastes ; radiation during nuclear fission ; ocean dumping of toxic wastes ; effects of pesticides (including herbicides), and insecticides ; and number of similar consequences of the manipulations by man. Natural hazards are being replaced by more damaging technological hazards. Every day a section of population and additional land is exposed to risks of all sorts. Even the most suitable areas for human settlements are not free of hazards.

The aim here is not to jump into the so-called 'environmental bandwagan'. As mentioned earlier, geographers have been studying environment traditionally. On the other hand, such studies in geography are not limited to environmental hazards, problems and disasters only. There are other aspects as well which are no less important. Aspects of environmental perception, education, law, diversification and mapping have been considered to be important for desirable management practices and awareness about the environmental quality. The issues for the study in geography also include the relationship of man with the physical environment, settlements, health, crime, water and air. Some of the issues are of local significance but of national concern. These are social forestry in Kumaon, ecosystem in Dal Lake, and economic development in Mirzapur district. Probably, the basic issue in such studies is the working out a balance between the environment and development. Can we work out such a balance at all ? Can we ignore people's demand for better economic prospects or heavy industries, large cities or higher regional or per capita incomes ? Can we allow to deteriorate the quality of the environment for such development ? How to work out a strategy when there is a demand from the people of the same region for both, that is protection of the environment and economic development ? Thus, the challenge for the geographers is considerable, because the issues involved are of national, regional or local significance which can be obviously better explained by the geographers. The challenge is also great because the research techniques have become more refined. The language in the scientific literature has changed enormously which would however be apparent from some of the chapters of this volume. The challenge is also to reassert the place of geography in environmental studies.

The volume revolves around the work of Professor S.L. Kayastha. His contributions in geography and environmental studies have been valuable (Appendix X). Here an attempt has been made to expand his work, especially the conceptual development of different environment-related issues. In order to do so, the group of editors decided to organise this volume in 1984. The scope of the volume was restricted to the issues and challenges relevant to the Indian situation. It took some time to get in touch with the potential authors for chapters on the field of their specialization. We are grateful to them for finding out time from their pressing teaching and research engagements to prepare their chapters. The editing work was organised from three places : Varanasi, Calcutta and Almora. During the last six months, editors met several time, apart from the places mentioned above, in Jamshedpur, Mathura and Delhi. This volume could not have been brought out without the support of the Concept Publishing Company. We are extremely grateful to Shri Naurang Rai for having accepted our proposal and completing it so magnificantly within the target time.

25 December, 1985 EDITORS

# CONTENTS

# LIST OF AUTHORS

Shri Adhikary, Pradip Kumar, National Atlas and Thematic Mapping Organisation, Calcutta.

Dr. Asthana, V.K., Department of Geography, Kurukshetra University, Kurukshetra.

Mrs Ayangar, N.L. Subhadra, Department of Geography, University of Madras, Madras.

Km. Bano, Hanifa, Department of Geography, Aligarh Muslim University, Aligarh.

Dr. Dhabai, R.K., National Atlas and Thematic Mapping Organisation, Calcutta.

Mr. Dutt, G.K., National Atlas and Thematic Mapping Organisation, Calcutta.

Mrs. Dutta, Anima, National Atlas and Thematic Mapping Organisation, Calcutta.

Dr. Ganguly, S.R., Central Institute of English and Foreign Languages, Hyderabad.

Dr. Gopalakrishnan, K.S., Department of Geography, Madurai Kamaraj University, Madurai.

Prof. Jariwala, C.M., School of Law, Banaras Hindu University, Varanasi.

Dr. Kumra, V.K., Department of Geography, Banaras Hindu University, Varanasi.

Prof. Mukerji, A.B., Department of Geography, Punjab University, Chandigarh.

Dr. Mukhopadhyay, S.C., Department of Geography, Calcutta University, Calcutta.

Dr. Nag, Prithvish, National Atlas and Thematic Mapping Organisation, Calcutta.

Prof. Pandey, G.N., The Indian Terpentine and Rosin Co. Ltd., Bareilly.

Prof. Pathak, C.R., Department of Architecture and Regional Planning, Indian Institute of Technology, Kharagpur.

Dr. Prasad, Haushila, Department of Geography, Banaras Hindu University, Varanasi.

Prof. Ramesh, A., Department of Geography, University of Madras, Madras.

Dr. Rao, M. Sambasiva, Department of Geography, Madurai Kamaraj University, Madurai.

Prof. Raza, Mehdi, Department of Geography, Aligarh Muslim University, Aligarh.

Prof. Sen, Probhat Kumar, Department of Geography, Burdwan, University, Burdwan.

Dr. Singh, Alok Kumar, Department of Geography, Banaras Hindu University, Varanasi.

Dr. Singh, Anil Kr., Kumaun University, Campus Almora.

Prof. Singh, H.H., Department of Geography, Banaras Hindu University, Varanasi.

Dr. Singh, J., Department of Geography, Banaras Hindu University, Varanasi.

Prof. Singh, J.S., Department of Botany, Banaras Hindu University, Varanasi.

Dr. Singh, S.K., Department of Geography, Kumaun University, Campus Almora.

Dr. Singh, S.N., Department of Geography, Banaras Hindu University, Varanasi.

Dr. Singh, S.P., Department of Geography, Kumaun University, Naini Tal.

## SECTION I

# CONCEPTS AND ISSUES

G. K. DUTT

# ONE

# ISSUES AND CHALLENGES

ENVIRONMENT is indeed a matter of general concern. This is a basic attribute on which the whole life and economy system is built up and hence is of vital importance. During the last decade environment has shot into prominence all over the world. Everywhere, an awareness about the environment has developed ; its importance has been realised and efforts are being made to understand it with a view to evolve ways and means of its betterment and preservation. The basic concept of the discipline of geography contained in its obsession with the study of land, man and his economy, has provided the geographer a certain outlook, insight and approach towards understanding the environment. He has essentially a synthetic approach and is capable of providing due weightage to all the components of the environment. His contribution in this sphere is, therefore, unique in many respects. The environmental studies have to incorporate the natural hazards as the significant attribute because at local level the interaction between the physical and human elements does influence the environment of the region or the locality (Burton, Kates and White, 1978).

Erroneously at times pollution is equated with environment. In fact, pollution is one of the aspects of environment. Environment with all its components has to be viewed and considered in its totality. There are, indeed, a wide range of physical, biological and man made components that interact in building up an environment. This has made the subject complex and a matter for multi-disciplinary study.

Environmental conditions get substantially influenced by local factors. Factors, like regular recurrence of natural calamities—floods, drought, land-subsidence and the like or features like intensive exploitation of natural resources e.g., mining, deforestation, or a highly effective health phenomena in the form of predominance of an endemic disease, or a typical socio-economic condition engulfing a large percentage of the local population, all exert profound influences of different natures and dimensions on the environment that evolve in a particular area. The intensity of the major factor determines the essential character of the environment under such circumstances. Indepth studies of such problem areas reveal interesting features exposing intricate linkages between environmental conditions and the pattern of human behaviour and response. Some possible solutions towards meeting the challenges of nature may also be struck which, if adopted, may bring significant changes for the better.

The geographers would also be interested in making some case studies of areas or regions with well-defined physical attributes or natural settings like mountainous or hilly region, a river basin or a lacustrine basin. There are other such areas where transformations have taken place because of implementation of multipurpose projects like Damodar Valley Corporation. It is obvious that human occupance and the resultant interference with nature have modified to a greater extent the natural environment as it were, of those regions. The approach for making a study of those cases would be to assess the present position of the environment vis-a-vis the human occupance and his economic activities, the extent to which the balance of nature has been maintained or disturbed, the future trend of the activities and their effect on environment. The facts, thus analysed thread bare, could then be made use of in planning actions towards improving the environment by restoring balance of nature (Harvey, 1979).

## Assessment and mapping

One of the tasks for which the geographers are most eminently suited, both by training and aptitude, is that of assessing the state of the environment. As mentioned in the previous paragraph those facts from the set of basic informations on which the entire planning for development has to be made. The approach and methodology adopted is to be systematic so that the data is structured and presented in a uniform pattern for all the areas of studies, making them comparable with one another and universally useful and intelligible to specialists of all disciplines interested in planning for improvement of the environment (Dutt, 1984).

In pursuing such a project the first concern of a research worker would be the collection of data covering all the components of environment as far as possible. Obviously, a judicious selection of components has to be made depending upon the peculiarity of the situations varying from one region to the other. At this stage co-operation of bioscientists, public health specialists, foresters and host of other scientists may have to be sought in deciding about the parameters of the data on different aspects that are to be collected. Our experience in the matter of collection of data, makes us cautious as a research worker that this is one of the most difficult job that one may come across. It is a fallacy that most of the data that one may like to have in some detail are collected and recorded regularly and systematically by one agency or the other of the central, state or local government but they remain confined in the files of the respective departments and are never published because of the massive volume of the data and are not easily accessible (Dutt and Nag, 1985 B). Once the respective authorities are made to realise, by pursuations and discussions, the importance and usefulness of the data collected by them that are gathering dust in their files, they may allow to make use of the information. There would still be a gap in the information content that would have to be filled up by field investigations. Micromorphological parameters are one such aspect where precise information has to be generated by observation and field work.

Having collected the data, the immediate problem would be that of their proper presentation. The components of environment, singly and collectively, relate to a point or area on the spatial frame and the totality of the composite picture of the environment can only be brought out by a graphical representation of the data pertaining to the components (NATMO, 1982). Data without a reference to the spatial frame carry little sense and become fully expressive and relevant only when placed side by side in its appropriate position on the topographic map. The map thus presents a synthetic picture of the relevant components of the environment and acts as the most important tool in the matter of environmental planning. A map depicting various components of the environment singly or collectively is essentially a thematic map (Dutt, 1984). It has to evolve and adopt suitable methods of cartographic representation of data so as to be able to portray an easily comprehensible and yet composite picture of the environment. It may be realised that whatever data be available or collected regarding the components of the environment from a variety of sources, the ultimate and optimal utilisation of such data can only be made when they are plotted appropriately on large scale maps. The maps are the net results of any effort towards making an objective assessment of the environmental condition of an area. Maps are able to provide a comparable set a pictures of environmental conditions of an area/areas under spatio temporal dimensions. In fact, thematic mapping is almost the exclusive domain of a geographer and because of that he commands a very important position amongst an inter-disciplinary group of scientists engaged in environmental study or planning (Dutt, 1982).

## Pollution

Pollution of the air and water caused by human habitat and his economy has, of late, created an alarming situation. The pressure of population along with increasing exploitation of our natural resources has aggravated the situation. The normal laws and capabilities of nature of absorbing the pollutants have been upset. This has resulted in pollution of air and water much beyond human tolerance capacity. People and the administration were complacent or even oblivious until recently about the quality of air and water which prevailed. Mass deterioration in public health, increased observations on the quality and composition on these two basic needs of life and publicity given to these phenomena by the media have made the people and the government aware of the deplorable condition of the elements at least in some parts of the country. It is revealed that laws enacted decades ago towards controlling pollution caused by emissions and discharge of obnoxious substances by industrial establishments have stayed only in papers. In case of pollutants discharged in the rivers, the urban settlements have been identified as the chief culprits who contribute a major share of the river's organic pollutants. The case in point is the holy river Ganga ! Curiously enough, the technology and equipment for nullifying the toxic and harmful elements in the affluents have been developed by our own National Environmental Engineering Research Institute at various periods, but they have been rarely installed and utilised by the erring agencies because of sheer apathy or reluctance to incur a little extra expense. The observations of scientists of various disciplines and the efforts of different agencies set up by the government at the all India and state levels, a mass of information are now available on the very poor level of air and water conditions in the country ; but having a factual knowledge of the situation is not an end in itself ! Actions have to be initiated to enforce the laws already introduced and formulate further laws to meet the present situation. It is heartening to find that some actions in the right direction have been initiated by the present administrations.

**Rural environment**

Our obsession with the phenomena of pollution which is predominantly associated with urban and industrial areas make us loose sight of the environmental conditions of our rural areas where eighty per cent of our population live. The conditions prevailing in India have widened the gap between the rural and urban situations and therefore rural India needs specialised attention for tackling its environmental problems.

The most basic problem is the absence of essential amenities in the villages. In the decade of 80's, declared as the decade for *health for all*, various surveys have revealed that a large percentage of our villages do not have supply of potable and safe drinking water. This has, in its turn, given rise to water-borne endemic diseases covering very wide regions of the country where people have to suffer all through their lives beause of the compulsion of drinking contaminated water. The goitre prevalent area of the Tarai region of Uttar Pradesh, the flurosis affected areas of a major part of Rajasthan, the guinea-work disease in its endemic form prevalent over a large area of Rajasthan, Madhya Pradesh, Andhra Pradesh etc., all are caused due to intake of water contaminated with harmful chemicals or bacteria. The recurrence of gastro-enteritis quite often in various parts of the country during the dry spells of summer is all too common. Malaria has started reappearing again particularly in the canal irrigated areas. Recurrence of floods or droughts is all too common. Salinity of land, land subsidence and soil erosion are other features to add to the list of rural maladies (White, 1974).

Contrary to general belief, precise information regarding existence of the above features are not available. Casual list of villages that lack potable water supply or suffer from a particular endemic disease may be available but efforts towards meeting those challenges are half-hearted or ineffective because of lack of information or planning. Maps on large scale incorporating the actual position regarding the components of the environment could be of immense help in planning and monitoring environmental improvement. Once again maps emerge as important tools ! The average villager is economically backward, yet he is a very important force in the country's efforts towards betterment of the environment. He is often blamed as he culprit for destroying the forest and tree cover of the country without realising the fact that he is compelled to do that to meet the dire necessities of fuel and forage for his domestic animals. Can we reorient his efforts towards *fuel-farming* so that the community could grow *fuel plantations* on available poor quality common land and 'harvest' rather than 'exploit' their common plantation to meet their daily needs leaving the forests alone ! Perhaps the geographers could do a service by locating in his large scale maps on rural environment the rough pastures and poor lands which could be utilised for such purposes. This will make plans and actions positively realistic.

**Environmental education**

Obviously, the present day situation can be tackled only by coordinated and sustained efforts, long-term and short-term, by the government as well to the public in general. To meet the immediate and urgent needs of checking deterioration the administration at various levels will have to be equipped with factual information, maps etc. and to be persuaded to initiate action on that basis. The people will have to be made aware of the gravity of the situation and infuse in them a conciousness about the essentiality of a clean environment for the benefit of the present day and future generations. Their active participation and support can only bring success in any effort initiated by the government agencies.

As a long-term measure, the children of today are to be given some exposure to systematic environmental education. Let this not be another subject confined to a few lessons in the text books but let it be more application oriented bringing the students more close to the nature, getting him aquainted to the realities in the field and equipping him with the necessary aptitude to observe, plan and react. In this effort again large-scale environmental maps would be the essential tools in his hand which would provide him existing facts as well as the perspectives for the future.

**Knowledge vis-a-vis action**

Environment has attracted the attention of a large number of scientists, geographers, geologists, chemists, biologists, ecologists and a host of other social scientists. All of them had been studying and analysing different aspects of the environment in their own particular fields. The contributions made by the bio-scientists had been quite substantial because by applying their methods of observation and analysis they could obtain a large amount of factual data. The scientists of different disciplines had been quite active in this sphere who have produced a wide range of information and have established methods of monitoring and continuous recording of changes. Unfortunately, the geographer had lagged behind so far in this field and has failed to make his rightful place in this group of scientists. He has not utilised his special aptitude and training in synchronising and synthesising the different variety of data, establish their relation with the spatial frame and thereby produce a composite picture of the environment. He has been hesitant in applying his knowledge of the physical aspects of land to explain relationships that exist between the physical and biological phenomena. He has confined himself to merely making some academic statements, expressing his views as to what should be done or should not be done for preserving the environment. This is not enough. He has to come forward with positive contributions to prove his worth. The new administration at the Central Government has undertaken specific programmes towards improving the environment—*Ganga Project* and the *Waste Land Development Authority* are some of the cases in point. The geographer is competent to play his part in portrayal of the present situation of pollution level of Ganga or existence and quality of waste-lands in the country. These documents could be permanent sources of references for action plan and monitoring. His synthetic analysis of the facts may open up new avenues of thinking and approach. We wish very much that geographers in this country come forward to contribute towards the national efforts and establish their capabilities.

## REFERENCES


Burton, I., R.W. Kates and G.F. White (1978), *The Environment as Hazard*, Oxford University Press, New York.

Dutt, G.K. (1982), Mapping the Environment, *International Journal on Environmental Studies*, 19, 25-33.

Dutt, G.K. (ed.) (1984), *Atlas Cartography*, National Atlas and Thematic Mapping Organisation, Monograph 3, Calcutta (Preface).

Dutt, G.K. (1985), Thematic Cartography : Problems and prospects in India, Souvenir of the XVII All Rajasthan Geographers' Conference, Central Arid Zone Research Institute, Jodhpur.

Dutt, G.K. and *et al.* (ed.), *Proceedings of the International Seminar on Environmental Maps and Atlases*. National Atlas and Thematic Mapping Organisation, Calcutta (forthcoming).

Dutt, G.K. and P. Nag (1985 A), *Life and Work of Professor S.P. Chatterjee*, National Atlas and Thematic Mapping Organisation, Occasional Monograph 1, Calcutta.

Dutt, G.K. and P. Nag (1985 B), *Scenario of National Atlas and Thematic Mapping Organisation in 2001 A.D.*, Department of Science and Technology, Calcutta.

Harvey, David (1979), *Explanation in Geography*, Edward Arnold, London, 115.

Kates, R.W. (1978), *Risk Assessment of Environmental Hazard*, John Willey, Chichester and New York.

Nag, P., (1983), Perception of the Environment : Issues and Challenges, *National Geographical Journal of India*, XXIX (Nos. 3 and 4), 217-31.

National Atlas and Thematic Mapping Organisation (NATMO) (1982), *National Atlas of India*, Department of Science and Technology, Calcutta.

White, G.E. (ed.) (1974), *Natural Hazards* : *Local, National, Global*, Oxford University Press, London.

C. R. PATHAK

TWO

# MAN, ENVIRONMENT AND SETTLEMENTS

THE study of Geography has gone a long way from the days of environmental determinism. The days of Allen Sample and Huntington have gone by. The progress of human civilization along with the inventions and application of new technology to harness and develop physical resources has brought about the supremacy of human beings over nature. The man's role in changing the face of the earth has been of paramount importance. The world, as the home of man has been shaped up and modified according to the growing need of the mankind. The living environment of today is not the same as in the past. It has been created by 'man' and if need be arisen, he will further modify it. Hence, the environment today contains man at its core or as an integral part of it. The distinction between physical and man-made environment is blurred. The physical world has been constantly modified at the advent of mankind. In order to study the aspects of environment today and their spatial variation we need to study man and his relation to the environment in an integrated manner (Lewthwaite, 1966).

## Symbiotic relationship between man and environment

Nowadays we study the environmental aspects of human habitat and it varies over space in the context of man-environment inter-relationship. Whenever, we find that there is a symbiotic relationship between man and environment we call it "ecological balance" and whenever such environmental factors cannot support human needs and aspirations because of exterior deterioration and over-exploitation of such environment it is called "ecological imbalance". The balancing factors in environment and human relationship are multi-dimentional and complex in nature (Raza, 1980). At a time when environment cannot absorb the vagaries of human exploitation, environmental pollution occurs. Environmental pollution is not necessarily the result of industrial and economic development alone. It may also occur because of ill-development or under-development. Besides, ecological environmental quality is not necessarily good in a low level of technology or low level of development. Natural hazards, such as cyclones, floods and droughts or large-scale soil erosion create physical environmental problems. Human beings can take precautionary measures against such environmental hazards by adopting modern-technology like providing transport and communication facilities to such areas and by undertaking flood control or soil conservation measures or by supplying artificial water at the time of droughts. It is for sure that human beings may not be able to eradicate such natural hazards, but by adopting technological measures could reduce their intensity.

Environmental problems resulted from economic development are also multifarious. They could be observed in several scenarios of human settlement. If you could lump the environmental problems in hilly and tribal areas and other lowly developed areas, such scenarios could contain aspects of under-development, low level of technology and wasteful utilisation of natural resources. Apart from the above scenarios, the living environment can be regionalised into two sub-scenarios of sedentary-human settlements : (a) rural environment, and (b) urban environment. When geographers study human settlements, they classify them into two groups i.e. rural and urban and describe the pattern of settlement in terms of population distribution, migration, and socio-economic opportunities and bring out the distinguishing factors of various size classes of human settlements.

**Rural settlement scenario**

The standard classification of rural settlements into size classes or according to their physical layout may not serve our purpose of environmental study. Such settlements need to be studied according to their economic development levels, land holding pattern and occupational structure. In rural settlements there is inverse relationship between man and land and it is the man-land ratio including productivity that should be the first indicator of environmental standard (Pathak, 1975 &1982). Of course intensity of cropping, intensity of irrigation, intensity of landuse and diversification of cropping may constitute other indicators. Per capita income in a rural settlement will not indicate living standard if the land distribution is very inequitious. Our common knowledge of having low environmental standard in the villages reveals the existence of large number of marginal and small farmers and landless labourers. Hence, the socio-cultural environment in rural settlements is vitciated by the presence of poverty, illiteracy and inequitable distribution of natural resources.

Even though a few wealthy land owners reside in such villages and constitute vocal minority, they usurp governmental subsidy or investment for rural development unscrupulously; they do not improve the living standard of the majority, instead create social tension and economic deprivation. The community development programme with the provision of minimum level of living such as potable water supply, sanitation and medical care, transport and communication facilities connecting nearby higher order settlements, rural electrification and other infrastructure, is a big step forward towards the improvement of living standard. Although they constitute minimum needs for human beings, they will hardly improve the quality of life of the poor majority who could hardly pay for such facilities. Of course, denial of such facilities would never open up rural environment for development purposes. Hence, the pollution of rural environment is the result of under-development and low level of living which further deteriorates human environment and creates vicious circle of poverty, under-development and environmental pollution.

### Urban settlement scenario

Environmental issues which emerge with the growth of urban settlements vary widely from one size and functional class of urban places to the other. Such urban environment is related primarily with the living condition of the people. Although urban places are distinguished from the rural ones in occupational structure and in density of population the built environment is the most distinguishing criterion, whereas, such built environment varies significantly in living standard from place to place.

Geographical spread of environmental issues in urban places can be identified in terms of their occurrence from one size and functional class of settlement to another and also as they vary within a single urban place. The latter type of variation is more prominent as the size of the city increases. Large cities are differentiated internally in terms of population density pattern, functional landuse, transport network and other infrastructure. Hence in such larger cities, while we observe reasonably good living standard in some parts of the city, in other parts, slums or working men's colony or unauthorised occupations having very low level of living standard (Pathak, 1983). A number of shanty mohallas or neighbourhoods indicating low living standard and human occupance grow in such large cities and metropolises. The environmental problems in cities and towns are results of human action or lack of it. The small and medium size towns suffer from insanitation, poor housing condition, low employment opportunity and low per capita income. There is no clear cut zonation of functional landuse. The bazaar located at the central part of such towns is being surrounded by congested houses of various standard varying from mud wall and thatched roof to one or two storey pucca buildings without having any centralised sewerage, water supply and other community facility. Roads are hardly passable and are seldomly clean. Municipal services are almost non-existent because the municipalities have very limited investable fund for providing such services and are having very poor managerial capabilities. The tax payers not only belong to low income group, the revenue they pay is necessarily small and hence, unless there is tremendous subsidy and investment from the government, the environmental standard of such towns could hardly be raised to the liveable one.

The problems of environmental pollution are more complex in character in large metropolitan cities. They have grown disproportionately to the services laid out in the past. Where cities are expanding very fast the services such as water supply, sewerages, road network, electricity and other municipal services cannot be extended without redeveloping the central part, whereas such redevelopment of the central part requires huge financial investment and large organisation for development management.

## Conclusion

Environmental issues are very, many and vary in intensity, in complexity and in nature from rural to urban and from small to large human settlements and pose challenges of various types. The methods of tackling such environmental problems would also vary according to the geographic, economic, social and technological levels of development. No single set of principles would be applicable universally nor such issues are universal in character (United Nations, 1974).

Any plan and programme of environmental improvement with the provision of community facilities and infrastructure improvement both in rural and urban settlements must be preceded by or at least simultaneously executed programme of economic and social development such as creation of employment opportunity, increasing per capita income, generation of gross domestic product and equitable distribution of such resources, so as to achieve symbiotic relation among physical, social and economic development in any village, town, city or region. Restoration of environmental ecological balance and its maintenance in any human settlement or a region are challenging tasks. The geographic analysis of such environmental problems would help in formulating rational strategy for environmental planning and development.

## REFERENCES

Lewthwaite, G.R. (1966), Environmentalism and determinism : A search for classification, *Annals of the Association of American Geographers*, 56, 1-23.

Pathak, C. R. (1975), District development planning in India, *Indian Journal of Regional Science*, VII (2), 188-94.

Pathak, C. R. (1982), Rural industrialization for rural area development (Rijvad), *Indian Journal of Regional Science*, XIV (1), 81-90.

Pathak, C. R. (1983) Environmental hazards and the low income groups in the slums of Calcutta Metropolis, *Indian Journal of Regional Science*, XV (1), 89-95.

Raza, M. (1980) Reflections on geographical view of man in ecosystem, *Transactions of the Institute of Indian Geographers*, II (2).

Raza, Mehdi and S. L. Kayastha (1984), Environmental studies in India, in

*Trends in Geographical Researches in India*, Country Report, 25th International Geographical Congress, Indian Science Academy, New Delhi, 47-55,

United Nations (1974), *Human Settlements : The Environmental Challenge*, A compendium of United Nations paper prepared for the Stockholm Conference on Human Environment, 1972, Macmillan.

S. P. SINGH
AND
J. S. SINGH

THREE

# ECOLOGY AND DEVELOPMENT

ABOUT 53 per cent (175[2] million ha) of India's total land is subject to serious environmental degradation ; of the cultivated area of about 140[2] million ha, nearly 60 per cent require soil conservation measures ; the annual flood damage on average is over Rs. 1,000 crores (area subject to periodic floods was 40[2] million ha in 1980, i.e. about 28 per cent of the total cultivated area) ; sedimentation is likely to reduce the life of twenty-one dams to one-third of what was expected ; water-logging accounts for at least one-fourth of the 40[2] million ha of irrigated land and salinity due to irrigation has increased alarmingly ; the crop fields lose more than 10 million tonnes of N+P+K annually ; 47 per cent of the cultivated land is Zn-deficient ; soils of 228 of the 250 districts examined are low in N (and probably also in organic matter) ; 70 per cent of the available water is polluted ; before the Yamuna enters Delhi its water contains seventy-five coliform organisms per ml and in Delhi the coliform organism count reaches 240,000 orgnism per ml ; forested area accounts for a little over 10 per cent of the land, but as the data of the Central Himalaya indicates (forests with crown density 60 per cent or more account for only about 4 per cent of the total area) the good forest cover is likely to be negligible ; the Himalayan forests are in much worse condition than when conservation began about 100 years ago ; despite an investment of Rs. 7,510 crores on major and medium irrigation works, the average yield for irrigated land is only 1.7 t of grain/ha against the target of 4-5t of grain/ha ; the transmission losses of hydroelectricity in Jammu account for up to 62 per cent per year : air pollution levels in major Indian cities are going up exceeding permissible limits in contrast to the trend in western countries ; the urban population has doubled in the last 20 years and slums population has increased at a still higher rate ; about a third of the urban population has no access to latrines of any kind ; traditional building materials are becoming increasingly scarcer, in Karnataka, for example, a poor person has to buy bamboo at Rs. 1,200 per tonne to build his house, whereas, the paper industry gets it at the concessional rate of Rs. 15 per tonne ; although per capita per day grain consumption has remained constant during the last 30 years (431—438 g), the pulse (the only source of protein for the poor) consumption has gone down from 70 to 40 gm ; the population increase is unabated ; many diseases have come-to-fore due to the polluted water or scarcity of water ; in the Central Himalaya women walk about 7 hours a day to collect the fuel-wood ; more than 10 per cent of the India's rich flora (15,000 species) and fauna (75,000 species) are on the way to extinction ; of the 134 most threatened species 99 are Himalayan ; the river Ganga originating from the Central Himalaya, carries 340 million tonnes of sediment each year ; the present rate of erosion in the catchment area of the Himalayan rivers (100 cm per 1,000 years) is five times higher than the rate prevailing in the past 40 million years (21 cm/1,000 years) ; the construction of 44,000 km-long road in the Himalayan region generated 2,650 million $m^3$ of debris ; each year 550 $m^3$ of debris per km of the road is produced by landslides and rock falls, causing 24 million $m^3$ of sediment to slide down the slopes killing vegetation and choking mountain streams. More details are given in Anonymous (1982), Singh *et al.* (1984), and Valdiya (1985).

The above symptoms indicate that the ecosphere of the country is far from healthy and the "developmental activities" have worsened it instead of improving it. Further, the various economic activities though have reduced the people's miseries in some cases, they have also added new ones and deepened the many old problems.

The problems of the human future range far beyond ecology, yet ecology is an integral part of them. Ecological knowledge can make possible the "appropriate" harvest of ecosystems, contribute to planning such that the use and the disturbance of environment are balanced, and clarify the meaning and potential dangers of pollution and retrogression (Whittaker, 1975).

In the current ecological concepts which place the man inside the ecosystem, another dimension relating to the human perception of changes in environment, has been added. The fact that the man's adaptational faculty is under continuous stress due to the changing attributes of the environment. The future course of development will undoubtedly be determined by the outcome of how the human civilization reacts to the present phase of large-scale changes in the environment.

**Conflicts between Economic and Ecological Values**

Although the ecological and economic systems have a myriad of intricate interconnections, the most simple and the most obvious is the ecological systems which provide raw materials to the economic systems and absorb the waste generated by the economic systems Fig. 3.1. Therefore, the economic system will always be constrained by the production and the waste absorption capacities of the ecological system. The moment one or both of these capacities are exceeded, ecological backlashes are bound to occur. Economic system can indeed, be predicted as having a more or less parasitic role in the biosphere (Wilen, 1973). Even

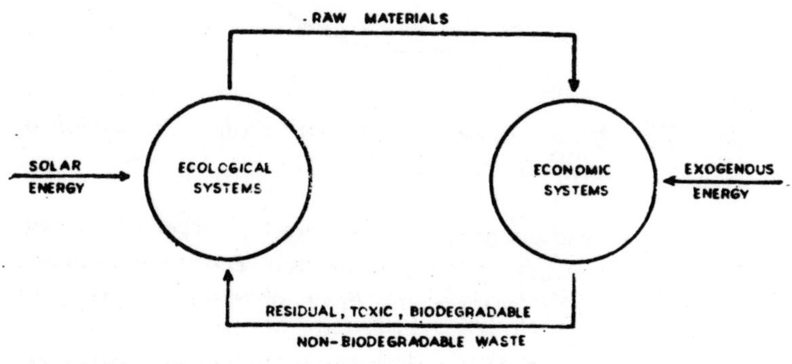


FIG. 3.1 : *Broad linkages between ecological and economic systems*

the seemingly innocuous, non-fossil fuel-based agriculture in the Himalaya entails an expenditure of 7 units of energy from adjoining forests in terms of fuel, fodder and leaf manure, for each unit of agronomic energy that it yields (Singh *et al.*, 1984).

Thus, while the ecologist lays down an environmental imperative that requires an end to economic growth or sharp curtailment of it as the price of biological survival, the economist may counter with a socio-economic imperative that requires the continuation of growth as the price of social welfare (Heller, 1972). In fact, the ecological, socio-cultural and economic systems are inter-connected and inter-dependent Fig. 3.2 and they have to be considered with a holistic and not sectoral view point.

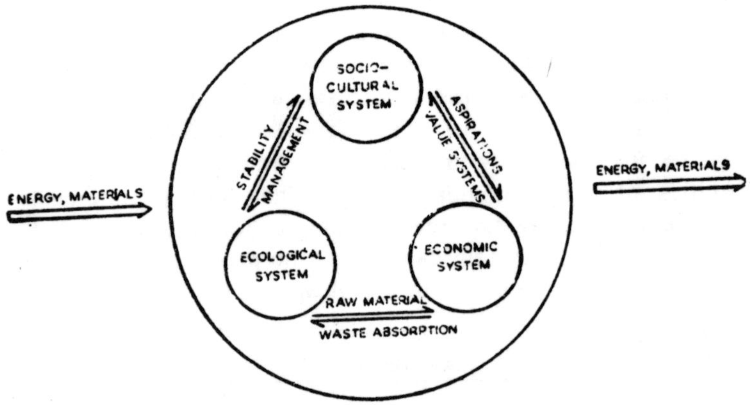


FIG. 3.2 : *Broad linkages between ecological, social and economic systems*

Both ecological and economic values can be served in a variety of ways, but combining ecological and economic values adds geometrically to the complexity of a development programme (Caldwell, 1984).

It has to be kept in mind that an important characteristic of our multi-loop, non-linear feedback social system is the occurrence of a fundamental conflict between the short-term and long-term consequences of a policy change. A policy which produces improvement in the short run is usually one which degrades the system in the long run, and those policies and programmes which produce long-run improvement may intially depress the behaviour of the system (Forrester, 1971).

The ecologist is not opposed to development, he only pleads that any development plan be examined for its environmental impact before implementation and if it is found compatible with the ecosystem properties of the place, safety mechanisms be built in the plan for protection and repair of the environment, and from implementation onwards a continuous monitoring be made of the more sensitive environmental parameters and if any deterioration in these parameters is detected, the development-generated activity be stopped or modified accordingly. The ecologist also pleads for preservation of unique ecosystems as environmental heritages in the spirit of an obligation to future generations, because he knows that man can never reproduce such ecosystems and that these can act as standards for assessing the impact of man on nature in furture, and as gene reservoirs (Singh and Chaturvedi, 1981).

### Aims of Development and Ecological Approach

Man's urge for development is old. The reason ascribed for the clearing of European forests and woodlands in the middle ages, for example are the beliefs that the earth was created as an abode for man (Piaget, 1929) and that man has to put finishing touches to the incompleted world (Darby, 1956 ; Rorabacher, 1973) so as to make it more habitable and more comfortable. Development to technology helped the process of landscape modification, ushering in ultimately the era of profound urbanization and industrialization.

Without much fear of contradiction we could perhaps state that the aim of ecologically sustainable development is to maximise human well-being or quality of life without jeopardizing the life-support system. Although there is no unique definition of quality of life, Swartzman (1972) has enumerated four groups of variables, viz., economic, ecological, socio-cultural and political, which might be considered as indicators for the quality of life. Thus the quality of life depends on economic as well as non-economic resources. It is possible that the increased demand for economic resources is a maladjustive reaction to the scarcity of non-economic resources, and an improvement of the social environment will reduce the threat to the physical environment caused by excessive demand for material resources (Foa and Foa, 1973).

Van Dyne and Innis (1972) have emphasized the interconnections and natural dependence of ecosystem attributes, determinants of human well-being and societal attitudes as follows :

| Ecosystem attributes (water, food, energy, air, minerals) | ⇌ | Determinants of human well-being (cultural resource system, social organization, spatial distribution, energy, water, air, biota, minerals) | ⇌ | Societal attitudes and values |
|---|---|---|---|---|

Each major determinant of human well-being has its own complex structure, e.g., air has CO, $O_2$, $SO_2$, $NO_2$, $CO_2$, $O_3$, etc., as major sub-components.

The use of ecology in development planning has the aim, both enhancing the goals of development, and anticipating the effect of developmental activities on the stability of ecosystem attributes, in general, on the natural resources in particular, and on the processes of the larger (i.e. beyond the local) environment as well. Thus, an ecological appraisal of a particular type of development will focus on the area of the resource to be exploited and also on larger environment and human inhabitants (Dasmann *et al.*, 1974).

More often than not, a regional development results in an increase in size of population and/or its activity. It is important that this increased level of population or its activity be in conformity with the carrying capacity of the environment. Carrying capacity is the measure of individuals of any species that a particular environment can support without getting deteriorated. According to Dasmann *et al.* (1974), carrying capacity has following levels :

*Absolute carrying capacity*

The maximum number of individuals that can be supported by the resources of the environment at a subsistence level.

*Security carrying capacity*

The level at which a species population is normally held by the influence of other species living in the same environment.

*Optimum carrying capacity*

A level which is generally considered more desirably by those concerned with the health or productivity of the species involved. At this level individuals in the population will have adequate supply of all essentials.

The carrying capacity of the environment can also be considered in respect of its waste-absorption capacity when an economic system is superimposed on an ecological system.

## Urban-Industrial and Natural Ecosystems

Because of the complexities of the inter-relationships in the 'holocoenotic view' of environment, ecosystem based management-development is finding more and more support today. From the stand-point of the primary functional role, Odum (1983) recognized four kinds of interrelated ecosystems in a given landscape. In this scheme, as given below, the urban-industrial heterotrophic environment is supported by the remaining three types of life-support environments.

| System | System attributes |
|---|---|
| 1. Non-biological systems (urban-industrial environment) — | Heterotrophic city and industrial system dependent on large areas outside it for energy, fibre, water, etc. High energy-consuming hot spots, larger and more poisonous output of waste products. |
| 2. Mature systems (protective life-support environment) — | Old-growth forests, climax grasslands, oceans, more protective than productive ; stabilize substratas, buffer air and water cycles, moderate extremes in physical factors. |
| 3. Dissipative systems (waste assimilative environment) — | Natural or semi-natural ecosystems that bear the brunt of assimilating the wastes produced by urban-industrial and agricultural environments, e.g., inland and coastal water ways, wetlands, etc. |
| 4. Growth systems (productive, life-support environment) — | Early successional or growth-type ecosystems, such as croplands, pastures, tree plantations and intensively managed forests. |

Of the above systems, the mature systems and most of the dissipative ones are 'natural', the growth systems are 'domesticated' and the urban-industrial systems are 'fabricated'.

Let us look at the compartmentalization of the environment in the continental United States (U.S.), a developed country (Table 3.1) ; and in India, a developing country (Table 3.2). Both

TABLE 3.1. **Land use in the Continental United States, 1980**

| Landuse | % of the total area | |
|---|---|---|
| Fabricated | | |
| Urban-industrial | 6 | |
| Total | | 6 |
| Agroecosystems | | |
| Croplands | 18 | |
| Pastures | 6 | |
| Total | | 24 |
| Natural ecosystems | | |
| Range | 33 | |
| Forest | 30 | |
| Parks, wilderness and wildlife refuges | 4 | |
| Water | 3 | |
| Total | | 70 |

(*Source* : *Odum, 1983*)

TABLE 3.2. **Land use in the Continental India**

| Landuse | % of total area | |
|---|---|---|
| Fabricated | | |
| Urban-industrial | 5.5 | |
| Total | | 5 5 |
| Agroecosytems | | |
| Croplands | 43.6 | |
| Pastures | 14.6 | |
| Total | | 58.2 |
| Natural ecosystems | | |
| Cultivable but wasteland | 12.2 | |
| Uncultivable wasteland | 6.4 | |
| Forests | 10.7 | |
| Total | | 29.3 |
| No information available | | 7.0 |

(*Source* : *Anonymous, 1982*)

are similar in having relatively large land areas and granaries and have democratic set-up. However, they also differ in many ways. The U.S. has comparatively very large technological base, low population density and while India's population is growing rapidly than that of the U.S. which is almost stabilized (Singh, 1978). In the United States, agroecosystems and urban-industrial systems (including rural transportation corridors) make up 30 per cent of the area, leaving 70 per cent in natural ecosystems variably affected by grazing, exploitation of timber, recreational activities and pollution load.

In comparison, the fabricated (urban-industrial) plus domesticated system (agroecosystem) constitute about two-thirds of India, leaving only about a third, for natural ecosystems. However, proportion of land under the urban-industrial category alone is similar. A sizeable portion of the natural systems of India includes waste-lands, supporting the amount of vegetation far less than the potential level. Some of them are themselves subject to one or the other destabilizing forces such as soil erosion that has driven them back to a extremely primitive stage of succession. In the United States about 4 per cent area consists of parks, wilderness and wildlife refuges which is highly protected from the human disturbance. Such a category is represented in a negligible proportion in the densely populated country, like India. Moreover, the existing forests of India are mostly 'second-growth'. Saxena *et al.* (1985), on the basis of population structure over large area, has concluded that the Central Himalayan forests are dominated by a few successional species which have less protective role than the old growth forest (Odum, 1983). Data on the Himalayan waterbodies indicate that a majority of them have little additional life support capacity (Zutshi *et al.*, 1972 ; Singh *et al.*, 1982). Situation in waterbodies of plains, such as the river Ganga or Yamuna is still worse. Thus, the comparison shows that ecologically the Indian condition is far more grave than that of the United States. However, the area of natural ecosystem required to support the urban-industrial system will also depend on how many times its power level is greater than that of the natural systems, greater the difference the larger would be the area of natural ecosystems required to dissipate the disorder output of the urban area (Odum, 1983). The urban-system (the hot spots) may flow from hundred to thousand times more energy or unit area (upto 3 million Kcl/m²/yr for fuel-powered urban-industrial system compared to 2000 Kcl/m²/yr for solar-powered natural ecosystem) than the natural ecosystems (Odum, 1983).

Although data on energy consumption density (Kcl/m²/yr) are not available for our cities, indirect sources indicate that the cities of the United States are exceedingly "hotter" than our cities. The per capita consumption of fuel energy in India is about 1/50 times less than in the United States. The U.S. agriculture is far more fuel energy-consuming than the Indian agriculture. In Manhattan (the New York city centre) $4.8 \times 10^6$ Kcal/m²/yr energy is consumed by the people, which is higher than the solar energy that reaches the earth ($1\text{-}2 \times 10^6$ Kcal/m²/yr) (Odum, 1983). Thus, an urban area of a given size would require several times greater area of domesticated and natural ecosystem in the US than in India, in order to meet the productive, dissipative and protective demands. Further, the life-support capacity of the natural environment can vary according to the degree of stress already imposed on it. For example, a polluted lake would have far less additional life-support capacity than an oligotrophic lake.

It is difficult to determine objectively how large natural environment must be preserved within a country. It is possible that the high ratio of 2.3 : 1 between natural and man-made environment in the U.S. may not represent a better situation than about 0.6 : 1 ratio of India. Keeping in view the current and still low level of energy flow in the cities of the latter country, however, this should not make us complacent. Odum and Odum (1972) state, "even the simplest models clearly demonstrate that high powered systems such as cities require an abundant life-support from nature. If large areas of natural environment are not preserved to provide the needed input from nature then the quality of life in the city declines and the city can no longer compete economically with other cities that have an abundant life-support input". Evidently, for maintaining the urban-industrial systems, management and preservation of protective and dissipative systems are essential.

Situation might have been poorer than the above two countries in some European countries and Japan, which are not only industrialized but are also densely populated. But, these countries have been able to reduce the pressure of their urban systems on the remaining systems by utilizing the services of what Borgstrom (1969) calls "ghost acres", which are the areas outside the country providing food and other general life-support goods and services. Being a major exporter of food, the U.S. provides ghost acres for several developing countries. However, the U.S. also utilizes natural resources of a number of developing countries. India of today is almost exactly self-sufficient in food but provides "ghost acres" for certain other developing countries by exporting tea, fibre and other natural products.

### Technology development and social system

Developing countries like India and China with their large forces of scientists and technologists have the potentiality for attaining high-level technology. In case they realize this potentiality, they may save their land from becoming the "ghost acres", and balance among various ecosystems may not deteriorate further.

However, a high energy-consuming technology will require the support of large areas under natural ecosystems. Keeping in view the population density and the large proportion of the land devoted to croplands, it would not be possible to save large areas of natural ecosystems. We do not know whether or not it is possible to develop sophisticated technology network without increasing the energy flow on unit area basis. This needs to be worked out in detail.

If our progress on the path of high technology level were to be partial and the so-called "change to the twentyfirst century technology" were to be slip-shod and rough, numerous sociological and ecological problems are likely to be generated. In the initial stages, at least the import of technology would be at the expense of natural resources. This would add to the stress on the natural protective and dissipative systems, some of which, in addition, would have to be converted into productive systems. In spite of all the technological development in agriculture during the last decades, the per capita pulse consumption has gone down from about 70 gm per day at the time of Independence to mere 40 gm per day in the present. In the Himalaya, where the people had reasonably balanced diet in the past (Berreman, 1977) the availability of pulse presently is merely 14 gm per day per person. Problem, as this may submerge us in numerous health, sociological and environmental catastrophes, leaving little option for the further progress on the envisaged path of high technology atta nment in a democratic set-up as our's.

Dangers as this and those enumerated in the beginning of the paper, highlight the significance of "holistic approach", one of the major ecological concepts. In recent years the emphasis on environmental conservation in India has been considerable but meaningful linkages between conservation, development programmes and technology used are too often ignored. For example, in the Himalaya hundreds of agencies are involved in planting trees and made several attempts to repair the damaged mountain systems. Even felling of trees has been banned in higher elevations. Simultaneously, agriculture is being strengthened and new roads are built up to open the remaining inaccessible areas. Singh *et al.* (1984) and Singh and Singh (1984) on the basis of studies on energy relationships between the forest and agro-ecosystems have shown that the Himalaya cannot be revegetated as long as the present form and size of cultivation of crops continues, and Valdiya (1985) has emphasized the relation between road construction and accelerated landslides or rockfalls and deforestation.

Caldwell (1984) has recently analyzed the political aspects of ecologically sustainable development ; a few points are recalled here. The implementation of such development is beset with many problems. Reordering of human behaviour is an important aspect. The social invention has lagged behind the growth of sophisticated technologies. Governments are called upon to provide quick and temporizing solutions to matters which cannot be addressed without recourse to more basic measures for which the public generally is unprepared. Certainly, ecologically sustainable development requires a high order of common purpose and understanding between people and their government and depends upon effective political support. Educational efforts are necessary to move the world into an ecologically sustainable mode of behaviour.

In order to divert the focus from 'man outside nature' approach and in order to emphasize the importance of socio-economic and cultural influences on ecological systems, a "noosystem concept" has been proposed ; and a plea has been made to recognize applied ecology as that integrative science which encompasses the principles, concepts and methodologies of both ecology and environmental science in an attempt to educate personnel across the total environmental spectrum (Barrett, 1984), leading finally to a transdisciplinary approach to education.

## REFERENCES


Anonymous (1982), *The State of India's Environment : A Citizen's Report*, Centre for Science and Environment, New Delhi, 189.

Barrett, G.W. (1984), Applied ecology : an integrative paradigm for the 1980s, *Environmental Conservation*, 11, 319-22

Berreman, G.D. (1977), Ecolcgy, Demography and Social Strategies in the Western Himalayas : A case study in Himalaya. *Ecologies—Ethnologie* : Centre National de la Recherche Scientifique, Paris. 401-26.

Borgstrom, G. (1969), *Too Many : An Ecological Overview of Earth Limitations*, New York Collier Books, 400.

Caldwell, L.K. (1984), Political aspects of ecologically sustainable development, *Environmental Conservation*, 11, 299-308.

Darby, H.C. (1956), The clearing of the woodland in Europe, in Thomas, Jr. W.M.L. (ed.), *Man's Role in Changing the Face of the Earth*, University of Chicago Press, Chicago, 183.

Dasmann, F.R., J.P. Milton and P.H. Freeman (1974), *Ecological Principles for Economic Development*, John Wiley and Sons Ltd., London, 252.

Foa, U.G. and E.B. Foa (1973), Measuring quality of life : Can it help solve the ecological crisis ? *International Journal of Environmental Studies*, 5, 21-6.

Forrester, J.W. (1971), Counterintuitive behavior of social systems. *Technology Review*, 53-68.

Heller, W.H. (1972), Coming to terms with growth and the environment, in Sehurr, S.H. (ed.), *Energy, Economic Growth, and Environment*, John Hopkins University Press, Baltimore and London, 3-29.

Odum, E.P. (1983), *Basic Ecology*, Saunders College Publ., Philadelphia, USA, 613.

Odum, E.P. and H.T. Odum. (1972), Natural areas as necessary components of man's total environment, Proc. 37th N.A. Wild and National Research Conference, Washington, D.C., Wildlife Management Institute, 178-89.

Piaget, J. (1929), *The Childs Conception of the World*, Rouledge and Kegan Paul Limited, London, 231-357.

Rorabacher, J.A. (1973), Geo-historical approaches to environment : the evolution of environmental thought in western civilization, *Journal of Geography*, 31-41.

Saxena, A.K., S.P. Singh and J.S. Singh (1985), Population structure of forests of Kumaun Himalaya : Implications for Management, *Journal Environment Management*, (In Press).

Singh, J.S. and O.P. Chaturvedi (1981), Ecology, environment and development, in Singh, J.S., S.P. Singh and C. Shastri (eds.), *Science and Rural Development in Mountains*, Gyanodaya Prakashan, Naini Tal, 47-73.

Singh, J.S. and S.P. Singh (1984) (eds.), An Integrated Ecological Study of Eastern Kumaun Himalaya with Emphasis on Natural Resources, (HCS/DST/187/76), Department of Botany, Kumaun University, Naini Tal.

Singh, J.S., U. Pandey and A.K. Tiwari (1984), Man and Forest : A Central Himalayan case study, *Ambio*, 13, 80-7.

Singh, S.P. (1978), Population, affluence and ecological demand, in Singh, J.S. and B. Gopal (eds.), *Glimpses of Ecology* (Prof. R. Misra commemoration Volume), International Scientific Publication, Jaipur, 541-62.

Singh, S.P., M.C. Pant, A.P. Sharma, P.C. Sharma and R. Purohit (1982). Limnology of shallow water zones of lakes in Kumaun Himalaya (India), Gopal B., R.E. Turner, R.G. Wetzer and D.F. Whigham (eds.), *Proceedings of Wetlands : Ecology and Management*. International Scientific Publishers, Jaipur, 39-49.

Swartzman, G.L. (1972), *Optimization Techniques in Ecosystem and Landuse Planning*, U.S. I.B.P. Grassland Biome Technical Report No. 143, Colorado State University, U.S.A.

Valdiya, K.S. (1985), Accelerated erosion and landslide prone zones in the Central Himalayan region, in Singh, J.S. (ed.), *Environmental Regeneration in Himalaya*, Central Himalayan Environment Association, Naini Tal.

Van Dyne, G.M. and G. Innis (1972), *Macro-level Ecosystem Models in Relation to Man : A Preliminary Analysis of Concepts of Approaches*, U.S. I.B.P. Grassland Biome Technical Report No. 162, Colorado State University, U.S.A., 61.

Whittaker, R.H. (1975), *Communities and Ecosystems*, 2nd ed., MacMillan, New York, 385.

Wilen, J.E. (1973), A model of economic system—ecosystem interaction, *Environmental Planning*, 5, 409-20.

Zutshi, D.P., V. Kaul and K.K. Vass (1972), *Limnology of High Altitude Kashmir Lakes*, Verh Institute Verein Limnol, 18, 599-604.

PRITHVISH NAG

FOUR

# ENVIRONMENTAL PERCEPTION

"THE study of man's relationship with the environment has been increasingly concerned with explaining why man behaves as he does ; why, among other things, he grows one crop rather than another, or chooses particular transport routes, or builds factories in specific places" (Knowles & Waring, 1976). The environment is assessed based on how it is being perceived. Farmers have their own estimation of floods or droughts or other types of hazards. Based on their perception, strategy for acceptable returns, risk-assessment and damages are being made. Perception of the environment which depends on the individual image of the world, influences decision-making and finally the behaviour of the individuals. Environmental preferences can be identified in terms of landscape preferences or residential desirabilities.

Decisions are not always governed by economic principles. Neither all the human activities can be explained by theories such as those of Von Thunen, Isard, Ullman and others. Decisions are based on non-economic factors as well, which involve subjective elements in man-environment relationship. Behavioural geography is interested in the interpretations of spatial reality that depends on human nature, on the conscious subjects. As the name behavioural geography indicates, it is interested in human action and behaviour in relation to environment (Paasi, 1984). Environmental perception and behaviour are being studied because these aspects lead to decision-making and ultimately the transformation of the environment.

"Environment is taken to refer to anything external to the perceiver which influences, or might influence the perception process" (Pocock and Hudsin, 1978). Man and environment paradigm, as a major offshoot of human geography, includes 'the notion that individuals and groups relate to their environment through their perception or cognition of them (Blunden, Haggett, Hamnett and Sarre, 1978). Whyte has also expressed similar views on man-environment interaction. According to her, "Man reaches decisions and takes action within the framework of his perceived sets of elements and links rather than any externally defined objective set" (Whyte, 1977). However, Brookfield differentiates between perceived environment and environmental perception. Attaching more importance to the former, he explains perceived environment as 'the whole monistic surface on which decision is based, including natural and non-natural, visible and non-visible, geographical, political, economic and sociological elements in terms of its scientific classification'. His idea of environmental perception is 'perception of the whole environment, or specific selected elements within it, or even of space in the abstract : essentially environmental perception is a property of the mind rather than a construct of that mind' (Brookfield, 1964).

**Basic approaches**

It may be interesting to evaluate the different types of environment according to man's behaviour. Sonnenfeld, in behavioural classification, identifies a hierarchy of environments, each of which includes elements important to man (Sonnenfeld, 1978). According to him 'the broadest level is the geographical environment, constituting both proximal and distal elements of Man's universe. This reduces in stages to the operating environment, which is that environment impinging on man with which he is likely to be involved in one way or other, to the perceptual environment of which he is aware ; and finally to the behavioural environment, which elicits a behavioural response from the individual. In the present context we are more concerned with the perceptual environment which is again classified into : (a) sensory environment, and (b) symbolic environment. The basic difference between the sensory environment of universal elements and the symbolic environment of idiosyncratic elements is that while the sensory elements have meaning derived from the nature of their composition or energy, the elements of sensory environment require a cultural and personality transformation before they achieve meaning' (Sonnenfeld, 1969). Floods, droughts, frost, avalanche, earthquake and cyclone are the examples of sensory elements of the environment while aesthetic, social and political environment or highly valued landscape are symbolic elements of the environment. Like Sonnenfeld and Brookfield (Brookfield, 1969), Whyte has also classified the concept of the perception of the Environment into inside view and outside view (Whyte, 1977).

Of late, stress has justly been placed upon spatial perception, by which is meant what given environmental exposures communicate to specific human groups. Environment communicates; indeed, all communication has to reach its human target through this sensory (environmental) field (Wagner, 1972). Downs has summed up topic under study using the term Geographic Space Perception and reviewed the past studies into following typology: (a) structural approach; (b) evaluative approach; and (c) preference approach (Downs, 1970). The structural approach refers to 'mental basis for the storage of relevant information from the environment' or to 'mental or cognitive organisation of space' on which man is 'able to implement decisions as behaviour'. The evaluative approach is 'concerned with the evaluation of the environment via spatial images, and seems to relate the evaluation to decision-making and therefore to behaviour.' Hence, this is one step ahead of the previous approach. Such an approach involves the study of peculiar environmental phenomena and their chance of occurrence, such as, perception of floods or droughts. The preference approach incorporates the previous two approaches. The concept of preference grades the spatially differentiated objects on a 'scale of preference with relation to some specific behaviour objective'. All these approaches have been taken into account for perception research.

An example of the application of above approaches may be mentioned here. White's edited volume on 'Natural Hazards: Local, National, Global'' contains 32 articles primarily on geophysical hazards. In this introductory article, White puts forward five major hypothesis to study natural hazards. For the present study the following portion of the hypotheses is of greater interest to us :

Variation in hazard perception and estimation can be accounted for by a combination of the following :

1. Magnitude and frequency of the hazard ;
2. Recency and frequency of personal experience, with intermediate frequency generating greatest variation in hazard interpetation and expectation ;
3. Importance of the hazard to income or locational interest ;
4. Personality factors such as risk-taking propensity, fate control and views of nature.

For individuals, the choice of adjustment is a function of :

1. Perception of the hazard ;
2. Perception of the choice open to them ;
3. Their command of technology ;
4. The relative economic efficiency of the alternatives ;
5. The perceived linkages with other people (White, 1972, 1974).

Keeping in view the above factors, the perception of hazards was studied as a part of hazard research under the aegis of Commission on Man and Environment of the International Geographical Union (established in 1968). The approach to the actual study of perception, based on the above hypothesis, deviated with the progress in field work. For example, in the same volume by White, there is diversity of interest among the authors of six articles regarding the following components of perception of flood hazard : (a) chance of occurrence (frequency) ; (b) information system ; (c) damages (crops and households) ; (d) identification of flood zone according to the intensity ; (e) adjustments ; (f) educational and socio-economic class of the respondents ; (g) relief measures ; (h) idea of fate, God and insurance ; (i) comparison with other hazards ; and (j) embarkment, bridges and other government development projects (Nag, 1983).

The impact of behavioural sciences on geography has resulted in an attempt to understand the significance of the individual's experience of the world. Lowenthal has claimed that "personal geographies" form part of the common world view."

"The whole structure of the shared picture of the world is relevant to the life of every participant ; and any one who adheres to a consensus must personally have acquired some of its constituent elements' (Lowenthal, 1961). The introduction of the geography has lead to more anthropocentric human geography. The behaviour approach to human geography is more concerned with the study of men and not objects. Much of this recent research is subsumed under the rubric of environmental perception (English and Mayfield, 1972).

The perception research requires a con-disciplinary approach in a cross-cultural perspective. Researches in this field have shown an inclination towards psychological, behavioural, quantitative and fieldwork oriented techniques. Some of these techniques are discussed here in detail. However, perception research is still in its infancy and a sound methodology is yet to be cultivated. Anne V.T. White in her Technical Note, Guidelines for the Field Studies in Environmental Perception, has tried to bring together methods and techniques, and has evaluated them for use in the field in widely differing environments and cultural settings (Whyte, 1977). Other attempts have also been made to develop a research methodology to study diverse problems. These attempts are piecemeal and represent individual efforts in this field.

## Psychological components of perception research

The Oxford University Dictionary (1961) explains "perception" as the process of taking cognizance of a sensible or 'quasi-sensible object' or as the 'intuitive recognition of a moral or aesthetic quality'. Similarly in psychology, it has been defined as the 'process of becoming immediately aware of something', and is the stage prior to recognition or identification or apperception of an object (Drever, 1966). Apperception includes not only perceptual thinking but also imaginative thinking. "An image may be defined as our apprehension of an object or objects in the absence of the object or objects themselves which originally determined our sense-perception (Ross, 1972). The image is considered to possess designative, appraisive and prescriptive components (Pocock, 1978). In geography the terms perception and imagination are used in a similar sense. Lowenthal in his article explores the role of 'imagination' in 'personal geographies' (Lowenthal, 1961). However, geographers and psychologists have shown an increasing interest in environmental images and their various connotations, such as mental images, mental maps, mental models, cognitive maps, spatial impress, space preference, spatial scheme, perceived environment or perceptual maps. The aim of the Research Group on Environmental Perception of the Psychological Institute of the University of Bonn concerns with the relationship between the "tacit dimension" and explicit knowledge." Specifically they are interested in how visual perception is opt to deceive us about what can only be known by means of linguistically mediated reasoning and how it can be used to reveal what is hidden behind or has escaped attention within the common use of language which may not always be well understood (IGU Working Group News Letter, 1982).

In psychological studies perception is treated more in the physical sense of the term, with a tinge of neurology. Such studies pertain to the continuous functioning of sense organs; association of sense impulses with the central nervous system; or receptiveness of sense organs. The activity of the various sense organs in response to stimulation from the environment is called sensation (Crow, 1972). Though perception is explained as 'sensation with fringe meaning (Ross, 1972) or simply meaningful sensation, psychologist place emphasis on sensory perception. Brightness and colour contrast, optical distortion, moon illusion, subjective horizon, slope perception, visual and induced illusion and the like have occupied a significant place in the perception research in psychology. The trend towards perception research in psychology and geography seems to be paradoxical. The psychologists are more inclined to lay stress on physical or geographical components while geographers have realised the significance of imagery components in perception research. This does not signify that research trends in both the disciplines are not at all complementary. Similar to geographers, successful attempts have been made by psychologists to prepare perceptual maps of towns based on the report of the inhabitants and showing the supposed relationships of the main roads and landmarks (Halan E. Ross, 1971). Perception of the urban environment has been fascinating for researchers in this field, probably owing to considerable variation in a relatively small spatial unit. Lynch's work, based on psychological concepts, is one of the earliest examples. Mental images of the inhabitants of Boston, New Jersy and Los

Angles were traced, comprising of sequential elements such as nodes and paths, spatial elements such as districts, edges and landmarks (Lynch, 1960 and 1976). Appleyard elaborated the idea of mental image of the urban environment by 'comparing field surveys of form, visibility, use and significance patterns with the inhabitants' maps, as it was apparent that they were using each of these attributes to structure the city (Appleyard, 1970). Similarly, Strokes experiments with 143 students' images of the city of Hamilton and summarises the variations in the images with respect to districts, routes and landmarks (Strokes, 1976). However, such researches are not limited to the direct sensory experiences of the urban environment but researchers in this field have found an additional vista : the 'mental maps of areas as perceived through cognitive processing of personal experience (Whyte, 1977)' Sanders and Porter's comparison of the students of Tanzania and United States regarding the outline map of Africa (Sanders & Porter, 1974) ; or Gould's study of the perception of residential desirability in the United States of the students of State University of California (Berkeley), Minnesota, Pennsylvania and Allabama (Gould, 1966) ; or Christopher's investigation of settlers' perception of Southern African environment through promoters, official literature, other settlers, travellers and natural historians are few examples of the perception through cognitive processing (Christopher, 1973).

Pocock and Hudson have rightly mentioned that 'there has been a feedback from geographers studying environmental images to psychologists, reawakening interest in relationships between people and non-laboratory environment' (Pocock and Hudson, 1978). The inclination towards more humane components in geographical research and environmental components in psychological research is an example of cross-disciplinary approach in perception research. Both the approaches have contributed to a sub-discipline called "Environmental Psychology". Yi-Fu Tuan regards this sub-discipline as a field of study which contributes to the general theory of learning in its explicit recognition of the role of the physical environment and in its concern with the problems of planning and design (Yi-Fu, 1972, 1977).

**Behavioural components of perception research**

The behavioural revolution in social science has had a considerable influence on the discipline of geography. However, geographers have shown more concern with the spatial pattern of behaviour rather than the behavioural process (Davis, 1972). Spatial behaviour is understood to be a product of personal views, experiences and preferences. English and Mayfield have discussed the objectives of spatial behaviour research. According to them, the objective is to discover those shared characteristics which provide clues in developing methodological and theoretical insights through analysis of the perception-evaluation, learning and response of individuals to their environment (English and Mayfield, 1972).

Downs has emphasised the need for a more anthropocentric view of spatial human behaviour. Previously, man was treated as black box, that is 'an unknown constant in the study of the environment/behaviour relationship' (Downs, 1970). The behavioural revolution has rightly placed man in the environment-spatial behaviour equation. Hence, the black box is replaced by white box. Erickson has further elaborated the idea of black box and analysed the relationship between environment and behaviour. He mentions two basic and conflicting models of man : behaviourist view and phenomenologist's view, based on outside stimuli (stimulus-response framework) and inside consciousness of man's action, respectively. However, these approaches have been discussed earlier. Besides the stimulus response theory and cognitive theory there is another intermediary approach which may be referred to as centralist approach or cognitive behavioural approach (Bluden, 1978). However, in practice, this approach lays more emphasis on cognitive process 'in order to better understand and predict patterns of human behaviour'. Erickson remarks that the behavioural geography should more appropriately be called cognitive geography and to call the psychological revolution in geography 'behavioural' is to completely distort the revolution's principles and practice. On the other hand, like many other researches Gould attempts to place environmental perception and spatial behaviour under the rubric of behavioural geography, which includes cognitive as well as psychological components (Gould, 1969).

The individual preference and choices compose an environment on which decisions are based. This environment may be subsumed as behavioural environment, a subset of perceptual environment. Sonnenfeld defines it as 'that of the environment of which the individual is aware which also elicits a behavioural response or towards which a behaviour is directed, such as results in a conscious utilization or transformation of environment (Sonnenfeld, 1978). In brief, the individual perception and preference contributes to the decision-making process in the behavioural environment. Further, 'choice and behaviour, together with decision-making and policy, represent the later stages of the perception process which lies between cognition and action' (Whyte, 1977). The behavioural model (IPDE) for accident prevention (e.g. automobile etc.) can also be applied for perception research The four components of the model are identification of the relevant cues (I), prediction of their consequences (P), decision upon a course of action (D) and execution of the decision (E). These components are similar to perception, evaluation or assessment, decision-making, and action (Thygerson, 1977).

An attempt has been made by Fine to relate risk scores with decision and action. He provides a method to quantify risk score (R) for a hazard situation comprising of the following factors : potential consequences (C), exposure (E), and probability of occurrence (P). The arbitiary risk score is the product of the above-mentioned three factors or $R=C\times E\times P$ (Fine, 1977). This technique represents a useful method for identifying major hazards before an accident occurs. Casual factors with high severity but very low frequency can be detected quickly by these methods, whereas a large data base and a long time period would be necessary to detect them statistically (Thygerson, 1977). Cost justification (J) of a hazard can also be calculated by incorporating the above formula, which is $J=R/CF\times DC$. With the help of risk score (R), cost factor (CF in dollars) and degree of correction (DC= estimate of the degree to which the proposed corrective action will eliminate or alleviate a hazard) the estimate cost can be justified. In addition to the formulae, an action sheet related to the risk score was prepared by Fine.

On the other hand, researchers have preferred to study environmental behaviour which has two aspects : the study of behaviour itself ; and the study of the consequences of the behaviour (Knowles and Wareing, 1976). With the introduction of the perception research the focus of the study has shifted from post-behaviour consequences to pre-behaviour consequences. This takes into account cognition, risk assessment and finally decisions which result in behaviour. The decisions are taken on the basis of perceived space or environment and operated on behavioural space. Hence, decision-making is a very significant process in between perception and behaviour. This also indicates that man acts on the space which he perceives rather than the objective space, contrary to the economic theories related with the myth of economic man. The theories of industrial, agricultural or settlement location of Weber, Von Thünen and Christraller, therefore, do not take into account the related behavioural components. However, recently behavioural components have accounted for a more viable explanation of the interaction of factors of the environment. For example, Harvey has considered the role of non-economic factors, such as psychological elements, in examining the landuse patterns (Harvey, 1976 and Neumann and Morgenstern, 1944). Further, the perception of push-pull factors at both the ends of migration have an important role. There has been an attempt in migration research to quantify the perception cost associated with the push-pull factors. The push-pull model is proposed to understand the 'uniqueness of human spatial behaviour resulting from stimuli perception and judgment (Fuller, 1978). Moreover, other social scientists have stressed the significance of perception in migration and spatial behaviour. The more interesting case study is how decisions are made in uncertain conditions. Game theory has been taken into consideration in simulated conditions to examine the human behaviour. Competitive strategies are formulated in a pay-off matrix under various possible combinations of controllable and non-controllable variables. The decisions are taken for minimizing risk and for maximising profit. The game of strategy as a defensive measure against the unknown action or behaviour of a person, firm, country or nature can be applied in cards, matching coins, military strategy, landlord-tenant problems in rented farms, producing and marketing cattle-feeds, the gamble of dry land agriculture and the time. Parthasarthy has utilized game theory for securing assured agricultural income in drought prone areas in pure and mixed strategies (Parthasarthy, 1978). Gould paired strategies ; in a 'two-person-five strategy-zero-sum game' to investigate the farmers' choice of crop(s) for maximum pay-off

(Gould, 1963).

The agricultural activities are not controlled by large scale resource managers in developing countries. The poor and illiterate farmers have their own experience and estimation of the nature which accounts for decision-making in agricultural practice. During the IGU Workshop on Perception of the Environment (1978), attention was drawn upon farmers' knowledge of the environment, pests and pesticides, rural development and risk assessment. Research methods and their application regarding community environmental knowledge were discussed in detail. Ethno-science or folk science waş thought to be outcome of rich cognitive systems of symbolic and practical value. The Repertory grid method questionnaire method or interview are employed to study the cognitive process involved in the knowledge of the environment. Such experiments have been conducted in West and East Africa.

In the real world situation, the farmers do not have total knowledge or perfect ability. Behavioural theory has been developed to explain the deviations from the theoretical norms in the real world and to take account of human motivation and behaviour (Knowels and Wareing, 1978). Researchers have laid much stress on the variation of perception of technical and scientific personnel on the one hand and the resource user on the other (Burton and Kates, 1969). The theories in social sciences can be improved by taking into account the behavioural components involved. The normative theories of economists must take into consideration such concepts as learning theory and the models of behavioural scientists (Henshall, 1970).

**Issues and challenges**

During March 1973, Man and Biosphere (MAB) panel felt that a comprehensive interpretation of the concept of perception was necessary in order to establish how man understands the environment he lives in and the alternative choices in environmental management available to him. The MAB Project 13 is aimed to stimulate research on "Perception of the Environmental Quality" under the chairmanship of Professor Ian Burton, Institute for Environmental Studies, University of Toronto, Canada. The stress was on man/biosphere relationships (including the complex interrelationships of eco-systems) instead of man-environment interaction.

The MAB Panel expressed its concern over the man's increasing capacity to modify and alter the biosphere which may either lead to desirable or undesirable results. It was realised that there is serious gap in knowledge which needs qualitative and quantitative understanding and clarification of discrepancies between the expert and the user perception of the environment. The expert panel assigned high priority to the following research projects :

(a) Perception of typical man-made landscape of historical, aesthetic and ecological importance.
(b) Perception of national parks and other relative unmodified natural environments.
(c) Perception of the environment in isolated or peripheral ecological areas.
(d) Perception of quality in urban environments.
(e) The development of systems models for comprehensive environmental planning and assessment.

The first four projects recount to the study of specific or unique environments which can be placed in technology-environment or technology-biosphere equations. The fifth project explores the accepted model of technology and tests 'alternative environmental quality policy decisions and thus probable wide range impacts' (MAB Panel on the Perception of Environment Quality, 1973).

Under the same chairmanship, International Geographical Union (IGU) Council at its Moscow meeting in August 1977 decided to establish a similar working group entitled "Perception of the Environment". This working group, in collaboration with UNESCO-MAB, however, emphasised on (a) development or research designs for studies of perceptions of highly valued landscape and environmental hazards; and (b) development of suggestions for research methods and instructions.

The combined MAB/IGU Working Group held a preparatory workshop in Ibadan along with the African Regional Conference of the International Geographical Union, Nigeria. International group of scholars were involved to contribute and share their experiences in perception research and a report of the workshop was produced to inform a large circle of interested persons (IGU,

MAB Working Group, 1978). The working group identified three main areas in which it would like to develop and carry out collaborative research :

(i) Perception of highly valued or unique landscapes.
(ii) The management of pests and pesticides.
(iii) The interrelationship of environmental hazards and development including (a) studies and communications links for hazards research with either a hazard focus or a development process or project focus : (b) farmers' and decisions-marks perceptions of climate variability and climatic changes ; and (c) perceptions of soil erosion and soil degradation or loss.

For the perception of highly valued landscapes a working definition was worked out, however, the term 'value' lacks conceptual clarity. 'A specific part of an ideal type of environment which is regarded as attractive' was termed as highly valued landscape. This definition can be interpreted in terms of : (a) aesthetic attributes of the landscapes ; (b) recreational possibilities ; (c) desirability for residential purposes ; (d) personal childhood reminiscences ; (e) economic or monetary value. Apart from attractiveness, behavioural drives were considered to be significant and a rank-order was prepared accordingly. Based on this rank order, values attached to landscapes could also be ranked. However, such values are 'from the agreement of a person's perception with socially and culturally derived value systems'. Fourteen topics were listed for future research in this field. Moreover, it was felt that by a series of case histories it would be possible to demonstrate how the varying cultural and social importance attached to landscape has an impact on spatial or locational decisions, keeping in view the cross-cultural components (Lowenthal, 1978).

The interesting study of pests and pesticides hazards reveals how the environmental hazards created by technological hazard, the pesticides. In addition, pesticide application is a fine example of a "hazard chain, that is, where efforts to control one environmental nuisance (pests) lead to another environmental nuisance (illness genetic damage to crops, increased genetic resistance among insects and steady shift towards monoculture vulnerable to other pests and climatic variations)". On the other hand, there cannot be universally accepted pesticides. The research proposal in this field aims to consider (a) how companies set standards for risk protection when preparing their product, (b) what scrutinizing mechanisms are available to see that these products are acceptably safe, (c) the extent to which farmers actually obey the instructions for use printed on labels or communicated to them, and (d) what improvements can be made in each of these management areas to ensure that the pesticides induced hazard is minimised. Perception gap between farmers and agricultural advisers was also realised here. Sufficient scope for-stock of indigenous knowledge or techniques for pest prevention practices was felt important. The forthcoming introductory seminar is expected to review literature on the perception of pests prevention techniques and evaluation of research methodologies in this field. Environmental Perception Research Group has brought a working paper for providing guidelines for research in this field (Tait, 1981).

The third sub-group on environmental hazards and development appears to be a step forward to the two preceding subgroups. As indicated earlier, sufficient work has been done on perception of hazards. However, a sound methodology is to be evolved for research in this field. It was felt that communication links between the participants in this sub-group will be helpful in this regard. The focus of this was on the multiplicity of hazards in relation to development processes.

The other topics discussed in were the perception of current climatic changes and variability and perception of soil loss. Both the topics are related to agriculture. It is interesting to know how the process of agricultural decision-making is effected with the better information of climatic variability. For example, how the information about the climate, collected by MONEX-79, influence the perception of resource managers or policy makers for agricultural prices, on one hand ; and relatively less literate Indian farmers, on the other, against the natural factors. The soil erosion is owing to human agencies, such as, agricultural practices, irrigation, overgrazing, desertification, deforestation, use of fertilizers, urbanization and the like. The perception of the soil loss varies according to the resource managers operating at different levels. For an average Indian farmer, having operational holding below 5 acres, the soil erosion does not matter much. But the resource managers at micro-level perceive the continuous soil erosion. In relatively under-populated country like Zambia the bush fire ultimately leads to the loss of the soil. 'Soil loss thus presents a hazard which may be imperceptible to those closest to it and often involves many individual managers none of which is able to respond effectively'. The other issues for research in environmental perception are :

(a) Post-disaster surveys
(b) Response to disasters
(c) Environmental perception and planning
(d) Perception of forest scenic quality
(e) Public perception of the desert
(f) Everyday environmental cognition
(g) Perception of air quality (Dworkin and Pijawka, 1981).

**Environmental perception research in India**

Of late, this field of research has attracted some Indian geographers. However, a few attempts were made earlier for the study of human behaviour in space. Recently, a large number of studies made related to the perception of hazards. Kayastha and his team of research workers were especially interested in flood hazards in Lower Ghagra Plain (Kayastha and Yadav, 1978, 1980). Perception of pollution has been another major field of interest. Karan's study on the Chotanagpur industrial area is one of the earliest in this field (Karan, 1977). As mentioned earlier, the perception of the pests and pesticides have been identified as a thrust area in the perception research. The study of environmental pollution in Bhopal city raised some additional issues for consideration, i.e., safety measures not only for the use of pesticides, but also for the production and storage of pesticides and its ingredients (Nag). However, the evacuation of the city more than once and resettlement of the population thereafter could have been another important field for perception research. Similar studies on perception and evacuation have been made elsewhere during natural disasters, such as earthquakes or floods.

Kumra has made a study of the perception of air pollution in Kanpur city (Kumra, 1981 and Kumra and Singh, 1983).

Kayastha and Ram Bilas have studied another vital problem and its perception in Varanasi city, i.e., river pollution along the ghats of River Ganga (Kayastha and Ram Bilas, 1983). Further down the River Ganga, Nag made an assessment not of pollution, but of a technological hazard in Calcutta (Nag, 1980). It appears that urban centres, similar to the western experiences, have been the focus of attraction for perception research. The studies of Desai. Saha and others are also urban-oriented (Desai, 1979, 1980 and Saha, 1982).

Over seventy per cent of the country's population lives in rural areas which have specific environmental problems. On the other hand, rural scene is subjected to change at a very faster rate. Hence, the perception of floods, droughts, soil erosion and the like is equally significant as the perception of development hazards or shocks or even rapid urbanization. Perception or region-specific problems may lead to greater adjustments in man-environment relationship. Such changing relationships are likely to be perceived at different levels by the planners. administrators or resource managers on the one hand, and the farmers or rural population on the other.

India is a developing country where the scarce resources are not available in abundance for environmental protection. The funds are diverted for the technological advancements and their applications. As a result, the technological hazards, such as in the case of Bhopal tragedy, will have a greater impact in terms of damages of all sorts. Natural hazards are being replaced by technological hazards. Every day a new section of population and additional land is exposed to risks. With the advancing technology, even the most suitable environment for human settlement is not free of hazards. There is a great threat to the rich civilization of the country with the inclusion of land, settlement and population in the advancing risk zones (Nag, Saha and Prasad, ......). The perception of such risk in the mind of the people should be a matter of great concern in a democratic set up. Researches in this direction is of a vital importance. Sincere efforts should be made in order to find out the national priorities and to make national policy for the management of the environment in tune with the perception of the people of the country.

## REFERENCES

Appleyard, D. (1970), Styles and methods of structuring a city, *Environment and Behaviour*, 2, 100-17. Also published in Blunden, John, P. Haggett C. Hammett and P. Sarre (eds) (1978), *Fundamentals of Human Geography*, Harper and Row, London, New York, Hguation, San Francisco and Sydney, 7-16.

Brookfield, H.C. (1969), On the environment as perceived, *Progress in Geography*, Edward Arnold, 1, 53-80.

Burton, Ian and Kates, Robert W. (1964). The Perception of natural hazards in resource management, *Natural Resource Journal*, 3 (412), 412-41.

Christopher, A.S. (1977), "Environmental Pollution" in Southern Africa, *The Southern African Geographical Journal*, 65 (1), 14-22.

Crow, L.D. and A.C. Crow (1973), *Educational Psychology*, Eurasia Publishing House, New Delhi.

Davis, Wayne K.D. (1972), Geography and behaviour, in Davis, Wayne K.D. (ed.), *The Conceptual Revolution in Geography*, University of London Press Ltd., 331-40 (editorial comments).

Desai, Anjana (1979), The choice of residential space : As perceived by the students of the Gujarat University, *Transactions of the Institute of Indian Geographers*.

Desai, Anjana (1980), The environmental perception of an urban landscape : the case of Ahmedabad, *Ekistics*, 47 (283), 279-85.

Downs, Roger M. (1970), Geography space perception : Past approaches in future prospects, *Progress in Geography*, Edward Arnold, 2, 67-108.

Drevir, James (1966), *A Dictionary of Psychology*, Penguine Reference Books.

Dworkin, J.M. and K.D. Pijawka (1981), *Air Quality and Perception : Explaining Change in Toronto, Ontario*, Environmental Perception Research, Working Paper 9, Institute for Environmental Studies, University of Toronto.

Erickson, Neil J. (1967), Behavioural Geography : A bulb for the black box, *New Zealand Journal of Geography*, 61, 19-29.

Erickson, N.J. (ed.) (1980), *Environmental Perception and Planning in New Zealand*, Environmental Studies Unit, University of Waikato, Hamilton,

Fine, William T. (1971), Mathematical evaluation for controlling hazards, *Journal of Safety Research*, 157-66.

Fisher, G.H. (1971), The Perceptional Map Newcastle upon Tyne, Department of Psychology, University of Newcastle upon Tyne.

Fuller, Grey (ed,) (1978), A Curriculum Guide for Population Geography, East-West Population Institute, Honolulu. 55-7.

Gould, P.R. (1963), Man against his environment : A game theoretical framework, *Annals Association of the American Geographers*, 53, 290-7.

Gould, Peter R. (1966), *On Mental Maps*, Michigan Inter-University Community Mathematical Geographers, Discussion Paper 9, Also Published in English and Mayfield (eds.), *op. cit.*, 260-82.

Gould, Peter, R. (1969), Methodological developments since the fifties, *Progress in Geography*, Edward Arnold, 1, 2-49.

Harvey, D. (1976), Theoretical concepts and the analysis of agricultural land-use pattern in geography, *Annals Association of the American Geographers*, 56, 361-74.

Harvey, David (1979), *Explanation in Geography*, Edward Arnold, 18-23.

Henshall, Janet D. (1970), Models of agricultural activity, in Chorley, R.T. and P. Haggett (eds.), *Socio-Economic Models in Geography*, London, 425-58.

IGU/MAB Working Group (1978), *Report of the IGU/MAB Workshop on Perception of the Environment*, Working Paper 5, Ibadan.

Karan, P.P. (1977), Perception of environmental pollution in Chotanagpur Industrial Area, India, *National Geographer*, 12 (2), 17-24.

Kayastha, S.L. (1980), Environmental perception and problems, *National Geographical Journal of India*, (Special issue), 26.

Kayastha, S.L. and R P. Yadav (1977), Human perception and adjustment to Environmental Hazards : A case study of three villages of Ganga Flood Plains, in Robert C. Eidt. *et al.* (eds.), *Man, Culture and Settlement*, National Geographical Society of India, Varanasi.

Kayastha, S.L. and Ram Bilas (1983), Public awareness towards river pollution along the ghats of River Ganga at Varanasi: A case study of environmental assessment and perception, paper presented in the National Conference on River Pollution and Human Health, Varanasi.

Kayastha, S.L. and R.P. Yadav (1978), Flood hazard in Lower Ghagra Plain (U.P.) India : A study in perception and impact on socio-economic development, IGU/MAB Workshop on Perception of the Environmental Quality, Ibadan (Nigeria).

Kayastha, S.L. and R.P. Yadav (1980), Aspects of flood hazards in the Middle Ganga Plain : A case study of selected settlements in Eastern Uttar Pradesh, Paper presented in the Symposium of the IGU Working Group on Perception of the Environment, Sapporo (Japan).

Kirk, W. (1975), Historical geography and the concept of the behavioural environment, *Indian Geographical Journal*, 25, 152-60.

Knowles R. and T.J. Wareing (1976), *Economic and Social Geography : Made Simple*, W.H. Allen, London, 12-24 (Chapter on Environmental Perception and Behaviour).

Kumra, V.K. (1981), Kanpur City : A Study in Environmental Pollution, Tara Book Agency, Varanasi.

Kumra, V.K. and R. Singh (1979), Perception of air pollution : 'A case study of Kanpur City, *Indian Journal of Air Pollution Control*, 2 (1), 50-4.

Lowenthal, David (1961), Geography, experience and imagination : Towards a geography of Epistemology, *Annals Association of the American Geographers*, 51, 241-61.

Lowenthal, David, (1978), Finding valued landscapes, *Progress in Human Geography*, Edward Arnold, 2 (3), 373-418.

Lynch, K. (1960), *The Image of the City*, Cambridge, Massachusetts and Harvard University Press.

Lynch, K. (1976), Managing the sense of a Region Cambridge, MIT Press.

MAB Panel on the Perception of Environmental Quality (1973), *Nature and*

*Resources*, UNESCO, VIX (2), 11-3.

Nag, P. (1980), Perception of technological hazards in an urban environment : The case of power crisis in Calcutta (India), *National Geographical Journal of India*, XXVI (Pts. 3 and 4), 174-84.

Nag, P. (1983), Perception of the environment : Issues and challenges, *National Geographical Journal of India*, XXIX (Pts. 3 and 4), 217-31.

Nag, P., Bhopal aftermath : Some more issues for consideration, paper submitted the *Geographical Review of India*, Calcutta for publication.

Nag, P., D. Saha and H. Prasad, Mapping of environmental risk zones paper accepted for publication in the proceedings of the International Seminar on Environmental Maps and Atlases. National Atlas and Thematic Mapping Organisation, Monograph 4, Calcutta (in Press).

Neumann, J. Von and O. Morgunstern (1944), *Theory of Games and Economic* Behaviour, Princeton.

Norton, William (1984), Historical Analysis in Geography, Longman, London, New York, 46-7 and 62-86.

Paasi, Anssi (1984), Connections between J.G. Grano's Geographical thinking and behavioural and humanistic geography, Fennia, Geographical Society of Finland, 162 : 1, 21-31.

Parthasarathy, R. (1978), Game theory model for securing assured income in drought prone areas in N.B.K. Reddy (ed.), *Drought Prone Areas of India*, Proceedings of the All India Symposium, Rayalaseema Geographical Society, Sri Venkateswara University, Tirupati, 457-66.

Pocock, Douglas and Ray Hudson (1978), Image of the Urban Environment, Focal Problems in Geography Series, The MacMillan Press Ltd.

Ross, Halan E. (1974), *Behaviour and Perception in Strange Environment*, Advancement in Psychology 5, George Allen and Unwin Ltd., London.

Ross, James S. (1972), *Groundwork of Educational Psychology*, Oxford University Press.

Saha, Durgadas (1982), Process of Mental Mapping among Cartographers in S.P. Das Gupta (ed.), *Essays in Thematic Cartography*, National Atlas and Thematic Mapping Organisation, Monograph 2, 38-43.

Sanders, R.A. and Porter (1974), Shape in revealed mental maps. *Annals Association of American Geographers*, 64, 258-67.

Sonnenfield, J. (1978), Geography, perception and the behavioural environment, in Bluden *et al.* (eds.), *op. cit.*, 26-32.

Strokes, Evelyn (1976), Student image of Hamilton, *New Zealand Journal of Geography*, 16, 30-36.

Tait, E.J. (1981), *Perception and Management of Pests and Pesticides—Guidelines for Research, Environmental Perception* Research, Workers' Paper 8, Institute for Environmental Studies, University of Toronto.

Thyguson, Aiton L. (1977), *Accidents and Disasters : Causes and Counter Measures*, Prentice Hall, Englehood Cliffs, New Jersy, 67-77.

Wanger, Phillip L. (1972), Cultural landscape and regions : Aspects of communication in English, Paul Ward and Robert C. Mayfield (eds.) *Man, Space and Environment*, Oxford University Press, 282-304.

White, Gilbert F. (1972), Collaboration in natural hazard research, (Geographical Record), *Geographical Review*, 62 (2), 280-1.

White, G.F. (ed.) (1974), *Natural Hazards : Local, National, Global*, Oxford University, Press.

Whyte, Anne V.T. (1977), *Guidelines for Field Studies in Environmental Notes* 5, UNESCO.

Wright, J.K., (1947), Terra incognitae : the place of imagination in geography, *Annals Association of the American Geographers*, 37, 1-15.

Yi-Fu, Tuan (1972), Environment psychology, *Geographical Review*, 62 (2), 245-56.

Yi-Fu, Tuan (1977), *Space and Place : The Perspective of Experience*, University of Minnesota, Minneapolis.

C. M. Jariwala

# FIVE

# ENVIRONMENTAL LAW

THE environmental pollution has become a cancerous disease and the modern technological developments are further posing a great threat to the very existence of living and non-living organisms. This disease is not confined to a particular region, nation but it has crossed transnational frontiers and reached even in the space. Here comes the role of the law—national and international—to control and regulate the development process to preserve the living environment. The technological developments are at a supersonic speed and the law, though not moves with that speed, has made some efforts to control and prevent the environmental pollution. Following are some of the achievements in that regard.

## Transnational law

It is said that the starting point of International Environmental Law is the Stockholm Conference. It was in the year 1972 that 113 members of the world community gathered together under the auspices of the United Nations at Stockolm to deliberate upon the issues of protection of human environment. The conference realised that environmental pollution was a major issue affecting the life of world population and the developmental process throughout the world. The conference resolved that it was the duty of all governments to preserve and improve the human environment for the benefit of all people and their prosperity.

The world community in order to attain the clean environment, made many attempts to arrive at some consensus to control certain pollutions. To name a few : the 1973 Brussels conference on marine pollution concentrated on the measures to prevent pollution of sea from ships. The United Nations conferences on the law of sea 1958, 1973 and 1981 were some of the great attempts where the member states focused their attention on subjects, *inter alia* pollution of sea water and suggested a co-operative approach to prevent sea pollution. The Habitat Conference in Vancouver, British Columbia, 1976 and the World Water Conference of Argentina, 1977 are some of the other attempts to regulate water pollution. Similar attempts were made to focus the attention of the world community on the problems of pollution of Antarctica, outer space, and pollution through use of nuclear weapon. In all these matters the progress to control environmental pollution is at a tortoise speed whereas the developmental process is moving with a supersonic speed. In this race the International environmental law is still in an embryonic stage and it has yet to equip itself to cope up with the fast degradation of the international environment.

The major problem, in controlling environmental pollution, is inequality amongst the world governments developed, developing and underdeveloped. The first type of States either do not want or are incapable of controlling environmental pollution. It is the developing States which are trying to a certain extent balance the development process with the protection of environment. On the other hand, the underdeveloped States are unable to meet the cost of controlling environmental pollution. Under these circumstances, to have a common approach of the world governments to protect earth-space environment is, if not impossible, a difficult task before the world community. However, whatever efforts are made by the international communities are commendable.

**Indian environmental law**

Now coming to the environmental law in India, generally the academics try to trace the history of environmental law from the British period. They point out some laws existing at that time controlling environmental pollution directly or indirectly. But this should not mean that the Indians were ignorant of the environmental pollution in ancient time.

Our *Dharma Sastras*, have rich material on this subject. For example, the trees were considered as the abode of God and a sacred place. The rivers were considered as pious and given the motherly status. Even the animals did not lag behind. Some of the animals were treated as the Chariots of Gods. The religious texts also tell us that in order to keep the environment clean, *Yagnas* were performed at the local and national levels.

Coming back to the discussion of the law controlling environmental pollution in India, the Constitution (Forty-second Amendment) Act, 1977 is a historic document in the world constitutions which introduced detail provisions for the protection of environment (Jariwala, 1980). On the one hand it imposes in Article 48A a duty on the state "to protect and improve the environment and to safeguard the forests and wild life of the country". On the other hand, Article 51A(g) imposes a duty on the citizen of India, "to protect and improve the natural environment including forests, lakes, rivers and wild life and to have compassion for living creature." There is no specific fundamental right in this regard but following the Supreme Court of India opinion in *Maneka Gandhi* v. *Union of India*, where the right to live and personal liberty guaranteed by Article 21 was considered a residuary right, the court might include the right to live in a clean environment as a fundamental right under article 21 of the Constitution of India (AIR, 1978).

The 42nd Amendment Act also made some changes in the legislative powers of parliament and the state legislatures. The subjects of 'forests' 'protection of wild animals and birds,' and 'population control' were transferred from the state subjects to concurrent subjects over which now parliament as well as the state legislatures get the power to pass a law. Apart from the above provisions, subjects directly or indirectly related to environmental pollution are also distributed between parliament and the state legislatures separately and also jointly. But, environmental pollution as such does not find any place in the distribution of legislative powers.

It was in the year 1860 that the Indian Penal Code for the first time made it an offence to corrupt or foul the water of any public spring or reservoir punishable with imprisonment extending to three months or with fine extending to rupees five hundred or both. In case of making atmosphere noxious to health the punishment was to the extent of rupees five hundred. It also made a punishable offence in case of rash driving, negligence in handling of poisons, explosive substances on machines or spreading infectious diseases dangerous to life. There were laws which regulated other aspect of environmental pollution which included, for example, protection of fisheries (1897), preservation of forests (1927), protection against traffic noise (1939), regulating the environment in factories (1948) and the like. In spite of these laws, the people were unconcerned about the environmental pollution, and the machineries were inactive to regulate environmental pollution. For these reasons in British India the legal control on environmental pollution was shelved in the cold storage.

The democratic Republic of India for two decades tried to regulate vast areas of environmental pollution including protection of wild birds and animals, fisheries, parks and open fields, river and sea water. Most of the regulations were at the state level and no uniform control was possible. It was on 23rd March, 1974, that a uniform law for the prevention and control of water pollution was passed (Act 6 of 1974). This law was enacted by the parliament against a joint mov of the States of Assam, Bihar, Gujarat, Haryana, Himachal Pradesh, Jammu and Kashmir, Karnataka, Kerala, Madhya Pradesh, Rajasthan, Tripura and West Bengal. This law enables to establish a board at the central and state levels for the 'prevention and control of water pollution and maintaining or restoring of wholesomeness of water.' Here water pollution has been defined as follows :

> Such contamination of water or such alteration of the physical, chemical or biological properties of water or such discharge of any sewage or trade effluent or of any other liquid, gaseous or solid substance into water (whether directly or indirectly) as may, or is likely to create a nuisance or render such water harmful or injurious to public health or safety, or to domestic, commercial, industrial, agricultural or other legitimate uses or to the life and health of animals or plants or of aquatic organisms ; the penalty for polluting water is imprisonment for a period extending from three months to seven years or fine which may extend to one thousand rupees or with both.

Another attempt of prevention and control of pollution was made on 29th March, 1981 when parliament, in order to give effect to the decision taken at the United Nations Conference on the Human Environment held in Stockholm in June, 1972, enacted the Air (Prevention and Control of Pollution) Act, 1981. For the prevention control and abatement of air pollution a provision was made for establishing boards at the central, state and joint levels. The function of the boards includes preservation of the quality of air and control of air pollution. It defines air pollution to mean "the presence in the atmosphere of any air pollutant" which means "any solid, liquid or gaseous substance present in the atmoshere in such concentration as may be or tend to be injurious to human beings or other living creatures or plants or property or environment." The penalties for polluting air is imprisonment for a term which may extend from three to six months or fine to the maximum of rupees ten thousand or both. Further, in case the offence continues then it shall be punishable with an additional fine which may extend to one hundred rupees for every day, during which such offence continues.

It was on the recommendations of the Tiwari Committee, a separate Union Department of Environment was set up in November 1980 to find out ways and means for the protection of environment. It is too early to evaluate the working of the department, but let us hope this department will act sincerely to check environmental pollution. Apart from the above penal laws, the law of torts makes provision from compensation in case of negligence, nuisance etc. Can compensation really compensate those who lost their lives in the polluted environment ? Will the Union Carbide compensate for the immediate and remote losses and the adverse effect on geography of Bhopal ?

There are water boards at the central and state levels to control and prevent water pollution. But the sad story is that today the majority of the rivers in India are so badly polluted that it is difficult to take a dip in the holy water. We can very well imagine the fate of all those who drink such water or use the water for irrigation or other purposes. These boards are also looking after the air pollution. Can the board retain the confidence of the people of India after the Bhopal tragedy ? The local authorities, instead of making the local environment clean, have been the important contributors to environmental pollution. The municipalities are allowing sewage to be dumped in the rivers. They are silently watching the increasing rate of water, air and noise pollutions.

Now coming to the judiciary, in the beginning it did not pay any serious attention to the problem of environmental pollution ; however, in the light of the legislative response the courts, particularly the Supreme Court of India, started entering in a field which at one time was considered as an exclusive area of technocrats. Justice Krishna Iyer in *Ratlam Municipality* v. *Vardhichand* set the ball rolling (AIR, 1980). In this case a locality within limits of Ratlam Municipality was stinking due to fifteen open drains and public excretion by nearby slum-dwellers. When this matter was moved to the Magistrate's Court the Municipality took the shelter of financial inability. The Supreme Court set aside the plea of financial inability on the ground that the Municipality with its primary duty to look after the public health, could not "run away from its primary duty by pleading financial inability," and that "the court will not sit idly by and allow municipal government to become a statutory mockery." The dynamism of the court required the municipality to provide "drainage systems not pompous and attractive, but in working condition and sufficient to meet the needs of the people." The consequences of allowing the plea of the municipality would, according to Justice Krishna Iyer, be that, the municipality "will legally defy its duty under the law by urging in self-defence a self-created bankruptcy or perverted expenditure budget". How far the *Vardhichand's* dynamism will open the eyes of other municipalities is difficult to say but it opens a new chapter in the Indian Environmental law.

**Future directions**

Today there are a number of laws passed by parliament and the State Legislatures regulating and preventing environmental pollution but unfortunately the said disease remains a disease ; if not incurable, most difficult to cure. The law provided for severe punishment and compensation but it has yet to effectively control the environmental pollution. Mere imposition of severe punishment or award of huge compensation cannot solve the problem. The law has to maintain a reasonable balance between the development process and clean environment. This can be done if positive efforts are made at all levels.

The major hurdles in the legal control are the ignorance of the people and state inaction. We have to educate the masses to arouse the public opinion against the assault on natural environment. This in turn will activise the state machineries which are responsible to maintain a clean environment. If we want the pious hope of the Directive Principles and the Stockholm Conference to maintain a clean environment a reality, a right to live in a clean environment must find a place in the fundamental right at the national and world levels. This will make the duties of the State and all those who pollute environment more meaningful and effective. But it must not be forgotten by those who claim the right that they have a fundamental duty to protect and improve the natural environment. It is pity that today we have forgotten that it is our *Dharma* to maintain a clean environment.

## REFERENCES

All India Reporter (AIR) (1978), *Supreme Court*, 597.

All India Reporter (AIR) (1980), *Supreme Court*, 1622.

Jariwàla, C.M. (1980), The Constitution 42nd Amendment Act and the environment, in *Legal Control of Environmental Pollution*, 1.

United Nations (1974), *Human Settlements : The Environmental Challenge*, A Compendium of United Nations Papers Prepared for the Stockholm Conference of Human Environment. 1972, Macmillion.

SECTION II

# ENVIRONMENT DEVELOPMENT AND PROBLEMS

ANIL K. SINGH
AND
J. SINGH

SIX

# DEFORESTATION IN UTTAR PRADESH : An Environmental Threat

THE herculean feats of economic gains and development have enhanced many times the severity and magnitude of deforestation. To be sure, the developmental activities of the day and in vogue with more modern and advanced technology without regard to man and his environment are proving ecologically destructive and there is no doubt that such developmental networks have presented various problems of environmental management. It has been really felt that modern developmental functioning has dual character of development and destruction. Marriot has rightly compared today's 'development' with 'over development' and according to him 'over development' is ". . . . the pressing of techniques upto and beyond the point of an optimum relation between man and his environment. In an over-developed area, too many techniques are too exhaustively applied by too many people to too little land" (Marriot, 1957). And certainly deforestation is one of the outcomes of various developmental activities having the character of extensional development. Thus, there is considerable truth in saying that deforestation is at havoc wreaked by thoughtless destructive activities of development (Singh, 1984).

Centuries back when the population was too small, needs were too little. Hence, the problems of environment were not so acute as nowadays. The history of forest invasion starts with the advent of man. In Uttar Pradesh (U.P.)—the study area, forest exploitation starts with the Aryan migration which took place sometimes around 1500 B.C. "These people were mainly agriculturists. Some of them were also engaged in pastoral pursuits. Forest clearings were made not only for laying out their villages but also for cultivation and pasturage. They cleared forests for the purpose of their settlements. As the population increased more and more forests were cleared for cultivation, habitation and pasture. They used wooden vehicles such as carts and chariots. They also constructed paths by clearing forests. The villages (grams) of Aryans consisted of thatched cottages. The doors, windows, post, rafters, ridges and bottoms were entirely made of wood which were obtained from the forests. Felling of forest was resorted too for fuel" (Dwivedi, 1980,). During later Vedic civilization and Epic age ; there were many large forest tracts spread over the central, northern, southern and western parts of U.P. along with comparatively poor vegetal cover of Eastern U.P. The position of vegetal cover remained more or less the same during the Gupta and Mughal periods. Gradually with the population expansion, forests began to shrink and it continued till the advent of the Europeans. Britishers have started clearing of forests at commercial basis leading to a major threat to the environment.

## Deforestation

The population explosion and consequent pressure upon natural resources have taken the toll in the northern hilly region of the state—"too much of the actual and proposed development of the Himalayan region is of this character . . . . Himalayas have been regarded as an inexhaustible reservoir of exploitable and expendable resources without regard to the future of the fragile mountain environment . . . ." (Berreman, 1979). In the Himalayan region, there are many causes and consequences of deforestation. Among various factors leading to deforestation are the activities directly related to development viz. the construction of roads, execution of various river valley projects, extension in agricultural lands, establishment of industrial centres, construction of buildings for various offices and holiday homes for tourists and many more miscellaneous constructions. Table 6.1 clearly depicts the magnitude of deforestation.

Forest fire is also a major problem related to the deforestation. Forest fires have many causes, "known and unknown."

TABLE 6.1 : **Purpose of deforestation and area accrued**

*(in ha)*

| Period | River Valley Projects | Agricultural purposes | Construction of roads | Establishment of industries | Miscellaneous purposes | Total |
|---|---|---|---|---|---|---|
| 1951-52 } 1972-73 | 93,200 | 79,700 | 3,400 | 19,300 | 19,900 | 2,15,500 |
| 1973-74 | 41 | 178 | 78 | 62 | 13 | 372 |
| 1974-75 | 02 | 3,895 | 333 | 41 | 186 | 4,457 |
| 1975-76 | 312 | 01 | 786 | — | 375 | 1,474 |
| 1976-77 | 8,732 | 289 | 145 | 23 | 854 | 10,043 |
| 1977-78 | — | — | 69 | — | 8,403 | 8,472 |
| 1978-79 | — | 07 | — | 3,999 | 429 | 4,435 |
| 1979-80 | — | — | 160 | — | 123 | 283 |

*Source* : Forest Statistics, Uttar Pradesh; Conservator of Forests, Research and Development Circle, Uttar Pradesh ; 1983, p. 31.

Forest fires play an adverse role in disturbing the vegetation which further leads to micro-environmental changes, as fires destroy the biotic states of areas under influence. Table 6.2 presents causes and role of forest fire in deforesting the forest

TABLE 6.2 : **Causes of forest fires and area burnt**

*(in ha)*

| Causes of fire | Area burnt during | | | |
|---|---|---|---|---|
| | 1978-79 | 1979-80 | 1980-81 | 1981-82 |
| 1. Fire originating in departmental fire conservancy operations | 4,886 | 1,003 | 4,990 | 31 |
| 2. External fires crossing the fire traces | 1,388 | 562 | 643 | 465 |
| 3. Fire due to carelessness of outsiders | 21,618 | 24,271 | 21,020 | 10,119 |
| 4. Fire originating from intention | | | | |
| (a) Of turning out games etc. | 283 | 238 | 1,739 | — |
| (b) Of malice | 239 | 134 | — | — |
| 5. Fire due to unknown causes | 14,975 | 9,455 | 36,025 | 2,164 |
| Total | 43,389 | 35,663 | 64,417 | 12,779 |

*Source* : Forest Statistics, Uttar Pradesh, 1983, p. 101.

areas. The government efforts, regarding the attempts made by them to protect forests from fires, is given in Table 6.3. From the Table it is evinced that the role of government is quite satisfactory in protecting the forests from fire.

TABLE 6.3 : **Area protected from forest fires**

| Area protected/burnt (ha) | 1978-79 | 1979-80 | 1980-81 | 1981-82 |
|---|---|---|---|---|
| Area attempted to be protected from fire | 37,88,296 | 36,63,180 | 38,77,311 | 38,43,797 |
| Area actually protected | 37,44,907 | 36,27,51' | 38,12,894 | 38,31,018 |
| Area burnt | 43,389 | 35,663 | 64,417 | 12,779 |

*Source* : Forest Statistics, Uttar Pradesh, 1983.

There are several other activities which leads deforestation. The commercial exploitation of forest resources by the contractors is an important causative factor of deforestation. "More commonly and rapidly destructive is commercial exploitation by outside contractors who devastate huge areas under government licence for quick profit. They cut forests for lumber, paper-pulp and charcoal. I have seen large forest tracts cleared, the great logs piled high only to be burned to charcoal" (Berreman, 1979). The extraction of major and minor forest products by various government and non-government agencies have also adversely affected the forest wealth in various ways. Such commercial exploitations are accelerating the quantum of deforestation. These activities are extinguishing various species in different scales. Among the highly affected species, especially by the extraction of timber are sal, shisham, eucalyptas, teak, asna, jamun and khair, etc. (non-coniferous) ; and chir, deodar, fiur and spruce, kail and cypress (coniferous). Table 6.4 sets out the actual picture of species-wise extraction of timber.

From the Table it is evident that among the coniferous species chir, deodar, fur and spruce have been highly valued for commercial exploitation, whereas in non-coniferous species sal and eucalyptus are worst affected by such exploitation.

There are various other usage of forest resources as well. Among these fuel is most important locally as well as regionally. The various agencies engaged in extraction of fuel from forests are—forest department itself and purchasers, right holders,

TABLE 6.4 : **Extraction of timber by species (cu.m. in round)**

| Species | 1978-79 | 1979-80 | 1980-81 | 1981-82 |
|---|---|---|---|---|
| **Non-coniferous** | 626,014 | 690,260 | 551,533 | 574,034 |
| Sal (Shorea robusta) | 223,337 | 251,342 | 184,257 | 193,669 |
| Teak (Tectona grandis) | 3,260 | 5,260 | 8,110 | 9,627 |
| Asna (Terminalia alata) | 22,265 | 13,892 | 9,202 | 9,187 |
| Shisham (Dalbergia sissoo) | 14,886 | 13,630 | 14,924 | 15,645 |
| Semal (Salmalia malabarica) | 4,706 | 5,228 | 8,313 | 2,558 |
| Gutel (Trewia nudiflora) | 2,411 | 1,290 | 476 | 578 |
| Haldu (Adina cordifolia) | 9,146 | 6,072 | 3,145 | 4,543 |
| Jamun (Syzygium cumini) | 11,045 | 9,215 | 4,862 | 5,897 |
| Tun (Cedrela toona) | 1,682 | 1,625 | 1,961 | 765 |
| Khair (Acacia catechu) | 24,812 | 8,642 | 5,349 | 3,534 |
| Kanju (Holoptelea integrifolia) | 5,738 | 3,353 | 2,842 | 2,167 |
| Siris (Albizzia procera) | 276 | 1,614 | 262 | 606 |
| Jhingan (Lannea coromondelica) | 2,705 | 4,182 | 1,801 | 1,742 |
| Kharpat (Garuga pinnata) | 756 | 1,029 | 530 | 457 |
| Sandan (Ougeinia oojeinensis) | 1,687 | 1,535 | 1,499 | 1,770 |
| Eucalyptas | 161,697 | 206,181 | 188,186 | 114,692 |
| Miscellaneous | 135,605 | 156,170 | 115,814 | 206,597 |
| **Coniferous** | 523,400 | 271,727 | 244,682 | 373,078 |
| Chir (Pinus roxburghii) | 413,654 | 240,652 | 204,297 | 337,253 |
| Deodar (Cedrus devodara) | 9,657 | 7,052 | 7,814 | 12,465 |
| Fir and Spruce (Abies pindrow and Picea Smithiana) | 92,869 | 13,008 | 22,026 | 14,155 |
| Kail (Pinus wallichiana) | 6,849 | 9,933 | 9,705 | 9,089 |
| Oak (Quercus incana) | 369 | 1,016 | 772 | — |
| Cypress (Cupressus torulosa) | 2 | 66 | 68 | 116 |
| Total | 1,149,414 | 961,987 | 776,215 | 947,112 |

*Source* : Forest Statistics, Uttar Pradesh, 1983, pp. 53-4.

concessionists and free grantees. The maximum quantity of fuel wood is being extracted by the purchasers followed by right holders and concessionists, free grantees and departmental. Table 6.5 reveals the structure of fuel wood exploitation by various agencies.

The statistics given in various tables are quite sufficient to tell the saga of speedy exploitation of forest resources. However, such statistics do not explain the holistic picture because there are other forms of exploitations also e.g. lopping, clearing, cutting

TABLE 6.5 : **Extraction of fuel wood by various agencies**

*(in cu. m.)*

| Agencies | 1978-79 | 1979-80 | 1980-81 | 1981-82 |
|---|---|---|---|---|
| Department | 92 | 195 | 582 | 6,149 |
| Purchaser | 1,598,400 | 1,613,492 | 1,099,525 | 1,074,942 |
| Right holder and concessionists | 717,205 | 711,747 | 748,215 | 661,167 |
| Free grantees | 39,577 | 33,669 | 13,445 | 7,548 |
| Total | 2,355,274 | 2,359,103 | 1,861,767 | 1,749,806 |

*Source* : Forest Statistics, Uttar Pradesh, 1983.

and extraction of various forest products by local inhabitants illegally, and it is quite difficult to make an estimation of such practices. At least a minimum requirement of one tree per annum per family either for construction or for the fuel can perhaps give us some idea. It is also a way of exploitation but is often ignored in government statistics. The average lopping rate by a family per year in the U.P. Himalaya is approximately 15 qts. which is expected to go up in near future. Grazing is also an acute problem and there is a considerable truth in saying that. "If we regard pasture and forest as the hill farmer's capital resources which yield a yearly interest in terms of grass, fodder, leaves and humus, then over-grazing and over-cutting of leaves means that the farmers are not only spending their entire annual interest but are also making in-roads into their non-replaceable capital resources. This spells disaster, if continued" (Ashish, 1981). There is an important aspect of grazing which affects indirectly forests through the reduction in regeneration as "cattle hoofs cut the soil on steep hill sides, leaving it to be washed away in the monsoon rains. . . cause immense damage to tree seedlings and so prevent proper regeneration after felling." (Ashish, 1981). Even the position of grazing land in terms of cattle pressure is not satisfactory. Tables 6.6 and 6.7 clarify the situation.

From the Table 6.6 it can be seen that the area under grazing records about 81 per cent and of which about 97 per cent area remains open throughout the year for grazing. In terms of animal population the pressure on grazing land is steadily increasing (Table 6.7). The above facts concerning with various kinds of

TABLE 6.6 : **Forest area under cattle grazing pressure**

*(in ha.)*

| Areas/Types | 1978-79 | 1979-80 | 1980-81 | 1981-82 |
|---|---|---|---|---|
| Total area under the control of forest department | 4,080,551 | 4,058,517 | 4,067,999 | 4,068,718 |
| Total area open for grazing | 3,229,260 | 3,273,451 | 3,245,819 | 3,303,780 |
| Area open for grazing for the whole year | 3,234,825 | 3,166,981 | 3,169,142 | 3,218,502 |
| Area open to grazing for part of the year | 94,435 | 106,470 | 76,677 | 85,278 |

*Source* : Forest Statistics, Uttar Pradesh, 1983.

TABLE 6.7 : **Animal population and pressure on grazing land**

| Animals | 1978-79 | 1979-80 | 1980-81 | 1981-82 |
|---|---|---|---|---|
| Cow and bullock | 1,046,142 | 1,010,192 | 717,670 | 886,861 |
| Buffalo | 243,600 | 216,065 | 224,025 | 210,031 |
| Goat and sheep | 662,042 | 507,663 | 611,386 | 602,587 |
| Camel and horse | 1,238 | 1,987 | 2,968 | 15,756 |
| Others | 39,763 | 21,607 | 11,730 | 30,802 |
| Total | 1,992,783 | 1,757,514 | 1,767,761 | 1,746,037 |
| Per ha. pressure of animal on grazing land | 1.67 | 1.86 | 1.83 | 1.89 |

pressures on forest wealth are explicit in order to explain the quantum of deforestation.

**Deforestation causing threat to environment**

Deforestation leads to not only deterioration in natural environment but also in cultural environment, as all the cultural features and conditions are dependent on the natural surroundings. The natural environment feeds cultural environment and there is a feedback as well. Forests are one of the basic geographical elements and any alteration in forest resources and wealth means alteration in the environment. However, the present study has a definite scope as it is concerned only with the certain major problems related with the deforestation.

The forest helps in shaping our natural environment in various ways as it influences micro-climate (by maintaining the temperature, humidity, wind, precipitation and radiation), soil environment (affecting soil composition, structure, soil-water and moisture, chemical properties. etc.), flood conditions (through the functionings of evaporation, sub-surface and siltation of river and lake beds, etc.) erosion (by the movement of water or wind), human health (by protecting physical and chemical contaminations, noise, etc.) etc. (Dwivedi, 1980).

The variation in temperature within and beyond the forest zones is an important aspect related with the micro-climatic conditions, as "forests exercise a moderating influence on air temperature. The influence is the greatest in a dense forest with heavy foliage" (Dwivedi, 1980). In a simplified way deforestation means increase in temperature. Besides, depletion of different species has different levels of temperature effects. Dabral and his team of workers have concluded that there is a considerable difference in air temperature inside different species and open places (Dabral, *et al,* 1969). Table 6.8 gives a clear picture of the same.

TABLE 6.8 : **Average air temperature (in °C) measured at 14.00 hours**

| Height from ground surface | Chir plantation | Bamboo plantation | Open |
|---|---|---|---|
| Above 5 ft | 25.9 | 26.3 | 27.3 |
| 3 ft | 26.0 | 26.3 | 27.5 |
| 1 ft | 26.1 | 26.3 | 28.2 |
| 6 Inches | 26.5 | 26.4 | 28.5 |
| At ground surface | 26.5 | 26.9 | 29.6 |

From the Table it can be explained that the deforestation of chir forests in a local sub-region will lead to an increase at least to 3.1°C provided other locality factors remain constant. Such impacts of deforestation are clearly visible in various hill areas where the level of deforestation is on higher side. During summers, due to increase in temperature use of fans have become usual phenomena. In a recent survey conducted by the authors reveals that almost all the persons belonging to age-group of 70-80 year or more who have a long experience of their surrounding, have expressed that the usual phenomena of snowfall have become intermittent and summers have become painful. Not only by the increasing demands of growing population but also by increase in temperature, the consumption of electricity for domestic needs as well as irrigation is going up, thus affecting other resources of the environment.

Wind movement is also related with the nature of species and the density of forest. "Dense forests with heavy foliage reduces the velocity of wind that moves through the forest regions and this reduction of wind movement brought about by forests and shelterbelts adds not only to the comfort of the men and animals but also increases the production of food crops and vegetables grown in leeward side" (Dwivedi, 1980). There are considerable experiences of such effects in the hill districts of the state under study. Deforestation also affects the soil and air moistures, as these moistures are related with the temperature and wind movements. The precipitation is a related phenomenon, as well, but a little bit controversial. The common belief is that the forests augment the intensity of rainfall. In this connection Blandfords example may help to realise the impact of forest rains (Dwivedi, 1980). Blandfords has compared the data of fourteen stations of central province of India during the period 1867-75 (when the region was not forested) with the data of the 1876-85 period (when it became forested). He observed that mean annual rainfall has increased in the latter period by 173 mm or about 12 per cent. Various studies on such topics confer the role of forests in precipitation.

Forests also help in the formation and composition of soils by adding humus, temperature, rainfall etc. Soil erosion is also directly related with the forest coverage to a certain extent. The surface which has a forest coverage, the amount of run-off will be less and flow of water in terms of velocity will also be less than that of open areas having no forest coverage. The flow of water (in terms of amount and velocity) influences the amount of soil erosion. The role of forests in retaining water is also of higl importance especially in the hill areas as the springs are fed by such retained water and its gradual discharge (Singh and Rawat, 1985). Deforestation has adversely affected the spring environment in several ways. The concentration of springs in hill areas of the State is directly related with the rock structure of the region and the forest species lying over there.

**Deforestation vs Environmental Degradations : Experiences and Examples**

It is worth quoting "Once beautiful, forested hills are becoming bare ; ugly land-scapes. The Himalayan are, says Eric Eckholm 'forms in ecological Gibraltar whose fate looms over the well-being of hundreds of millions' about 50 million in the mountains, and about another 300 million in the Indus, Ganges and Brahmputra valley systems. The ecological clock is ticking away. The deserts of Afganistan are in the late evening ; Bhutan and Arunachal Pradesh may still be in the early morning and the Central Himalaya is in the late afternoon of the Himalayan days. Wild life is being fast exterminated, and both man and mountain are becoming poorer." This statement expresses the feeling of the experiences and happenings in the hills. Further, he has also expressed his fear "Man has been the maker of deserts in past ages, even with bare hands, and often in the name of progress. Now that he is multiplying fast, and has more powerful technology, his capacity to endanger the balance between Man and Nature ('Only One Earth') has been enhanced many times" (Moddie, 1981). Berreman, an anthropologist, who lived in the Himalayan region of Uttar Pradesh, has made a valuable remark regarding the development of Himalayas, to quote him, 'more inclined towards destruction rather than development'. The green revolution has affected positively in improving the well-being of the people. At the same time its negative impacts from the environmental stand point may not be separated from the degradation of the environment in general and on forests in particular. There is no doubt in saying that the 'Green Revolution, which has taken place during last decade, has proved as the eater of green vegetal cover' thus, the replacement of 'natures greenery' by artificial 'green colour of crops' has dual character of both the development and destruction. Thus, intensification is rather more imperative than extensification, keeping the environment loses of vegetal cover, and replacement of 'natural greenery by artificial colouring should be checked' (Singh, 1984). For the short sightedness, people have jeopardised their future. "The deforestation and expanding unscientific (not based on land capability) agriculture in the hills are causing massive environmental degradation and untold misery" (Singh and Chaturbedi, 1981). In the various localities of the Himalayan region "orchids are getting scarce and rare due to indiscriminate felling of forests" (Maheshwari, 1980). Disturbances in wild life States, alteration in herbal vegetation and microbial community, increasing noise and mechanical disturbance to nestling birds are major issues related with forest environment. "In the past few decades, the growing impact of human population in the form of deforestation, conversion of forest land into agricultural fields, invasion over steep slopes. . . .has brought inversible ecological changes—the soils of this zone are being exported to the down hill regions without the later paying compensation" (Bisht, *et al*, 1984). The statements mentioned above, are enough to portray the human feeling of 'inhuman activities of human beings' in the name of development.

Soil erosion is a severe problem throughout the State and deforestation is one of the causes of it. In the Himalayas, the problem is more severe on account of massive deforestation. In several localities of Kumaun and Garhwal Himalayas, the incidence of soil erosion are very frequent and common. Due to deforestation the badly damaged compartments of Syamkhet, Karil, Ruinsi and Mahrora of Nainital; and Amri, Paraban Palbelon of Pithoragarh are experiencing more erosion (10-15% higher) than previous levels. The same situation prevails in the various forest areas of Almora, Garhwal, Mussoorie, Chakarata and Tehri districts, where deforestation is comparatively on higher side. Inhabited zones lying along the highways are experiencing a great havoc from soil erosion. In the southern part of the State, the situation of soil erosion is also alarming. "The ruthless cutting of forests and its conversion into agricultural land has led to gradual increase in volume of run-off. . .led to increase in the extent of soil erosion" (Yadav, 1984).

In almost all the districts of Uttar Pradesh (leaving hill districts) the deforestation has invited a new problem on soil environment. Previously when the woodlands and forest patches were available in abundance throughout the region, the cattle dungs were utilized as organic manure because the fuel needs were fulfilled by the jungles available nearby each and every village. But nowadays, the situation is entirely different. Such woodlands and forest patches have disappeared due to speedy extension of agriculture in order to meet the increasing demands of rapidly growing population. Thus, to fulfil the fuel requirement, cattle dungs are being used frequently in the form of dried cakes. From the southern districts of the plateau region to northern districts of Tarai region soil is suffering a great loss consisting of organic manures. From a recent observation, it has been found that in few districts of Eastern Uttar Pradesh, about 80 to 90 per cent of total cattle dungs produced are diverted to kitchen fuel rather than field manure. The situation varies as per the availability of woodland. The condition in southern, we tern and Tarai regions of U.P. is comparatively better. Social forestry schemes can help much in solving organic losses accrued to the soil. On the one hand, we are providing little organic manures to our fields, while on the other, the population pressure on the land has increased. Due to the agricultural extension progiammes and intensive farming, the crop cultivation intensity has gone up and at the same time the use of cattle dung manure has come down to a great extent, thus the soils of the region are really under threat and this problem is directly related with the shrinking of forests and increasing diversion of dungs towards fuel requirement. Further, the growth rate of cattle wealth is much less than the rate of increase in agricultural lands. Subsequently, amount of organic manure per hectare has decreased considerably.

Due to the high rate of siltation (much higher the expected) the life of Ramganga Dam Project has been calculated 76 years against the previous estimated life of 100 years, certainly this is all due to the misery of erosion as a consequence of deforestation (Mithal, *et al*, 1984).

Soil erosion—a by-product of deforestation, retrogrades the wheel of agricultural economy by reducing the fertility of soil. When soil moves down, the economy based on soil also follows it. Thus, the role of forest wealth in maintaining soil environment is unquestionable and if deforestation occurs (in the study area deforestation is an acute problem) or continues, the recurring loss in soil system is inevitable.

In hill areas of Uttar Pradesh, forests are performing a big role in maintaining the springs. The water retaining quality of forest and further its gradual discharge directly governs the nature of springs in terms of quality, quantity and periodicity (Singh and Rawat, 1985). This is not only, even the periodicity as well as their number are also affected by different species in

Kumaun and Garhwal region (Tables 6.9 and 6.10).

TABLE 6.9 : **Concentration of springs**

| Region | No. of springs observed | Forest zones with oak | Forest zones without oak |
|---|---|---|---|
| Kumaun | 60 | 43 | 17 |
| Garhwal | 60 | 47 | 13 |
| Total | 120 | 90 | 30 |

TABLE 6.10 : **Nature of springs**

| Region | No. of springs observed | Extinct | Seasonal | Reduction in discharge |
|---|---|---|---|---|
| Kumaun | 60 | 10 | 18 | 32 |
| Garhwal | 60 | 12 | 22 | 26 |
| Total | 120 | 22 | 40 | 58 |

*Source* : Singh and Rawat, 1985, p. 46.

From the above Tables it is evinced that the concentration of springs is more common in and around oak forest zones than in non-oak forest zones. Besides, the gradual depletion of oak forests has also adversely thwarted the periodicity of springs. In many areas where the depletion of such forests is high, the springs generally day up, while in some cases (if not dried up) a significant decrease in discharge is found. Because of this, nowadays hill areas face acute shortage of water during summers. Even they could not manage adequate quantity of drinking water for themselves during this season. Thus, the springs have been responsible for shaping or reshaping the economy and habitat of this region.

Retreat of glaciers in the Himalayan region is also important aspect from environmental point of view and is related with the climatic conditions of the areas concerned. Though the impact of deforestation on glaciers has not been studied scientifically yet it has been argued that deforestation as the agent of change in micro-climate may be one of the factors leading to the retreat of glaciers (Table 6.11).

TABLE 6.11 : **Retreat of glaciers**

| Glaciers | Period | Year | Retreat in m |
|---|---|---|---|
| Milam | 1849-57 | 108 | 1,350 |
| Pindari | 1845-66 | 121 | 2,840 |

*Source* ; Vohra, 1981, p. 141.

Areas endowed with rich natural vegetation attract wild animals and so any loss in vegetal cover compels them to migrate due to ecological fugacity. With the depletion of forests in the State many species of wild animals and birds have become extinct. The gradual shrinking of forest areas has disturbed the movement of birds and animals. On the one hand, deforestation as a repulsive factor (by reducing the opportunities of security) disturbs the habitation of animals, on the other it also provides opportunities to the hunter by providing better conditions of inside mobility.

## REFERENCES

Ashish, M. (1981), The Effect of over-population and overgrazing in the Kumaun, in J.S. Singh, *et al*, *Science and Rural Development in Mountains*, Gyanodaya Prakashan, Nainital, 23-8.

Berreman, Gerald G. (1979), *Himachal*: *Science, People and Progress*, I.W.C.I.A. Document 36, Denmark, 21.

Bisht, L.S. *et al* (1984), Organizational structure of rural living space in Kumaun Hills (India) : An ecological perception, *Rural Systems*, II (3), 165.

Dabral, B.G. *et al* (1969), Some studies on air temperature and humidity inside *pinus roxburghii* and *dendrocalamus strictus* plantation at New Forest, *Indian Forester*, 95(8).

Dwivedi, A.P. (1980), *Forestry in India*, Jugal Kishore and Company, Dehradun, 3, 44-5.

Maheswari, J.K. (1980), Some aspects of the plant life of Himalaya with reference to conservation of endangered flora, in T.V. Singh (ed) *Studies in Himalayan Ecology and Development Strategies*, English Book Store, New Delhi, 112-6.

Marriot, McKin (1957), Technological change in over-developed rural areas, in L.W. Shannon (ed), *Underdeveloped Areas*, New York, Harper and Bros., 423-33.

Mithal, R.S., *et al* (1984), Environmental impacts of the Ramganga Dam Project, *National Geographical Journal of India*, 30 (2), 85.

Moddie, A.D. (1981), Himalayan environment, in J.S. Lall (ed), *The Himalaya*

*Aspects of Change*, Oxford University Press, New Delhi, 341.

Singh, A.K. (1984), Social Forestry : A way to revival and survival of village economy, unpublished paper.

Singh, A.K. and D.S. Rawat (1985), Oak deforests threatening springs : An exploratory survey, *National Geographical Journal of India*, 31(1), 44-8.

Singh, J.S. and O.P. Chaturvedi (1981), Ecology, environment and development, in J.S. Singh *et al* (eds.) *op cit*, 60.

Vohra, C.P. (1981), Himalayan glaciers, in J.S. Lall (ed), *The Himalaya Aspects of Change*, Oxford University Press, New Delhi, 85.

Yadav, R.S. (1984), Environmental problems and eco-development strategy in South Mirzapur, *National Geographical Journal of India*, 30(4), 262.

PROBHAT KUMAR SEN

SEVEN

# FLOOD PROBLEMS
## A Challenge to Environmental Setting of A Region

THE hydrological phenomena of floods in any drainage basin may experience floods of varying magnitude in the flood plains as an annual occurrence or one at greater intervals of time. Since the magnitude and frequency of floods vary primarily on the precipitation conditions in the catchment, there are floods of different magnitudes of discharges at different intervals of time. commonly identified by their expected recurrence intervals (R.I.), such as annual floods, 5-year, 10-year, 25-year, 50-year or even 100-year floods. There may be certain endemic zones in the river valleys which may have annual floods of moderately high discharge, occasionally being frequented by extremely higher discharge conditions at different expected intervals of time. Thus almost all the river valleys are subjected to varying degrees of flood incidences, either creating minor damages or devastating havocs at times. Such calamities of floods pose a serious problem to the whole environmental set up of the region bringing about some of the phenomenal changes of the physical environment ushuring in accelerated erosional, transportational and depositional activities of the river with the consequent effects on the soil. Vegetation and micro-relief of the area on the one hand and devastating effects on the human habitat and activities of agriculture, industry, settlement, sometimes causing loss of human and animal lives and of materials on the other. In spite of some of the beneficial effects of floods on the natural environment the loss in terms of lives materials is in most cases much more grave

Floods thus pose serious challenges to the natural, cultural and socio-economic environment of the region. Paradoxically enough, the flood-prone areas of the river valleys of the world are the most habited parts of the world with high densities of agricultural and flourishing human activities and the story of their survival in case of floods is one of adjustment, abatement or protection as encountered with the available technological or indigenous efforts of the area.

We will like to concentrate here on the flood problems, without detailed deliberations of their genesis which itself constitutes the most important aspect of the occurrence of floods because of either (a) specific types of hydro-meteorological conditions giving rise to rain storm or cyclonic rainfall or excessive snow melting in the catchment, (b) earthquakes, (c) tsunamis or to some man-induced causes, (d) deforestation in the upper catchment, (e) failure of dams, (f) excessive urbanisation etc. Thus the nature of floods that occur due to rainfall have been classified by Ward as (a) flash floods, (b) single event floods, (c) multiple event floods, and (d) seasonal floods with varying intensities of their impact (Sen, 1985).

Whatever may be the nature of floods, some being forecast and some other being completely unforseen, bring about varying degrees of hazards and the measures of overcoming them in entirely the ability of the society to counteract them. The interesting observations with regard to the response of the human society to floods as an environmental hazard vary to large extent on the perception about floods which can vary between persons depending upon their respective trades in practice (Chakraborty, S.C., 1985). There are undoubtedly varying degree of capabilities to overcome this abrupt and unforeseen environmental hazard of flood in different countries. From the traditional experiences, the gravity of the situation has been in fact considered as only temporal in character and thus the flood prone areas have in most cases encountered the floods as a temporal phenomena without rather little commensurate comprehensive devices to overcome them permanently. Probably the first global awareness to focus on the problems, prevention and management of floods was discussed in the United Nations Seminar on *Flood Damage Prevention Measures and Management* at Tiblisi in Russia in 1969.

Endemic flood zones cover a fairly large area, being confined mainly in the lower riverine plains, deltas, coastal areas and piedmont zones especially frequented by cyclones or tornadoes. Even if they may have annual floods or devastating ones at considerable gaps of return periods, the human habitation and activities have hardly shifted away from these areas. The basic truth of such a paradoxical situation is because of the richness of the physical environment of these areas being very conducive to agriculture, industry. communication and settlements. Thus on basic corollary of a flood-prone area is its richness in environmental systems. The environmental, challenge, therefore, imposed on flood prone areas is tremendous and of the triple devices of adjustment, abatement or protection, we have to consider whether we would like to completely control the floods or survive with them as one of the occasional environmental hazards in an area.

In this context, it becomes incumbent on the environmental planners to take into account the debit and credit sides of the floods. The beneficial effects of floods have even been established beyond doubt. It fertilizer the floodplain by siltation, causes flushing of the drainage arteries, increases soil moisture and initiates a renewed healthy ecosystem of the area. The fertilisation of land by floods have been stressed by Mukherjee who considered the flood-prone areas as 'the tract of civilization' in case of Bengal delta (Mukherjee, 1938). Thus many other deltaic zones of the world such as the Mekong delta, Irrawady delta, Nile delta, Mississipi delta and many others owe a strength of their natural environmental system from the annual floods. Even in the river valleys, the flood plains thus owe their sustenance to the incidence of floods. From this perspective of beneficial effects of floods, there is a strong ground to allow nature to play its role, though it may be whimsical in nature. Therefore, the question of survival of mankind and sustenance of human activities in the flood prone areas may be advocated by means of proper adaptation and adjustments with floods rather than its complete stoppage. In the primitive and underdeveloped societies and countries such a symbiotic process of living with floods is evident in many parts of India e.g. Ganga flood-plains, Bengal, Tripura, Bangladesh and many other parts of the underdeveloped world. But, in some areas the occasional devastating floods might have overwhelmingly obliterated the human and animal existence, not to speak of the little material belongings of material culture that they possess.

The devastating effects of floods, on the other hand, far exceeds the beneficial effects on the sedentary habitat and material culture of man along with some of the phenomenal changes in the channel pattern and physical conditions of the area. The physical changes usher in the bank-side erosion, breach of levees, sometimes shifting of the river courses, causing removal of vegetation and subsequent impacts on the soil and vegetation of the area. There are significant examples of such changes of natural environment along with the drainage system in the lower Ganga plains, which have shown remarkable changes since the oldest maps prepared by Rennel in 1771 or in the sub-Himalayan piedmont zones. The loss of human and animal lives and properties are immense of which probably little records have been maintained in the past. With the growth of technology and material culture synchronous with urbanization and industrialization the material loss is colossal in certain habited parts of the world frequented by floods. Therefore, the huge investments in modern agro-industrialised societies have provoked much more of awareness to the problems of floods as a major challenge in the environmental set up of the area. Compared to its gravity in the past with much less of material development and technological input and even non-sedentary character of mankind, the man-nature interface in case of floods thus pose greater challenges to more advanced countries with huge infrastructural developments. Now the basic considerations of protecting by the engineering devices such embankments, dams, barrages, reservoirs etc. are encountered in the light of the return periods of 50 or 100-year floods as the most important criteria rather than the magnitude annual flood only.

## The Indian problem

The flood problems and the challenge to the environmental setting in India is of gigantic magnitude especially because of (a) the typical seasonal hydro-meteorological conditions which give rise to the highly variable nature of the monsoons, and (b) the vehemence and erratic nature of the tropical cyclones and rain-storms—the two major sources of excessive precipitation and floods. On the other hand, the man-induced factors of extensive deforestation in the regions such as Chotanagpur plateau and Himalyan catchment zones coupled with the excessive soil erosion in the upper catchments tremendously augment the flood magnitude. Furthermore, the effects of urbanization and industrialization are even perceptible in some localised areas.

The geographical distribution of flood prone areas are commonly associated with areas of high rainfall zones, cyclonic tracts, typical topographically and geomorphologically characteristic areas such as piedmont and deltaic and topographically low and depressed zones of the river basins. Therefore the macro-flood zones of India may be broadly grouped into the following zones :

(a) Upper catchment and piedmont zones of the Himalayan river basins.
(b) Ganga river valley—upper, middle and lower deltaic zone.
(c) Brahmputra valley and its tributary catchments.
(d) Deltaic river basins of peninsula India—Mahanadi, Godavari, Cauvery, Krishna and others.
(e) West flowing rivers basins e.g. Tapti, Narmada and others.

Therefore, the areas of recurrence of floods have their own environmental characteristics. The floods in the piedmont zones of the Himalayas obviously owe to the heavy rainfall or even snow-melting in the upper catchment which debouches on the plains, the most classic examples are of the Tista, Mahananda, Son, Ghaghra, Gandak, Yamuna, Ganga and many others. The deforestation and other man-induced erosional processes further augment the gravity of the situation to a great extent. Besides the short-spells of abnormally high rainfall conditions during the monsoons, the cyclones and rainstorms have invariable association with the incidence of floods.

**The environmental challenge**

Within the small compass, it is impossible to dilate upon the nature of environmental challenge created by floods in each basin but certain basic generalizations and case studies would be helpful enough for this comprehension.

The magnitude of flood havoc in India is of huge dimension but the age-long practice of overcoming them is one of adaptation and of meagre devices such as raising of embankments at a local indigenous level. Though we have adequate testimony of the flood havocs in different parts of the country, the huge damages caused by them have not been adequately recorded or assessed. The average damage caused by floods in India in each year is about Rs. 1,500 million per year (Rao, 1974) during the last two decades and there are certain measures taken up to overcome this natural hazard. Before 1954, the flood control measures were considered as a State problem though we find that the need for developing integrated environmental approach has been emphasised by many authorities as the problem far overrides the boundaries of the State. A comprehensive drainage basin assessment and training programme is a key to the success of the prevention or mitigation of floods in river valleys. Even in 1940s the need of developing inter-provincial flood control commission such as Ganga River Commission in lines of the Mississipi Valley Committee has been emphasised (Mazumdar, 1932). It is only after 1954 severe floods that the National Flood Control was launched in the same year (1954). At the national level the Central Water Commission, Ganga Flood Control Commission, Brahmaputra Flood Control Board, North Bengal Flood Control Board and constitution of Rashtriya Bar Ayog in July, 1976 are some of the organizations to conduct in-depth study of the floods and work out means and devices for ameliorating them.

In this context, the nature of flood problems in the lower Ganga plains may be taken up to assess the nature of challenge to the environmental setting. The whole of the lower Ganga plains in West Bengal is a notoriously well-known endemic flood zone while further beyond the problems in Bangladesh, frequented by severe floods, are some of the worst flood prone areas of the world.

The Bhagirathi is the main distributary of Ganga in West Bengal which though experiences a chronic mitigation of discharge has a flood prone flood plain, especially because of a large number of tributary river systems that fall into it but originate from the Santhal Parganas and the Chotanagpur plateau e.g. Ajay, Mayurakshi, Darakeshwar, Damodar and its distributary channels. The Bhagirathi flood plains have relatively lesser frequencies of floods because of the poor discharge but the high discharge in certain sections cause localized floods and havocs of bank-side erosion to a great extent. The heavily built-up settlements from Baharampur southwards have thus spill-over zones and the gradual processes of erosion have devoured even within a decade some of the well-built river side areas such as Patuli and adjacent areas. Thus in spite of the mitigated flow, in high monsoons the survival of some of the built-up areas is a matter of grave concern in the upper reaches of the Bhagirathi. The lower reach below Katwa have settlements over the levee one and thus do not pose so serious problems of the floods.

In the tidal creeks of lower Bhagirathi-Hooghly basin the floods, of course, play a more vehement role. The tributary rivers, already mentioned have some of the worst affected flood prone zones which experience regularly the menace of floods with the devastating consequences. The lower Damodar is possibly the best example which in spite of the construction of 5 dams in the upper reaches, have experienced serious floods in 1959 and in 1978. Long known as the 'River of Sorrow', it used to inundate annually vast areas of the lower Damodar basin inclusive of its number of distributary channels but has considerably been benefited with the implementation of the dams. The need for Damodar Valley Project was primarily to protect the flood prone areas of lower Damodar in the integrated river valley project. The Burdwan town and the adjacent areas and in the lower reach covering large tracts of Burdwan, Hooghly and Howrah districts used to remain inundated for long periods destroying agriculture and paralysing normal activities. The 1943 flood is one of the worst which completely swept away the railway lines near Saktigarh. Even after the building up of the dams and reservoirs with flood cushioning capacity of 1.0 m acre ft in place of recommended 1.5 m acres ft, unprecedented flood damages were caused to the lower basin and even affected the industrial belt of Durgapur-Asansol to a great extent.

Without going into the details of their causes, we may just assess the problems of the floods as a challenge to environmental setting of the region. The floods have proved to be quite beneficial in the Damodar basin and so also in the lower Gangetic plains as in any other flood affected areas. We may specifically mention the bare outline of the beneficial effects of the floods as follows :

**Beneficial effects**

(a) The floods acted as the natural flushing process of the Damodar and its distributary channels. Thus most of the channels had their sustenance through this annual process.

(b) Floods are the most vital fertilizing agent of the flood plains. Thus with a silt (composed mainly of silt and clay) concentration of 1 mg/litre rich in salts, the siltation of the flood plains enrich the soil tremendously which boosted up the agricultural production in the subsequent years to a tremendous extent.

(c) Floods act as natural processes of mitigating the pollution of the channel and basin which subsequently proved as an important sanitary agent. In consequence the incidence of diseases considerably decreased in the subsequent years.

(d) A more efficient drainage of the land was established, overriding the man-made obstruction, following the topographic conditions of the area.

(e) The water table was substantially nourished by the floods which proved beneficial for the future groundwater extraction.

**Non-beneficial effects**

In contrast the grave consequences of flood hazards, especially on the man-built material culture far exceeds the beneficial effects, not only in Lower Damodar and Lower Ganga basins but in all flood-affected zones.

**Material effects**

(a) The devastating floods cause complete devastation of the agricultural production of the area.

(b) The transport and communication and as such, the normal activities are completely paralysed.

(c) A great damage or complete devastation of the built-up areas such as houses, bridges, communication arteries and the belongings may be caused.

(d) Loss of lives both of human being and livestock, by some of the worst floods caused an immense loss.

(e) All economic activities are seriously hampered because

of these floods.

The magnitude of the floods with different return periods is, of course, the major determinant of the degree of loss and extent of the havoc.

**Physical effects**

Floods have not only a temporal effect on the material loss but usher in some changes of the physical environment and drainage arteries. A bare outline may be furnished as below.

(a) Some of the floods have been the cause of shifting of the river courses. The Bhagirathi has considerably shifted her course to a great extent primarily because of the accelerated erosional and depositional activities, forming cut-offs, meander scars, sloughs, water-logged areas which can be very well be assessed from the oldest map of Bengal by Rennel in 1771 and the topographic forms. The shiftings have devastated many of the built up areas along the Bhagirathi river site and even today pose a serious problem in certain sections of the channel.

Some of the most classical changes of the river courses are of Tista which shifted eastward and flowed into Brahmaputra instead of Ganga where it used to fall previously. The river Kosi, in its piedmont section over the North Bihar Plains, have been a playground of oscillation of the river course which shifted westward by about 112 km in about 130 years.

The Damodar which flowed in an easterly direction have continually shifted her courses from Khari in the north thereafter through Gangur, falling near Neosarai at Kalna, Kana Damodar, Amta Channel (the present Damodar) and ultimately to Mundeswari which carries ninety per cent of the high discharge leaving the residual amount to flow through the true lower Damodar known as Amta Channel.

Such changes of the river courses must have their grave consequences on the settlements and economic activities but unfortunately little record is maintained of the nature of loss of the poor poverty striken villagers of the area.

(b) The consequences of the floods in the upper catchment is fraught with some grave erosional processes amounting to even landslides and slope failures on a massive scale as reported in the Appalachian area. Most of the upper catchments in the Himalayas, such as that of Tista are frequented by such problems.

The hilly catchments, normally prone to the erosional processes experience a phenomenal rise of the erosion with the result of loss of soil fertility, vegetation and perpetual increase in the erosional processes.

(c) Deposition of coarse sand in the sub-Himalayan river basins due to floods requires adequate labour for its removal for practising agriculture.

**Remedies of the challenge**

Undoubtedly the floods pose as serious environmental challenges and the awareness for preventation and mitigation is gaining with the development of material culture. We have already dilated on both the beneficial and non-beneficial effects of the floods, sometimes with the consequences of a catastrophy whenever the threshold of normal flood exceeded. The modern engineering and technological devices to encounter them are gradually gaining ground. The most common devices that have been followed are some of the river training devices and forecasting as follows :-

—Stoppage of erosion in the upper catchment.
—Development of dams and reservoirs at vulnerable sites.
—Dyking or 'jacketting' of the channels to prevent overflow.
—Building of settlements on raised grounds.
—Construction of multiple channels for distribution of the excess river water.
—Development of shelters in flood prone areas.
—Prompt evacuation measures and protection of properties.

Besides the developed countries, India has also launched upon a vigorus campaign of flood control measures. The Damodar Valley Project, the Mahanadi Project, Kosi Project have undoubtedly mitigated the hazards of floods.

As an environmental challenge, the preventing of floods have thus been considered to be of immense importance to the human habitat. There is growing consensus that the prevention or mitigation of floods should not be tackled as a localized phenomena of the flood-prone areas only but should be tackled with a holistic approach of the drainage basins, as the upper catchment may initiate the problems which propagate and cause the havocs of the floods in the lower reaches. The checking of the erosion in the upper catchment though afforestation, restricted grazing and stoppage of unwise land-use will considerably mitigate the sediment yield and the flood discharge to a considerable extent.

The main river channel and its tributaries would require considerable storage facilities for flood cushioning and trapping of the sediment by the formation of reservoirs and dams, proper dyking of the flood-plain zone, rivetting of the bank-sides and construction of spurs. The experience of the Mississipi river basin extending over forty-one per cent of the area in forty-eight states of North America (basin area 322455 km²) speak of high floods in the lower deltaic zone but the above measures, as taken up by the Mississipi River and Tributary Flood Control Project (MR & T Project) has achieved a considerable measure of success (Noble, 1973) though the 1973 flood brought about the worst flood inundating 6.7 million hectares of lands, making sixty-nine thousand people homeless and causing an economic loss of about one billion dollars in spite of the measures taken by the project. It is opined that had there been no MR & T Project, the economic loss would have amounted to fifteen billion dollars and inundated additional land.

The Indian story of floods is no less dismal in spite of the river training programmes. As an example we may point out the devastation caused by the 1978 floods in the lower Damodar and Ganga basins in spite of the dams built up on the river Damodar and other tributary rivers of Ganga like Mayurakshi, Ajoy, Kangsabati and other rivers. While the floods cushioning benefit Damodar dams as much less than originally anticipated, the urbanised section of the Asansol-Durgapur region had experienced the consequences of the floods, never experienced even before the construction of dams, because of the high sediment yield and discharge as an outcome of urbanization and even deforestation over the area. The total area inundated amounted to thirty thousand km² affecting about 15 million people during the post-dam 1978 flood with a peak discharge of 3.85 lakh cusecs at Rhondia.

Thus the environmental challenge of floods is perennial in character with its dual contrasting role of benefit and of damage or devastation. A progressive deterioration of flood land, where fertilized by annual deposition of silt in the deltaic section obviously act adversely in the natural process whereas the prevention or mitigation is a positive gain of our material culture providing security to people.

In face of our growing concern for the abatement or prevention of floods, the economic considerations are of considerable importance as the river-training programme cannot essentially ensure a 50-year or 100-year R.I. of floods. The cost-benefit of huge engineering constructions never becomes feasible for protection of such floods. Thus the various other devices of proper forecasting, evacuation, crop insurance and flood damage insurance are some of the essential measures that should be given topmost priority in encountering the challenge of floods.

The Indian flood protection measures in the past was completely through indigenous efforts with some benevolence shown sometimes by the zamindars. Little is on record of the vast devastations of men and materials that had taken place through ages in India. In the post-Independence period, a commendable measure of success has been achieved through the different river training programmes for which the government has spent a huge amount.

The flood problem thus pose an environmental challenge which ought to be taken up with more detailed study and safety measures in the future retaining the sanctity of natural environment.

## REFERENCES


**Chakraborty, S.C. (1985), On understanding floods in West Bengal : A geographer's view, in Sen, P.K. (ed), the *Concept and Methods in Geography*, Institute of Geography, Burdwan University, Burdwan.**

**Damodar Valley Corporation (1978, D.V.C.—Its flood control capacity, past, present and future, (A cyclostyled report).**

**Majumder, S.C. (1942), *Rivers of Bengal Delta*, Calcutta University Press, Calcutta.**

**Mukherjee, R.K. (1938), *The Changing Face of Bengal*, Calcutta University Press, Calcutta.**

**Noble, Charles C, (1976), The Mississipi river flood of 1973, in Coates, Donald R., *Geomorphology and Engineering* (ed) Dowden, Hutchinson and Ross. Inc., Stroadsberg, Pennsylvania, U.S.A.**

**Rao, K.L. (1979), *India's Water Wealth*, Orient Longmans Ltd.. New Delhi.**

**Government of West Bengal (1979), Report on the floods in West Bengal in the year 1978, Irrigation and Waterways Directorate, Calcutta (Cyclostyled copy).**

Sen, P.K. (1985), The genesis of floods in the lower Damodar catchment, Sen, P.K. (ed), *op. cit.*

Sen, P.K. (1979), The recurrence of floods in the lower Damodar basin, *Indian Journal of Power and River Valley Development*, Calcutta, January.

Sen P.K. (1978), Evaluation of the hydrological and morphological characteristics of the Bhagirathi-Hoogly and Damodar interfluve, U.G.C. Research Project, Department of Geography, Burdwan University (Unpublished).

Sewell, W. R. Derrick (1971), Human response to floods in Chorley, Richard J. (ed), *Introduction to Geographical Hydrology*. Metheun and Co. Ltd., London.

Singh, R.L. (ed.), (1971), *India—A Regional Geography*, National Geographical Society of India, Varanasi.

R. K. DHABAI

EIGHT

# DIVERSIFICATION AND DEVELOPMENT OF RURAL ECONOMY

Because of an increasing spatial disparities in the beds of development, the concept of diversification and development is getting momentum in developing countries. The rural economy of such countries is biased with agriculture having excessive pressure on agricultural land, which in turn has definite limits because of scarcity and it cannot ultimately eliminate rural poverty. In any case according to Rao diversification of rural economy has to be achieved, so that the rural people may have alternative channels of employment (Rao, 1974).

In India, efforts were made from time to time towards rural economic development through diversification. Dantwala committee report on block level planning, suggested about the diversification of agriculture and rehabilitation of rural industries. This pointed out towards much greater and wider economic integration between agriculture and non-agricultural activities (Erik and Alfred, 1969). According to the projection made by the National Commission on Agriculture, "although the proportion of rural population in the country is likely to come down from about 80 to 71 per cent, by considering the rate at which industrialisation is taking place in the country, but there seems no significant structural transformation and large scale transfer of population from agricultural to non-agricultural activities."

The rural economy will thus continue to support a vast population. Even after the rural non-agricultural activities which substantially depend upon the prosperity of agriculture, are to be diversified. It seems that the absolute number of dependents on agriculture will, in fact, be much higher than the prevailing conditions. If, therefore, diversification on the rural economy is not to be dependent merely on the secondary and indirect on industrialisation, most of the new industries are now concentrated in large towns, remote from the rural areas. Ahead of us is the task of creating and maintaining sufficient work opportunities for the development of resources and employment in various forms for the utilisation of locally available manpower. In other words, the villages should be economic units in which agriculture and other related occupations are organised with the object of securing the maximum output and employment. However, to be able to provide a diversity of occupations, it is necessary for the village economy to accept and develop new technique. Such technological change, should be conceived as an essential and integrated aspect of social and economic change (Jayaraman, 1976).

Hence, there is a need to reorient the rural economy to support vast population. An industrial progress cannot as yet absorb the rural labour surplus, therefore, it would be necessary to introduce planned land-use to meet competing demands of food and commercial corps, animal husbandry, rural industries, forestry, etc. To generate widespread productive employment, rural savings would have to be mobilised for reinvestment in the rural economy, and to provide necessary help and social security to the weaker section.

**Inter-sectoral diversification and the development process**

Economic development is the function of vertical shift in the work force from primary to non-primary sectors leading to inter-sectoral diversification. As communities become more economically advanced, the numbers engaged in agriculture tend to decline relative to the numbers in manufacture, which in their turn decline relative to the numbers engaged in services.

Based on the classical division of economic activities into three sectors—primary, secondary and tertiary a general theory of evolution in phases could be put forwarded (Amin, 1976). In the first phase, development of secondary sector takes place at the expense of the primary sector ; in the second phase the tertiary sector takes over from the secondary and its relative share in economic activity as a whole increases more and rapidly reducing the relative proportions of the primary and even of the secondary sector. In the development process the relationship between vertical shift i.e. agriculture to industry and industry to services in the work force, on the one hand and of its horizontal mobilities i.e. over space on the other has crucial significance. It is through this complex system of inter-actions that the changing organisation of space responds to the dynamics of the development of productive forces. This relationship is of particular significance in the case of the developing countries which have almost invariably inherited both an underveloped economy and a distorted organisation of space from the period of colonial rule (Raza & Kundu, 1979). The work force in the underdeveloped countries is largely attached to agriculture where its employment, at the existing level of technology yields at best a subsistence return. By way of contrast, the highly developed countries devote a relatively small part of their labour input in primary production of food stuffs. Majority of the workers are employed in the production of other goods and services for growing market sustained by demand generated by a raising level of real income and formation of domestic capital. Thus, economic development in its labour aspect is seen as a task of shifting workers from agriculture to industrial and commercial employments.

In case of India, it is of the essence to recognise that, in the light to the colonial heritage, the challenge of bringing about basic socio-economic transformation could not be met without breaking the constraints on the inter-sectoral diversification of the Indian economy and without accelerating the process of vertical shifts in the work force a process which had been throttled during the whole of the colonial period.

The inherited and product of the long drawn process of horizontal mobility unbacked by vertical shift in the colonial economy is characterised by serious distortions in the economic organisation of space. There is, therefore, hardly any other option available, at least some time to come, except to embark upon a strategy of combining the maximisation of vertical shift with the minimisation of horizontal mobility. In other words, the problems related to vertical shift within the rural sector itself acquire crucial significance. It is through the solution of these problems that the multi-layered system of appropriate technology gets intertwined with the hierarchical system of settlements to provide a spatial organisation of the economy which responds to the requirements of the present stage of development, on the one hand and prepares the ground for the future advance in next.

In India about 76 per cent population living in rural areas and have a share of only 25 per cent of the total non-household employment of the country. Further, if we put household and non-household employment together, it comes only 5.5 per cent. The percentage of household and non-household industrial workers in rural areas also shows a fairly high degree of inter-regional variations (CV=.64). The decline of the household sector was mainly in the rural villages and in the small towns, wherein household sector continues to be the major element of inter-sectoral diversification. The employment structure in household and non-household industries in rural areas is associated with the indicators of diversification. Their relationship indicates that a higher percentage of secondary employment in rural areas would reflect a healthy economic base of the region.

**Diversification in the primary sectors and the development process**

The development of the primary sector is possible only by optimising the use of ecological factors. India inherited distorted economy from "British Raj" after it became independent. In the post-independent period, new plans and policies were formulated to relieve the population pressure from agriculture by absorbing them in other sectors of the economy. At the same time the primary sector which has earlier confined only with agriculture had also developed through the available ecological factors. For the first time as a part of programme the intensive agricultural development programme was launched in the beginning of the Third Plan at the district level, introduced some of the supporting activities like, animal-husbandry, poultry, fishing. piggery and pasture development etc. (Caprihan). The efforts to diversify the rural economic base were further made after exploring the possibilities of diversification within the sector. This reveals from the survey report on the "Rural employment and special area programmes" of the National Commission on Agriculture which indicate that next to crop, animal husbandry programmes have got the largest employment potential, which could be developed in drought prone, hilly, tribal and other less favoured areas, where crop production on its own may not be capable of providing suitable job opportunities to the unemployed and underemployed particularly for small and medium farmers.

After animal husbandry, forestry provides gainful employment opportunities. Employment offered in this sector is often in the off season, therefore logging is generally carried out in the lean agricultural months. Secondly, as forests are located in backward areas, direct employment in forestry activities can benefit the backward communities, such as tribes. Besides, the activities in the forestry sector require a very high component of unskilled labour and, therefore, would benefit primarily the unemployed and underemployed agricultural labourers and weaker section of the rural community.

Forest development activities which contribute to the generation of employment opportunities, can be those associated with (a) production forestry, (b) social forestry, and (c) minor forest produce. As regard forest production, the related activities are logging, road construction and regeneration. The regeneration of forest both natural and man-made consists of cleaning of forest, collection of seeds, rising of nurseries—preparation of seeds etc. This provides employment to unemployed in the neighbouring villages. The construction of forest roads will open up forest areas for secondary and tertiary employment. To sum up the forestry sector as a whole, the additional employment generated will be of order of 5.5 million man years by 2000 A.D. In addition there is a large scope for employment in the rural areas in the transport of forest produce in forest base industries such as saw milling, pulp and paper, panel products, matches, katha, resin and lac etc.

Next to forestry, fishing sector opens avenues for diversification in the primary sector. However, this sector employes comparatively a very small percentage of rural labour force, which can be increased by expanding the tertiary sector including preservation, processing and marketing of fish products.

**Diversification in the agriculture and the development process**

The development of agriculture based upon the available infrastructural and technological advancement in a country provides a diversified cropping patterns. The farmers have been very much concerned about higher monetary rewards. They harvest from different plots of land, producing different crops yielding different returns. The farmer has been interested in maximising these rewards. Such advantages could be obtained only when the pattern of cultivation is altered according to the expected variations in inter-commodity prices. The changing pattern of cultivation is mainly due to these price changes. Given the technological and geographical possibilities, no farmer would like to produce inferior crops yielding less monetary returns to him.

Mukhopadhyay and Sarkar in their study Economic Aspects of Cropping Pattern maintained that with the introduction of irrigation farmers move from inferior crops to non-inferior crops and then towards non-inferior non-food cash (industrial) crops (Mukhopdhyay and Sarkar, 1972). Hence, monitorization of crops or commercial orientation becomes more important in developing countries.

Samir Amin in his book Accumulation on World Scale narrates that.

> "The central American type of economy is highly monetized, wholly directed towards the commercial production (Sugar in the West Indies, fruit in the main land). In some countries of South America, the economy is only slightly monetary, owing to the juxtaposition, without inter-penetration, of a more or less closed agriculture. In Africa the primitive indigenous agricultural economy has been more or less commercialized (In the two forms of plantations and petty commodity production by natives). In the Arab and Asian world, agricultural economy with a semi feudal structure is pretty well commercialized in North Africa, Syria, Iraq and Turkey, very wide commercialized in Egypt and very little commercialized in Iran and Arabian peninsula. A somewhat similar structure is found in Brazil and Chilli. In southern and eastern Asia we find a model more or less common to the whole of this area (But different from Brazil and Chilli), characterized by an agrarian structure of a markedly feudal type (so that the degree of commercialization is rather slight)." (Amin, 1974).

In India, the trend of industrial crops became more marked with the advent of twentieth century. In 1937 the agronomist Sir John Russel mentioned in the report published by the Imperial Council on Agricultural research that "the area under cotton, sugarcane and wheat have risen corresponding with the increased area under irrigation and such trend kept pace with the increase in population. On the otherhand acreage under food crops show much less rise". (Government of India, 1937). This trend was further accelerated "by the interest of government departments of agricultural research and experimental farms in improving the quality and introducing new varieties of cash crops, while food crops were neglected" (Wadia & Merchant, 1954).

The accent on cash crops was to have other effects, especially, when the basis of Indian agriculture remained on a technically premitive level namely, the exhaustion of the nitrogenous contents of the soil, with the concomitant fall in agricultural yields. The British could have been expected to bring benefits to Indian agriculture, but they never did. It is true that they developed a really magnificient plan for explanation and brought the desert areas estimated at 20 million acres under cultivation by means of irrigation. There is only one thing to be said against this vast agricultural programme that it did not affect the food supply of the people. Its benefits went exclusively to cash crops for export. On the contrary, it is in the comparatively backward areas where subsistance type farming with most of the crops being food crops for local consumption still dominate the scene of cropping pattern (Dhabai, 1979). Therefore, in the present context in India, the problems of diversification in agriculture are of crucial significance.

## Process of Diversification and Rural Economic Development

The process of diversification as we have conceived in our model is a sectoral bifurcation into primary and non-primary sector, within primary sector into crop and non-crop and within the crop sector i.e. agriculture into inferior and non-inferior food crops and non-inferior non-food cash crops or industrial crops, which in their turn raise the cropping intensity (Fig 8.1).

The development of rural economy is the function of overall development of different sectors, for example, in crop sector—if the area under non-inferior crops is higher in a geographical unit, then it can be put in the category of developing areas. Further, if

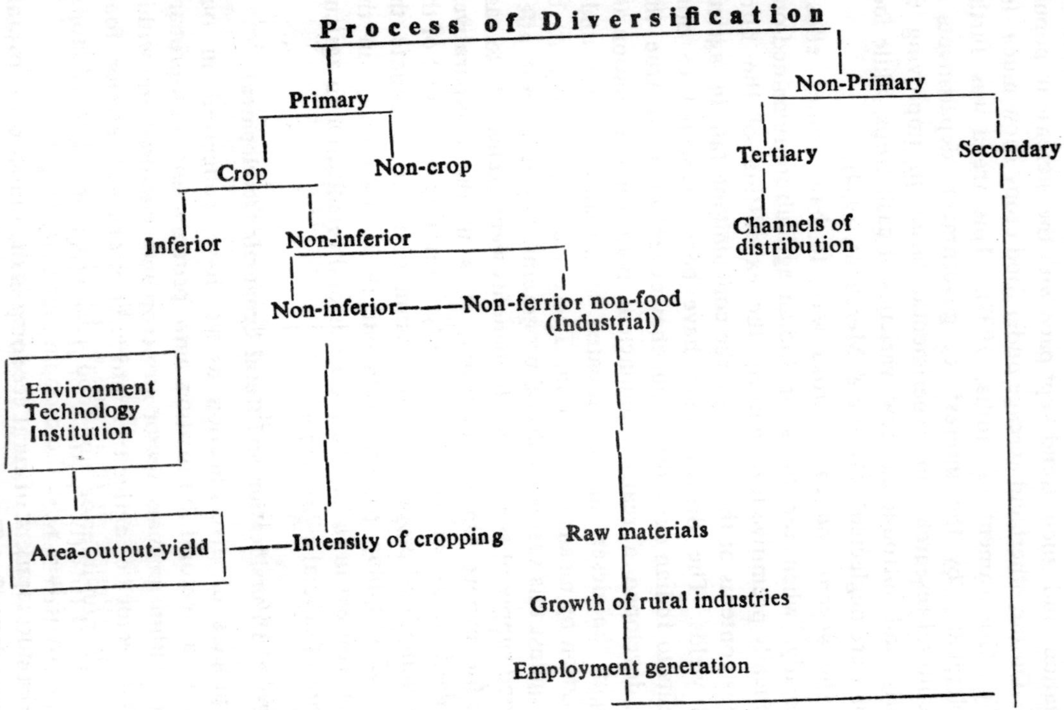


FIG. 8.1 : *Stage of Diversifications*

the area under industrial crops is higher than the area under non-inferior crops, then the region can be categorised as developed one, as far as crop sector is concerned.

Since, Indian rural economy is biased with crop sector by providing sizeable employment opportunities on one hand and high contribution to G.N.P., our attention is towards the development of this sector along with exploring the possibilities of employment in non-agricultural field for those 300 million rural poor, who live in 6,00,000 villages in our country. The development of agriculture is possible only by intensive agriculture meaning thereby practice of growing multiple cropping.

Now the nature and rotation of crops are to be determined for the best possible use of available land resource. In this connection our attention is to increase the net value added by growing short during crops on one hand and industrial crops on the other. The industrial crops would provide raw materials for agro-based industries. This would ultimately help in boosting up industrial growth which in turn would generate employment opportunities. In this way the development of secondary sector takes place on the expanse of primary sector more precisely on the development of agriculture. Thus, the proccess of diversification is inter-linked. The overall economic development cannot be achieved unless or until we strengthen our agricultural base.

## REFERENCES


Caprihan, S.P., *Fight Against Hunger in Developing Countries*, J.K. Jain Brothers., Bhopal.

Dhabai. R.K. (1979), Patterns of Crop Diversification—A Spatio-temporal analysis, in Mohammad. Ali. (ed), *Dynamics of Agricultural Developments*, Concept Publishing Company, New Delhi, 125-31.

Erik, T. and J.F. Alfred (1969), Relationship between agriculture, non-agriculture and foreign trade in development of Argentina and Peru, in *The Role of Agriculture in Economic Development*, N.B.E.R. Columbia University Press, New York, London, 165-218.

Government of India (1937), *Report of the Imperial Council on Agricultural Research*, New Delhi-5.

Jayaraman, K. (1976), Concept of Village as an Economic Unit, *Economic Times*, June 19,

Moonis Raza and Amitabh Kundu (1979), Aglomerated and Dispersed Patterns of Industrialisation : A Spatial Analysis, Paper presented in the

Seminar on National Development and Regional Policy at United Nations Centre for Regional Development, October 31 to November 4, Nagoya, 1.

Mukhopadhyay, A., and G.K., Sarkar, (1972), Economic aspects of cropping pattern, *Proceedings of the Symposium on Cropping Pattern in India.*, Indian Coucil for Agricultural Research, 584.

Rao, V.K.R.V. (1974), *Growth With Justice in Asian Agriculture : An Exercise in Policy Formation*, United Nations Research Institute for Social Development, Geneva.

Samir Amin (1974), *Accumulation on a World Scale—A Critique of the Theory of Under Development.*, (I and II), volumes I and II, Sussex, 167.

Amin Samir (1976), *Unequal Development : An Essay on the Social Formation of Peripheral Capitalism*, Sussex, and Harvester, 242-43.

Wadia, P. and K., Merchant (1954), *Our Economic Problem*, Bombay, 179-80.

A. RAMESH
AND
SUBHADRA IYENGAR

NINE

# ENVIRONMENT AND HEALTH
## Geographical Perspectives

THE earliest work in Geography relating to health and environment began under the banner of Medical Geography. Geographers in this specilized branch of Geography, in the mid-1800s, began to search for factors in relation to the incidence of certain diseases and their spatial pattern of occurrence and spread. In 1950s realising the importance of geographical assessment for the understanding of diseases ; the IGU (International Geographical Union) constituted a Commission on Medical Geography and its first report was discussed at the IGU meeting in Washington in 1952. Since then the analysis of health and diseases within the framework of Man and Environmental relationship has attracted the attention of geographers. It was again in the IGU Commission of 1976, at Moscow, the 24 year old term of Medical Geography was abolished and a more broader perspective of Geography of Health and Health Care was adopted.

### Historical perspective of health and environment

An historical prespective of research in Medical Geography and Geography of Health and Health Care has brought to the limelight, two important dimensions, namely, spatio-temporal analysis of Health and Health-care. In the early part of twentieth century most of the work was to assess and understand the causative forces of environment in the incidence and spread of diseases. Works pertaining to the assessment of ill-health conditions in natural regions in relation to environmental elements of topography, climate and vegetation were common in the West as well as in India. For instance, the works of McCleland (1859), Adams (1889), Rowntte and Subohm (1901), Hamston (1905), Rogers (1926), Hesterlow (1930) and Nityanadan Pillai (1941) were important contributions in this direction. In 1950s the work was focused on control of diseases through environmental managements like maintenance of rural sanitation, innoculation, water management, infrastructure, vector control, control of epedemics and endemics etc. The works of Rogers (1944), and Davery and Light Body (1956) in the field of research were significant.

From 1959 to this day, medical geographers and geographers specializing in health and health care are busy in assessing the environmental repercussions and associated health hazards of air pollution and respiratory disorders, water contamination and gastroentestinal problems. A large number of works in this aspect have been carried out by Stocks (1959), Allen Price (1960), Burkit (1962), Turner (1964), Leave and Seskin (1970), Ghosh (1971), Garud (1972), Ellis (1972), Dixit (1972), Howe (1972), Fraumeni (1973), Ghosh and Meade (1976), Graceus, Stone and Sutoto Sutejo (1976), Feachem, Mc Garry and Maha Duncan (1977), Weiss (1977), Mehta (1977), Gigss (1979), Wood and Margaret Lawrence (1980), Kumra (1980), Greco (1981), Singh (1981), Kumra (1982), John M. Hunter, Luis, Ray and David Scott (1982) and John M. Hunter and Marris D. Thomas (1984). At the same time works were also carried out in particular environmental regions at micro-level to assess both the direct influence of environmental elements and disrupted elements on the health status of the people. The great contributions in this field are the works of Sen (1957), Learmonth (1958), Sharma, (1958), Charg (1963), Man (1960), Sargent and Barr (1965), Hunter (1966), Fonaroff (1968), Justaz (1968), Shimkin (1970), Auliciens (1972), Chen Lincoln (1973), Bydyko (1974), Mcader (1976), Armstrom (1976), Vornov (1977), Liposkin (1978), Pyle (1978), Jelliffe (1978), Akhtar (1978), Learmonth (1978), Mukerjee (1980), Aiken Frost and Leigh (1980), Genhili (1980), Giles, Graham (1981), Akhtar (1982), Brownlea and Taylor (1984) and Ramachandran (1984). The study of control of diseases through maintenance of ecological balance and control of pollution were also the main thrust of research. The works of Mukherjee,

Chauhan and Singh (1976 and 1981), highlight the importance of this aspect.

In the works of contemporary geographers of health and health care the environmental determinants in utilization and optimization of medical and health units have gained greater importance. Assessment of location, allocation of health care units in relation to socio-economic and socio-geographic distances, assessment of urbanization process, health status in relation to health care units are the topics of research observation in the works of Noble La Montagne, Bellotic, Wechstller Weiss, Greenlick, Jones and Girt (1971), Knox and Hodgart (1978), Shonick (1979), Subramanian and Sambasiva Rao (1981), Etton (1982) and Robert (1983). Another contemporary theme in the works of geographers of health and health care in the various techniques of representation of medical and health data, which is well discussed in the works of Hopps (1969), Ramesh (1983) and Robert (1984).

**Approaches to the geograpichal analysis of health and environment**

The basic theme of geographical analysis is the assessment of spatial pattern and the process of man-environment interactions. Geographers have adopted a number of approaches to study the pattern of interactions and the consequences in geographical space. Some of the approaches applied in the field of geography of health and health care are—(1) Medical Cartography (as applied by Phillips) (1981) (Robert, (1984); (2) Ecological-Associative Approach (Robert, 1984); (3) Diffusion Studies (Robert, 1984); (4) Locational Analysis (Robert, 1984); and (5) Spatio-Temporal, Social and Behavioural Studies.

*Medical cartography*

Works in Medical Geography began in nineteenth century with the preparation of maps and atlases showing the distribution of diseases by medical persons as an important tool of analyse. The cartographic approach identifies the homogenous areas. It marks out the different disease regions and also analysis the available medical facilities in the geographical space. Aiken, McClleland and Hopps have adopted this approach in their work on analysis of health status of people in Thailand and Bengal. Aiken in his study has shown in his map the areas of incidence of **Dengue**

Hemorrahgic fever in relation to temperature and precipitation characteristics. McClleland has marked out the medical elements of disease and medical service in relation to topography, soil and climate in Bengal and North Western region. Hopps in his work has discussed the importance of computerized mapping of data on disease and environment.

*Ecological approach*

In this approach the thrust is on the spatial analysis of conditions of health of people through environmental interactions. The geographers through this approach assessed the incidence and spread of diseases in relation to topographic, climatic, biotic and socio-economic cycles. Geographers adopted this approach also to analyse the health hazards resulting from the interference of man in the environmental process. Examples can be cited from the works of Mc clleland (1859), Adams (1889), Hesterlow (1930), Sargent and Barr (1965), Fonaroff and Justuz (1968), Ashley (1969), Shimkin (1970), Girt (1971), Dixit (1972), Howe (1972), Dave (975), Gracey (1976), Meader (1976), Mehta (1977), Kraybill (1977), Voronov (1977), Learmonth (1978), Akhtar (1979), Mukerjee (1980), Kumra (1980), Chauhan (1981), Singh (1981), Alexander (1982), Akhtar (1982) and Brownlea and Taylor (1984).

*Diffusion studies*

This is an approach adopted for assessing the spread of disease contagin and diffusion of adoption of health care innovation. The notable works in this field are from Rogers (1944), Noble, La Montagne, Bellotic and Wechsler (1971), Weiss, Grenlick and Jones (1971), Girt (1971), Knox (1978), Hodgart (1978), Elton (1982) and Robert (1983).

*Locational analysis*

Contemporary geographers have adopted this approach to assess the locational attributes of an health care centre in relation to natural and socio-economic environment. The geographical models like gravity/distance decay, central place theory, allocation/location models etc. are all applied to assess the distributional pattern of health care units. The work of Robert (1984) is an important contribution in this field.

### *Spatio-Temporal, Social and Behavioural Approach*

The attempt in this approach is to assess the location-allocation of health care centres and opting of these centres by patients. The analysis is done through spatial, temporal, social and behavioural attributes, of providers and users. This is a newly emerging area of research, attracting the interest of all scholars.

### **Source, reliability and processing of health data**

The most important task of research is the collection and processing of the information. These are major problems in this as the data is not available for most of the region and what is available is also not reliable. Secondly reporting of cases and maintenance of recorded information is not standardized. This is especially so in case of data on death rate and causes of death. In many cases the information is overestimated and highly peneralized. Similar problems arise in the collection of environmental data, especially topographic and climatic, data, as there are only few stations and they are also unevenly located. Thirdly, the reports of the recording is not available for rural areas and areas at micro-levels. It is because of this that a true picture cannot be revealed.

From the above discussion it is very clear that there is a need to standardize data and also specifically organise the recordings of the information. In view of the new areas of research in the field of Geography of Health there is an urgent need for assessing the reliability of data and establishment of data bank at national level for better co-ordinated research.

## REFERENCES

Adams, A. (1889), *The Western Rajputana States : A Medico-topographical and General Account of Manwar, Sirohi and Jaisalmer*, London.

Aiken, S.R., Frost D.B. and C.H. Leigh (1980), Dengue hemorragic fever and rainfall in Peninsular Malasia : Some suggested relationship, *Social Science and Medicine*, 14D, 307-16.

Akhtar, R. (1978), Geography of Cancer in India, *Medicine Biology, Environment*, July-December 1978, 20-30.

Akhtar, R. (1979), Environmental factors and health in India, *Philippine Geographical Journal*, 23 (3), 109-14.

Akhtar, R. (1982), *Bibliography of Sources in the Geography of Health in India,*

Marwar Publications, New Delhi.

Akhtar, R. (1982), *Environmental, Agriculture and Nutrition in Kumaon Region*, Marwar Publication, New Delhi.

Akntar, R. (1982), India : A Study in Geocancerology, with special reference to mouth cancer, *IGU Working Group on Geography of Health*, Roma, 47, November, 1982.

Akhtar, R. and A.T.A. Learmonth (1986), *Geographical Aspects of Health and Diseases in India* ; Concept, New Delhi.

Alexander, David (1982), Disease epidemiology and earthquake disaster. The example of Southern Italy after the 23 November, 1980 earthquake, *Social Science and Medicine*, vol. 16, 1959-71.

Allen-Price : E.D. (1960), Uneven distribution of cancer in West Devon with particular reference to the divers water supplies, *Lancet*, 1, 1235-8.

Amin, M.A. (1977), Problems and Effects of Schistosomiasis in Irrigation Schemes in Sudan, *in Worthington E.B. op. cit.*, 407-11.

Armstrong, R.W. (1976), The Geography of Specific Environments of Patients and Non-Patients in Cancer Studies with a Malaysian Example, *Economic Geography*, 52, 161-70.

Ashley, D.J.B. (1969), Environmental Factors in the Actiology of Lung Cancer and Bronchitis, *British Journal of Preventive & Social Medicine*, 24, 58-60.

Auliciems, (1972), *The atmosphere environment—A study of comfort and performance*, Toronto University Press.

Azariah, J., S.R. Sekhar and Hilda, A. (1982), Potential Health Risks due to High Incidence of Faecal Coliform Bacteria in the River Cooum, Madras, India. *Geography Teacher*, (17), 107-15.

Bolin, B. (1979), Global Ecology and Man. *Proceedings of the World Climate Conference, Geneva*, 27-50.

Braden, Carrie, Jo ; M.S. and N.L.H. Braden, (1976), Environmental Challenges of Health and Health Care Administration. Chapter 3, in *Health Care—Delivery System Concept Community Health—A Systems Approach*, Prentice, Hall. U.S.

Bradshan, E.M., Schonland (1974), Smoking, Drinking and Oesophagel Cancer in African Males of Johannersburg, South Africa. *British Journal of Cancer*, 30, 157-62.

Brownlea, A.A. and C.M. Taylor (1984), Australia's environment and its influence on health. *Social Science and medicine*, 18 (11), 901.

Bydyko, M.I. (1974), *Climate and Life*, New York, Academic Press.

Burkit, D.P. (1962), A children's cancer dependent on climatic factors. *Nature* 194, 232-34.

Burton, Ian, Kates W. Robert and Gilbert F. White] (1978), *The Environment As Hazard*, Oxford University Press.

Bytchenko, B. (1966), Geographical distribution of tetanus in the world, 1951—1960, *Bulletin*, W.H.O. 34, 71-104.

Chang, K.S.F. *et al*,. (1963), Climate and conception rates in Hong Kong. *Human Biology*, 35, 366-67.

Chauhan, B.S. (1981), Desert Medicine : Ecological Approach to Health Problems, *Proceedings of the IGU Working Group on the Geography of Health*, Madras, 1-4, December.

Chen, Lincoln, C. ed. (1973), *Disaster in Bangladesh; Health Crisis in a Developing Nation*, Oxford University Press; New York.

Cleek, R.K. (1979), Cancers of the environment : the effect of scale. *Social Science and Medicine*, 13 (D), 241-47.

Cook (1971), Cancer of Oesophagus in Africa. *British Journal of Cancer*, 25, 853-80.

Dave, J.M. (1975), The Hazards of environmental pollution : *Proceedings of All India Seminar on Environmental Pollution*, Bombay, November, 5-7.

Davey, T.H. and W.P.H. Lighbody (1956), The control of disease in the tropics, *H.K. Lewis*.

Dazo, B.C. and J.E. Biles (1972), Schistomicsis in the Kainji Lake area, Nigeria, Report on a survey made in October-December 1970. Unpublished, WHO/SCHISTO/72.21.

Dixit, J.M. (1972), The Hazards of environmental pollution. *Proceedings of All India Seminar on Environmental Pollution*, Bombay.

Doll, B. Armstrong, (1981), Cancer, Western diseases, their emergence and prevention, *Edit: H.C. Trowell*, London, 93-110.

Ellis (1972), Mortality from heat illness and heat aggravated illness in the United States. *Environmental Research*, 1-58.

Elton Peter (1982), The measurement of distance when testing hypotheses of contagion, *Social Science and Medicine*, 16, 1065-66.

Farid, M.A. (1977), Irrigation and malaria in arid lands in Worthington E.B. *op. cit.*, 413-19.

Feachem, R. M. Mc Garry, Mana Duncan (1977), (eds.), *Water, Wastes and Health in Hot Climates*, Johnwiley, Sons, Toronto.

Fonaroft, L.S. (1968), Man and malaria in Trinidad : Ecological perspectives of a changing health hazard. *Annals of the Association of American Geographers*, 58, 526-56.

Fraumerni, J.E. (1973), Genetic determinants of cancer. Host Environment Interaction in the Etiology of Cancer in Man, *IARC*, Lyon, 42-55.

Gangadaran (1974), Epidemiology of cancer of the oesophagus in India. *Indian Journal of Surgery*, 36, 293-8.

Garud, M.K. (1972), Effect of Low level of air pollution in Kanpur on the growth of children. Unpublished. *M. Tech. Thesis IIT, Kanpur*.

Gentilli, J. (1980), Some climatic factors in Australian health. *Social Science and Medicine*, 14 (D), 85-100.

Gerald, F., Pyle and Robert M. Cook (1978), Environmental risk factors of California encephalitis in man, *The Geographical Review*, 68, 150-70.

Ghosh, D.K. (1971), Air pollution in Kanpur city and its effect on respiratory disease. *M. Tech. Thesis, IIT Kanpur*, 25.

Ghosh, D.K. and G.D. Agarwal (1974), Adverse effects of air pollution on human health with particular reference to respiratory diseases. *Journal of Institution of Engineers*, 55.

Giggs, J.A. (1979), Human health problems in urban areas, in D.T. Herbert and D.M. Smith (eds.), *Social Problems and the city—Geographical Perspectives*, Oxford.

Giles, Graham, G. (1981), Biometerological investigations of asthma morbidity in Tasmania using co-spectoral analysis of time series. *Social*

*Science and Medicine*, 14 (D), 111-21.

Girt, J.L. (1971). The Location of medical services and diseases ecology—some conclusions on the effect af distance on medical consultations in a rural environment. *University of Guelph (mimeo)*.

Girt, J.L. (1972), Simple chronic bronchitis and urban ecological structure in N.D. Mc Glashan (ed.). *Medical Geography : Techniques and Field Studies*, Methuen, London, 219-31.

Glick, B. (1979), The spatial auto correlation of cancer mortality by cause in Chile. *Social Science and Medicine*, 13(D), 123-30.

Goldsmith, J.R. (1967), Seasonal variation in mortality, in S.W. Tromp. W.H. Weiche (eds.), *Biometerology 2*, Oxford Pergamon, Press, 97-114.

Goldsmith, J.R., A. Landan (1968), Carbon monoxide and human health. *Science*, 162, 1352-9.

Gracey, M., D.E. Stone. Sutoto Sutejo (1976), Environmental pollution and diarrhoean disease in Jakarta, Indonesia, *Journal of Tropical Paed. Envir. Child Health*, 22 (18).

Greco, D. *et al*. (1981), Epidemiological surveillance of diseases following the earthquake of November 23, 1980, in Southern Italy. Disasters. *International Journal of Disasters Studies*. Pract. 5 (4).

Hamston, J. (1905), *Some Aspects of the General and Medical Topography of Ajmeer*. Rajputana Printing Press, Jaipur.

Hathcock, A.L., C.A. Greenburg and Dakan (1982), An analysis of lung cancer on a microgeographical level. *Social Science and Medicine*, 16, 1235-8.

Hayness, R. (1983). The geographical distribution of mortality by cause in Chile. *Social Science and Medicine*, 17, 355-64.

Hesterlow, A.M.V. (1930), Public health of Coimbatore District in relation to environment. *The Journal of Madras, Geographical Association*. V (2).

Higginson (1976), Importance of environmental factors in cancers, *Bulletin*, I.N.S.E.R.M., 52, 15-23.

Hodgart, R.L. (1978), Optimizing access to public services : a review of problems, models, and methods of locating central facilities. *Progress in Human Geography*, 2 (1).

Hopps, H.G. (1969), Computerized mapping of diseases and environmenta data. *Bulletin, Special Libraries Association (New York), Geography and Map Division*. No. 78, 24-31.

Hopps Howard C. (1971). Geographic Pathology and the Medical Implications of Environmental Geochemistry in Canon H.L. and H.C. Hopps (eds.). *Environmental Geochemistry in Health and Disease*, Geological Society of America, Boulder, 1-11.

Howe, G.M. (1972), *Man, Environment and Disease in Britain : A Medical Geography Through the Ages*, Newton Abbot, David and Charles.

Howe, G.M. (1977), *A World Geography of Human Diseases*, Academic Press, New York, 431-76.

Howe, G.M. (1982), Spatial, appraisal of ischaemic heart disease and lung-bronchus cancer in Glasgow Scotte. *IGU, Working Group on Geography of Health*, Rome.

Howe, G.M. and Loraine, J.A. (eds.) (1974), *Environmental Medicine*,

Heinemann Medical.

Hunter, J.M. (1966), River blindness in Nangodi, Northern Ghana : a hypothesis of cyclical advance and retreat, *Geographical Review*, 56, 398-416.

I.G.U. Working Group (1984), *Health in the Tropics : Geographical Perspectives*, Symposium Papers, Zambia, April 20-23.

Jelliffe, D.B. (ed.) (1978), *Child Health in the Tropics*, Monograph, 4th ed., 58.

John, M. Hunter, Luis Rey and David Scott (1982), Man made lakes and man made diseases, Towards a policy. *Social Science and Medicine*, 16, 1127-47.

John, M. Hunter and Marris D. Thomas (1984), Hypothesis of leprosy, tuberculosis and urbanization in Africa. *Social Science and Medicine*, 19, (1), 27-57.

Kayastha, S.L. and Ram, B. (1981), The importance of water supply to rural communities and related problems : An environmental assessment with special reference to Varanasi district, *Third Indian Geography Congress National Association of Geographers*, New Delhi.

King, P.E. (1979), Problems of spatial analysis in geographical epidemiology, *Social Science and Medicine*, 13(D), 249-52.

Knox, P.L. (1978), The intra urban ecology of primary medical care : Patterns of accessibility and their policy implications. *Environment and Planning*, A. 10, 415-35.

Kraybill, H.F. and Mehlman M.A. (1977), *Environmental Cancer*, Wiley.

Kumra, V.K. (1980), Environmental pollution and human health : A geographical study of Kanpur city, *National Geographical Journal of India*, 26, (I and II), 60-69.

Kumra, V.K. (1982), *Kanpur city : A Study in Environmental Pollution*, Tara Publication Varanasi.

Kutschenreuter, P.H. (1959), A Study of the effect of weather on mortality, *Trans. N.Y. Acad. Sci.*, 22, 126-38.

Lave, L.B. Seskin (1970), Air pollution and human health, *Science*, 69, 723-33.

Learmonth, A.T.A. (1958), Medical Geography in Indo-Pakistan : A study of twenty years ; Data for the former British India, *The Indian Geographical Journal*, 33 (1), 1-59.

Liposkin, H. (1978), Dietary, environmental and herediatary factors in the development of colorectal cancer, *Cancer Bulletin*, 30.

Mann, Ida (1966), *Culture, race, climate and eye disease, an introduction to the study of geographical opthalmology*, C.C. Thomas, Springfield, 373.

Marubini, E. (1982), Medical geography and trail to seek risk indicators of oncological diseases : Pitfalls and fallacies, *IGU, Working Group on Geography of Health*, Roma, 4-7 November.

Mayer, J.D. (1982), Relations between two traditions of medical geography : Health systems planning and geographical epidemiology. *Progress in Human Geography*, No. 2, 216-30.

McClland, J. (1859), *Sketch of Medical Topography and Climate on Soils of Bengal and North West Provinces*, London.

Mc Glashan (ed.) (1972), *Medical Geography : Techniques and Field Studies*, Methuer & Co. Ltd., London.

Mehta, J.C. (1977), *Urbanization and Environmental Health Habitat*, New

Asian Publishers, Delhi, 156-60.

Melinder, Meade (1976), Land development and human health in West Malaysia, *Annals of the Association of American Geographers*, 66, 428-39.

Misra, R.P. (1970), *Medical Geography of India*, National Book Trust, New Delhi.

Mukerjee, B.N. (1976), A simple method of containing health changed index and its application in micro-regional health planning (India), *Regional Studies*, 10 (1), 105-23.

Nityanandan Pillai, P.K. (1941), Public health and its relation to environment, *Indian Geographical Journal*, 10 (16), 4.

Noble, J.L.a. Montagne, M. Bellotic and Wechsler. H. (1971), Variations in visits to hospital emergency care facilities : ritualistic and meteorological factors affecting supply and demand. *Medical Care*, 9, 415-27.

Okafor, S.I. (1982), Policy and practice : The case of medical facilities in Nigeria, *Social Science and Medicine*, 16, 1971-7.

Pyle, G.F. (1979), *Applied Medical Geography*, V.H. Winson and Sons, Washington.

Ramachandran, H. (1984), Environmental factors, *The Indian Geographical Journal*, 59 (1).

Ramesh, A. (1982), Problems and issues in the study of human comfort zones in India, *IGU Working Group on Geography of Health*, 4-7, November, Rome.

Ripert, C. Ekobos, and Palner. D. (1977), Environmental assessment : The political health impact of the construction of 57 dams in the Mandara Mountains, North Cameroun, unpublished USAID document, Manda Mountains Water Reservoir Project, No. 631.

Robert, S. (1983), Distance and Utilization of Health Facilities in Rural Nigeria, *Social Science and Medicine*, 17, 563-70.

Rogers, L. (1926), Small pox and climate in India : Forecasting of epidemics, *Medical Research Council Special Report Series*, No. 106, H.M.S. London.

Rogers, L. (1944), Cholera incidence in relation to rainfall, absolute humidity and pilgrimages : Innoculation of pilgrimage as a preventive measure, *Transactions of the Royal Society of Tropical Medicine and Hygiene*, 38.

Rogers, L. (1984), Cholera incidence in India : Forecasting and control of epidemics, *Indian Medical Research Memoir*, 9.

Rowstree, B. Seebohm (1901), *Poverty, A Study of Town Life*, McMillan, London, 182-221.

Swaf, R.K. (1973), *Environmental Health*, Proceedings of Seminar held at National Environment Engineering Research Institute, Nagour.

Sargent, F. and Barr, D.M. (1965), Health and the fitness of the ecosystem, *Environment and Man*, Travellers Research Centre, Hartford, 28-46.

Sharma, S.N. (1956), A study into the incidence and ecology of cancer of lavrix and adjacent parts in Assam, *Indian Journal of Medical Research*, 46, 525.

Sen, U. (1957), Distribution of cholera on the soil of West Bengal, *Geographical Review of India*, 19, 45-6.

Shannon, G., R.L. Barkur, C.A. Meler (1969), The concept of distance as a factor in accessibility and utilisation of health care, *Medical Care Review*, 26, 143-51.

Shimkin, D.B. (1970), Man, ecology and health, *Arch. Envir. Health*, 20, 111-27.

Shonick, W. (1979), The public hospital and the local ecology, *International Journal of Health Service*, 9 (3), 359-96.

Singh, R. (1981), Moradabad city : A study of environmental pollution and human disease : *Third Indian Geography Congress, NAGI*, New Delhi.

Sistori, G. Gastini (1982), Demographical and epidemiological aspects of agency in relation with the geographic environment. *IN.R.C.A.*

Subramanian, A.R. and T. Seshani (1981), Effects of some climatic aspects on the human body in India, *Proceedings of the IGU Working Group on the Geography of Health*, Madras, 1-4, December.

Subramanian, A.R. and A. Sambasiva Rao (1981), Eco-climatic perspectives of health reports in Maharashtra, *Proceedings of the IGU Working group on the Geography of Health*, Madras, 1-4, December.

Subrahmaniam, V.P. and A.A.L. Sarma, (1981), A climatic study of human comfort zones in South India, *Proceedings of IGU Working Group on the Geography of Health*, Madras 1-4, December.

Turner, W.C. (1964), Air pollution and respiratory disease, *Proc. Roy. Soc. Med.* 57, 618.

UNDP/WHO (1979), Research on the epidemiology and methodology of schistosmiosis control in man-made lakes (RAF/11/217) Ghana and Egypt, Project findings and recommendation. Unpublished *WHO document*, 2.

Verharselt, Y. (1976), Geographical approach to geocancerology, *Medicine, Biology and Environment*, 4 (2), 39-48.

Verharselt, Y. (1977), Notes on geography and causes, *Social Science and Medicine*, 11, 745-8.

Verharselt, Y. (1981), Geography of stomach cancer in Belgium : An approach, *Geographic Medica*, Budapest, 11, 104-15.

Voronov, A.G. (1977), The geographical environment and human health, *Soviet Geography*, 18, 230-7.

Warner, P.M. (1981), Towards a geographical contribution to gerontology, *Progress in Human Geography*, 5 (3), 317-41.

Weiss, J.E. Greenlik, J.F. and J.F. Jones (1971), Determinants of medical care utilization : the impact of spatial factors : *Enquiry*, 76-92.

Weiss, W. (1977), Lung cancer mortality and urban air pollution, *Adm. Rev. Respir. Dis.*, 115, 250.

Wellmei, H. (1983), Dengue haemorrhage fever in Thailand ; *Springer — Verkey*, Berlin.

WHO (1972), Health hazards of the human environment, Geneva : *World Health Organization*.

Wilner, D. and R.P. Walkey (1963), Effects of housing on health and performance, Dhul, L. (ed.), *The urban condition*, New York, 215-28.

Wissman (1978), Spatial aspects of ageing, Washington, *Association of the American Geographers, Resource Papers*, 78-9.

Wood, D.C. and Margaret Lawrence (1980), Air pollution and human health in Greater Manchester, *Environment and Planning*, 2, 1421-39.

Worthingtin, E.B. (ed.), (1977), *Arid Land Irrigation in Developing Countries : Environmental Problems and Effects*, Pergamon Press, Oxford.

H. H. SINGH
AND
V. K. KUMRA

# TEN

# SLUMS : Threat to Urban Environment

IT IS correctly said that the 20th century has been called as the age of urbanization. In 1900 about 8 per cent of the world population lived in urban areas. By 2000 A.D. this percentage will rise to 50 and about three billion people will be living in the cities of the world. India is no exception to this process of urbanization, though our rate of urbanization is comparatively low but the size of urban population is very large. According to 1981 census it is 156 million—fourth largest urban population amongst the countries of the world. In number only U.S.A., U.S.S.R. and China have more urban population than India. By 2000 A.D. India would have 350-400 million people (about 40%) living in urban areas with 20 cities each with a population of over a million and 600 cities having 50,000 to 5,00,000 persons living in each of them. Delhi, Bombay, Calcutta and Madras would be amongst the 30 largest cities of the world with their population more than 10 million. The pattern of city growth in the 'Third World' in general, and in India in particular, is exhibiting an over increasing deterioration of the total material, social and cultural atmosphere. The ordinary man is experiencing various socio-economic strains and stress and a shameful inadequacy in civic amenities. In India a large proportion of urban population are living in slums in various shapes and forms, such as 'ahatas', 'chawls', 'bustees', shanty towns, 'jhoperpatti' and pavements. Thus, our cities are also responsible to generate the 'culture of poverty' and deteriorating phenomena of squatter settlements. In the coming years, as now, the urban landscape will continue to mirror, probably more glaringly, the contrasts and evils of society, economic disparities, social inequalities, cultural alienation and an ever-increasing deprivation for a great majority of poor city dwellers. The present trend of urbanization, if not checked, then naturally there will be more sky scrapers, more spacious and luxurious apartments for the privileged few and at the same time the number of those 'squat in squalor and drink from the drains' will also increase. The poor and the unskilled labours from the rural areas and from the small urban centres will continue to flock to the big cities in the hope of better employment and better life, only they will add to the growing number of unemployed and further deterioration in the quality of city life spreading more slums and blighted areas. The growth of slums in cities not only deteriorates the urban landscape but also causes spread of diseases, and ill-health, thus making the city centre for crimes, other evils, tension and conflicts. It is estimated that more than a quarter of the urban population of cities in India lives in slums or squatter settlements. The picture of unintended growth of squatter settlements has become an inevitable phenomenon causing threat to the city planners and authorities.

The conception of slums differs from region to region. The climatic condition, the level of poverty are also contributing factors to the varying degrees to which the slum lives can sink. Slum is regarded as a specific type of disorganized area and others, treat the term 'slum' and blighted area as synonymous. According to Bergel (1970), slums are characterised as areas of substandard housing condition within a city. A slum is always an areal unit and a single neglected building, even in its worst stage of depletion does form a slum. This reveals that a slum is always a areal unit with its specific character such as uglyness. According to United Nations Reports on 'Urban Land Policies' a slum is a building, group of buildings, or area characterised by overcrowding, deterioration, in sanitary conditions or absence of facilities or amenities, which because of these conditions or any of them, endanger the health, safety or morals of its inhabitants in the community (Report, 1952).

Further, a slum is a complex product of many factors, as is true of many other social phenomena but poverty is the foremost cause. According to Stokes (1970) 'slums are those areas of the city in which housing and resulting social arrangements develop by processes so different from those by which the general growth of the city proceeds, that they will destroy the city'. In other words, slum is the home of the poor and the stranger if nothing else which is different to the city population living in planned cities. The U.S.A. Housing Act of 1949 defined slums as 'any predominantly residential area where the dwellings by reason of delapidation, overcrowding, faulty arrangements of design, lack of ventilation, light and sanitary facilities or any combination of factors that are detrimental to safety, health, and morals'.

In India for the first time the scientific definition was developed by the Seminar on 'Slum Clearance' held in Bombay in May 1957. According to which 'slum is described as a chaotically occupied, unsystematically developed and generally neglected area which is overpopulated by persons and overcrowded with ill-repaired and neglected structures. The area has insufficient communication, indifferent sanitary arrangements, and inadequate amenities necessary for the maintenance of physical and social health and the minimum needs and comfort of human beings and the community. There is a general absence of social services and welfare agencies to deal with the major social problems of persons and families, in respect of substandard health, inadequate income and low standard of living, who are the victims of biological, pathological and social consequences of the physical and social environment'.

## Background of the work done in India and abroad

The problems of growing slums and blighted areas were realized after the First World War. But before that a number of workers raised the growing nuisance of slums in industrial cities. Zorbaugh (1929) in his monograph 'The Gold Coast and the Slums' mentioned that slums develop into an area of high land values but cheap rents. Burgess (1939) from his intensive study of U.S.A.'s cities mentioned that the slum develops within the zone surrounding the central business district. With the expansion of commercial and industrial ventures the neighbourhood becomes infilterated with industrial storage and wholesale operation. Due to less income of workers, they are bound to live in surroundings because of cheap rental value in congested localities. Further,

Lewis (1940) in his article entitled 'The Future of the Italian Cities' stated that the factory, the rail, the road and slums existed together. Filth was thrown on the roads, with great ease—an offence in many English towns in the 16th century. Engles (1944) in his article 'The Condition of the Working Class in England' described the possible conditions of the workers living in the industrial epoch as slums in the city of Manchester. Charles Abraham (1946) stated that the residents of slums cannot afford good housing because the private enterprises will not provide it at prices which they can afford. He analysed very critically their income structure and other related problems. The worst slum condition occurs when the physical slum is accompanied by over-crowding. This now seems to be the general tendency in most cities. After IInd World War due to shortage of dwellings and increasing overcrowding caused spread of slums. Lewis in his article 'Slum Life in Mexico City' has explained the actual life condition of slum dwellers. According to him, economic growth, full employment and lowering of racial discrimination in employment may raise the income of neighbourhood population, thus improving their socio-econmic conditions. Christopher (1970) very critically examined attitude of slum dwellers to live in congested and unhygienic localities.

In India very limited work is carried out by geographers on slums, their distribution and causes of growth. But other social scientists contributed much towards the genesis of slums. Malkani (1957) raised various problems of people living in slum localities and also suggested measures to improve their living condition with the help of Corporation authorities. Later on Mazumdar (1960) stated in his article 'Social Contours of An Industrial City—Social Survey of Kanpur', that due to industrial growth after First World War, problems of housing had risen up in the city which resulted growth of slums near to new industrial establishments. Nambier (1970) described that the poor living condition of slum dwellers was because of less income. Trivedi in the same year observed that slums are not confined due to the industrial growth only, they exist in the midst of the non-industrial area also. But the problem of industrial slums, is acute owing to the concentration of all the unhappy features of the unhygienic and insanitary life which exerts its own tale. Chatterjee (1961) stated that 'bustees' are product of rapid industrialization, constructed by huts which have a common structural arrangements. Due to the excavation of earth for constructing buildings shallow tanks were formed. These tanks not only serve as source of water but also help in increasing incidence of diseases. Ramchandran in the same year with his very useful study of slums given a new thought to the study of slums. The holder of low income migrants tend to occupy illegally both the public, semipublic and private lands and created slums and blighted areas for themselves. He has also suggested measures to check further growth of slums in big cities and to improve the physical condition of slums. The physical condition of slums cannot be improved only by providing amenities to them, but more effectively by educating them. Desai and Pillai (1970) made intensive study of slums of Indian cities. They stated that owing to lack of proper attention from the Town and Country Planning Department, the cities have been chronically incapable of accommodating the ever increasing flow of migrants from rural areas and small towns.

Geographers' contribution in evaluating the problems of slum dwellers and genesis of slums in cities was appreciated in seventies. Singh (1972) in his work on KANPUR, examined various factors of slum growth and also analysed various categories of slums and associated problems. Muslim majority areas have enormously swelled up and in consequences overcrowding, congestion, and insanitary conditions prevailed in many parts of Kanpur city. Dubey (1976) in his study of 'KAVAL' towns pointed out that a greater part of the low class residence represents the horrible slums of poor and insanitary state of affairs. They are menace to the health of the urban life and hence require immediate demolition and their inhabitants need to be provided with better accommodation. Badami (1977) in his monograph evaluated very critically the problems of slum dwellers and also suggested appropriate measures for improvement of living condition in slums and blighted areas of Kanpur city. Kumra (1980) did significant work on slums of Kanpur city. He has analysed several problems of slum dwellers and also examined their attitude towards shifting to another better site for living. He observed that most of the earnings of slum dwellers spent on food, medical expenses, and using alcohols which make their economic and social status very low. Further, Pathak (1983) critically evaluated the physical environment of Calcutta's slums, and provided suitable measures to improve them. Mehta and Kulkarni (1983) examined the factors responsible for choice among slum dwellers to live therein. The response of slum dwellers to their environment is best manifested in their separate physical entity. One can always remark the slum, pockets quite definitely in an any urban area.

**Causes and consequences of slums growth**

According to a report of National Building Organization the back-log in urban housing in Indian cities rose from 2.9 million houses in 1971 to 4.9 million in 1981 and it is further estimated that by the end of this century it will rise to 8.0 million. This target will be beyond the Government housing targets along with private agencies. And the housing needs of high income group and middle class people is largely taken care of by many of the private and public residential construction agencies, it is the poor who bear the brunt of acute housing shortage. Flimsy and digy dwellings, rotting garbage, open and overflowing drains are the main physical features of the many residential localities of the cities in which the poors are forced to live. The swell of population and lack of housing has transformed the face of the cities into all encompassing slums. It has been realized that slums are formed because their residents cannot afford the poorest housing in the city. According to the Director, National Institute of Urban Affairs the housing schemes of the Government in most of the big cities generally benefit people in the organ zed sector, those who have steady source of income to pay the instalments to the concerned housing boards. In a recent survey conducted by WHO, it is found that as many as two-thirds of the urban households of Madras, and Ahmedabad cannot afford even a house of Rs. 5,000, the cheapest that most recent technology can construct.

Industrialization in selected cities or urban centres have contributed greatly for the growth and genesis of slums in most of the cities of the world. The factories, mainly the cotton industry, the chemical and plastic industry, the iron industry, generally claim the best sites, near a water course for need of large quantities of water for processes of production. While factories are usually set near the rivers, or the rail-road lines that paralleled the rivers, no appropriate authority was exercised to concentrate factories in a particular area to segregate the more noxious industries that should be far away from the human settlements. But owing to lack of functional planning, and the jumbling together of industrial, commercial and domestic functions went on steadily in industrial areas, slums have started growing around the industrial establishments. In the industrial towns that, grew up on older foundations, the workers were first accommodated by turning one family dwelling into rental-barracks. In these made-over dwellings, each separate room may accommodate a whole family from Dublin and Glasgow to Bombay the standard of room family long held. Moreover, the combination of dark rooms and the dark walls formed an ideal breeding medium for bacteria, mosquitoes, especially since the overcrowded rooms afford the maximum possibilities of transmission through breath and touch, causing spread of dangerous diseases. Another consequence of slums growth is unhygienic conditions, because of absence of the most elementary traditions of municipal services. In the slum localities near to industrial establishments or on open spaces, all the dwellings are sometimes without water even from local wells. With this shortage of water for drinking and cleaning purposes, it is no wonder that the filth is accumulated. Open drains represent foulness, and any toxic municipal affluence. This will also contribute in causing spread of diseases like typhoid, cholera etc.

Slums also develop with the expansion of commercial ventures, the neighbourhood town becomes overcrowded because low income group including hawkers, labourers, rikshaw-pullers and daily wage earners and recently arrived poor regional ethnic, then move in and become the exclusive inhabitants of these areas. Owing to less rental income to the owners, causing decline in building condition and because of carelessness by dwellers, the neighbourhood becomes a slum. Thus, the slums develope into an area of high land values but cheap rents, a curious contradiction that results from the lands being held 'in pawn' so to speak on the assumption that the CBD will expand bringing into an area of new business firms, manufacturing establishments and high priced rental units like hotels and apartments. Due to lack of maintenance activities, the face of such localities converts into slums and blighted areas.

It has been observed that the pattern of land distribution in which slums and blighted areas are located in or near the main city represents a generalization fulfilled only in industrial cities, where commercial and industrial activities are dominant and does not apply to preindustrial cities. In such type of cities, formerly popular in Europe and still in a few developing nations of Asia and other parts of the world, the core areas are commonly inhabited by the elite with the slums located on the peripheries where the houses near to the city fringe are small, flimsily constructed, often one room, hovels into which families crowd. The disadvantages of distant locations are borne by the low class income group, who has to travel long distances to get city's facilities. Such pattern of slums growth is largely seen in the extensive squatters or shanty towns that have developed around the cities in Asia, Africa and Latin America to large number of persons migrated from rural areas. In a few instances, they are located closer to the cities on unoccupied or undesirable land. These slums grow like cancer spreading very fastly. 'Living conditions and levels of health and sanitation in these outlying slums are generally described as extremely low and municipal officials, seems powerless in their efforts to turn the tide, (Schnove, 1965) !

The slum environment is responsible for the degradation of mental outlook of residents, especially, its effect on growing children is most remarkable. Children daily witness scenes of violence, drunkenness, crimes and prostitution, which encourage them to do the severe offence. Because of the poverty and unhygienic conditions of the slums there is all-round deterioration in the environment with overcrowding full of all possible social and economic crimes.

**Morphology of Slums according to their types and condition**

The morphology of slums is concerned with the physical appearance as well as social implication. Physical appearance of slums indicates the structure of slums, their types and building material used. The appearance is generally one of structural average and decline. Social implications relate with the social structure of the locality. Like other areas within or adjacent to the city, the slum changes and move. It is perhaps the most sensitive part of urban settlements because it reflects quickly and in different manners, mostly negative, to the various social, economic, administrative and other aspects of community life. One fairly universal characteristic of the slum is the poverty of its inhabitants. Slums differ greatly with respect to the social organization of their dwellers. They range from the slum in which the inhabitants are stronger to one another to the family slum in which there is wide acquaintance betwen the inhabitants. It has been found that slums inhabited by immigrant groups may have a firm social organization.

The functional character of cities also affects the morphology of the slums. In Indian context specially in metropolitan cities slums are generally of three types : (*i*) Areas with single or multistoreyed buildings, built longago as per standards prevailing in those days and which are to-day in deteriorating conditions and require immediate action for demolition, (*ii*) The second type of slum localities consists of semi-permanent structures which are on both authorized and unauthorized lands. A few of these areas consists of semi-permanent structures built of corrugated iron sheets. They are called as 'chawls' in Bombay, and (*iii*) This type of slum is the hutment colony or the squatter colony. These huts are usually called as 'Jhoppad Pattis', built with an assortment of materials consisting hardboard, mud, zinc sheets, gunny bags, tats etc. Such hutment colonies are not many in city proper but the outskirts areas of the city and along railway lines, have a large number of them owing to the open areas available there. Crimes are at extreme in such slums and it was found that people have left regular jobs to do full-time business in illicit liquor preparation and sale.

The India Habitat (1976, Government of India) report clearly indicated that 'the process of migration in large number of cities has created problems of accommodation, utility and service to a magnitude that the cities cannot meet. Migrants occupied whatever land that was available and constructed huts and shacks from material that they could afford. Thatch, mud, bricks, tin, wood ressian, and canvas were assembled to provide shelter. The huts multiplied and grew into squatter settlements. The older part of the city became very congested and most of the old buildings were occupied by as many as families, which further created congestion and lack of facilities, giving birth to old slums'. Our metropolitan cities such as Bombay, Ahmedabad, Bangalore, Calcutta, Delhi, Madras, Kanpur etc. are experiencing the alarming growth of squatter like 'chawls' 'bustees', 'jhuggihopari', katras, and 'cheries' type of structure. Such squatter settlements have witnessed the multiple growth of population. Table 10.1 shows the type of slums and their general characteristics in big cities of India.

**Slums of Indian cities**

The slum population is 33 million (1984) and it needs immediate and urgent attention in the country. It is further estimated that nearly 25 per cent of population in cities live in slums or squatter settlements. It has been observed that while the large cities are bursting at their seams, the small and medium towns in the country are stagnating. Since 1951, Class V and Class VI towns have hatled in number. While Class I towns increased their share of the urban population from 44 per cent to 60 per cent in the last three decades, the share of town with less than 20,000 persons (Classes IV, V, VI) has diminished from 30 per cent to only 14 per cent. This reveals that urban population in big cities is increasing at soaring rate while population in small towns is decreasing. It has greatly affected the housing situation of big cities. In Bombay 77.6 per cent, Delhi 58 per cent, Kanpur 61.5 per cent, Madras 48.4 per cent people live in one room dwellings. Table 10.2 shows the percentage of population living in slums in selected cities of India (1981).

The dramatic increase in the size of Indian metropolitan cities has not reduced the pressure of population or poverty in rural areas. India is facing problems not only of 'teeming cities' but also of 'teeming countryside'. The huge migrating rural population, each and every year with a tremendous swelling of larger urban centres is multiplying the slums in the cities and reducing the area of the countryside. During the last thirty years, our urban population has increased by 74.0 per cent during the same period, rural areas lost 15,000 km$^2$ land to expanding cities. The metropolitan centres have recorded a horrible increase of slum dwellers. About 40 per cent of the 1.68 million urban population in Kanpur lives in slums according to 1981 census while in Kanpur the percentage is highest. It is lowest at 10.20 per cent in Bangalore. Numerically, Calcutta has the highest slum population of 3.24 million or 35.35 per cent in an urban population of 9.16 million. Bombay comes next with a slum population of 3.15 million or 38.38 per cent of the urban population of 8.22 million.

TABLE 10.1 : **Showing types of slums and their characteristics**

| Type | Location | Structure | General condition and available facilities |
|---|---|---|---|
| 1 | 2 | 3 | 4 |
| *Older Type* Old city slums | Around CBD | Old pukka buildings (multi-storeyed) but densely populated | House without repairs, absence of sanitary and other facilities, no proper drain and low standard of living. |
| 'Ahatas' and Katras | Mostly near the industries and congested commercial and residential areas | Big enclosures with one or two rooms flat | Lacking bath rooms, laterines, kitchen and open space, mainly occupied by low class industrial workers and unhygienic environmental conditions. |
| 'Bustees' | Inside the oldest part of the city | Mostly 'kachcha' house with tinshade used as cattle house | Prevalence of livestock dirt, cow-dung. Mostly occupied by milkmen, washermen, riskshaw pullers, hawkers, without sanitary facilities. |
| 'Chawls' and 'Jhuggi Jhopari' | Located in and sometimes in the periphery of CBD | Mostly kuchcha house with tinshed, brick walls, ressin, plastic sheets, mat etc. | Lack of urine facilities, no drains, public taps or wells for water supply, overcrowding and juvenile deliquency etc. |

TABLE 10.1 : (*Contd.*)

| 1 | 2 | 3 | 4 |
|---|---|---|---|
| *Old Rural Settlements* | | | |
| Abadis | Small villages, located at outer fringe of the city on low lands | Mostly kachcha houses with a few brick-wall houses, mostly occupied by labourers, and daily wage earners | No proper facility of drinking, water, using open latrines, no sanitary facilities. |
| *Newer Slums* | Located in and around colonies, most of them on unauthorized lands | Few are pukka with kachcha walls, some are in the form of Jhuggies | Rented to poor workers, without sanitary facilities and common taps. |
| *Temporary Hutments* | | | |
| | Along roads, railway lines and also in few open spaces on Govt. lands | Constructed with wood, scraps, tents, or with soil using tin shade as roof | Unhygienic conditions, no water supply, using house front as latrines. Occupied by mostly beggers, and orphans. |

TABLE 10.2 : **Showing slum population in selected areas of India (1981)**

| Cities | Percentage of population living in single room | Slum population as per cent of total population |
|---|---|---|
| Bombay | 77.6 | 38.30 |
| Calcutta | 64.8 | 35.35 |
| Madras | 48.4 | 31.87 |
| Ahmedabad | 64.0 | 26.16 |
| Delhi | 58.0 | 30.19 |
| Kanpur | 61.9 | 40.00 |
| Poona | 51.9 | — |
| Hyderabad | 27.4 | 21.28 |
| Lucknow | 50 8 | 38.83 |
| Jaipur | — | 15.62 |
| Bangalore | — | 10.20 |

(These figures are compiled by National Building Organization of Ministry of Works and Housing, December 10, 1984). The percentage of slum population in other cities are : Delhi 30.19, Madras 31.87, Hyderabad 21.28, Ahmedabad 26.16, Lucknow 38.83, and Jaipur 15.62.

Bombay's housing situation is the worst. It is learnt that in the problematic city of Bombay about 50 per cent of the inhabitants are compelled to live in 'chawls' and 'Johpadpattis'. In addition, there are a number of slums on land belonging to the government.

Table 10.3 shows the number of slums in Greater Bombay according to ownership of land occupied by them.

Table 10.3 indicates that there are 466 slum localities in

TABLE 10.3 : **Number of slums in Greater Bombay (1984)**

| Ownership of land | Number of slums | Approximate number of huts |
|---|---|---|
| 1. State Government | 189 | 42,867 |
| 2. Municipal | 80 | 47,291 |
| 3. M.W.B. land | 10 | 4,920 |
| 4. Private | 187 | 1,04,993 |
| Total | 466 | 2,00,071 |

Bombay having 200,071 'huts' absorbing about 1.3 million persons. In addition to this, a sizeable population of Bombay lives in squatter settlements with the surroundings of sky scrappers. According to a survey of the Bombay Municipal Corporation, 15 per cent of the people live in 1,000 buildings which are on the verge of collapse, approximately 5.00,000 persons live on pavements, which depicts the great shortage of housing. Every year a shortage of 55,000 dwellings is added to a previous shortage of 8,00,000 houses. Bombay's housing crisis has been compounded by the rapid deterioration of its housing stock (Bombay Municipal Corporation, 1984). Dharavi stretching from Matunga, Sion, and Bandra with half a million squatters is amongst the India's biggest slums. Practically the whole area belongs to the municipality, having been acquired by the Ex-Improvement Trust. The lowlying and marshy nature of land and the consequent considerations of the costs involved in reclaiming the same had been the main factors responsible for this area lying idle so far and for not being developed under the Improvement Schemes. The water-closets, washing places, water taps are too inadequate, being common for a number of structures and detached from the 'chawls'. The part of the city in which old buildings predominate soon got blighted and many of the areas lapsed into slums of worst type. The 'patra chawls' (single storeyed structures, corrugated iron sheet structures) located mainly in the northern outlying areas of the city, is one of the densiest inhabited areas of Bombay where man is living in worst hygienic conditions.

If Bombay depicts the evils of thriving Third World Metropolis, Calcutta represents all symbols of a deteriorating city. Calcutta's housing mess has its origin in its development as a centre of Britisher's commerce and industry, basically in jute, textile, and engineering goods. Due to lack of civic infrastructures, thousands of migrant labour were housed in shacks errected on tracts of fallow land in and around the city. According to a CMDA report about 66 per cent of its population lives in temporary buildings. The city of Calcutta has been confronted with the alarming pattern of migration since the peripheral areas have not been developed with the adequate production, income and employment prospects. The sky scrappers have grown in the city of Calcutta to its physical proximity, the thousands and thousands of squatter settlements have emerged. In Calcutta there are more than 1,000 slums which include 20,000 damaged houses, require immediate demolition. The city of Calcutta has not reserved any specific earmarking of land for service personnel or migrants, thus consequently the mushroom growth of squatter slums has become a great problem with a spectacular growth phenomenon. Most of the 'bustees' of Calcutta are located in low-lying areas around ponds and ditches. About 0.5 to 1.0 million people are living on pavements (Dogra, 1985). Garbages are allowed to accumulate on the impaved pathways. 'Khatala and Cowshades' are other nuisance inside the 'bustees' which choke the drains and sewers of the area and allow free breeding places for mosquitoes and flies. Thus, the 'bustees' of Calcutta are subjugated to a degenerating type of living environment. Almost the rooms of the 'bustees' are non-ventilated and receive little sunlight in any part of the day or fresh air in any part of the year. Very few 'bustees' possess water taps within their boundaries. In the majority of cases, municipal taps, far away, are to be used, the number of users ranging often from 200-300 per tap. It is estimated by the CMDA that there is annual demand of 60,000 houses every year to meet the essential needs of migrants for shelter. This can be only done if long-term planning measures are to be taken for slum improvement.

Slum conditions in Delhi prevail in all parts of the city except a few in New Delhi areas. Congestion, both of houses on land and of people within houses, obtains here in the most acute form. According to a report on "Jhuggi 'Jhopari' Settlement in Delhi" prepared by Town and Country Planning Organization, Ministry of Works and Housing, there were 199 squatter settlements which accommodated 12,749 households in 1951. In 1973, there were 1,373 such settlements which are six times more than that of 1951, occupied by 141,755 households. The annual growth rate of the squatter households has been calculated as 11.8 per cent during the period as compared to that of 4.5 per cent for the city as a whole. In the year 1951, there was one household over 20 households living in squatter settlements. Table 10.4 shows the increase in growth of 'Jhuggi—Jhopari' in Delhi.

According to a survey three-fourths of the slum families in Delhi are in-migrants families. More than 40 per cent of these families are headed by displaced persons from West Pakistan in 1947 (Report, 1957). According to place of origin, 45 per cent of all the in-migrant families, including the bulk of refugees, came from

TABLE 10.4 : **Growth of Jhuggis in Delhi**

| Year | Number of clusters | Number of Squatters | Growth index 1951=100 |
|---|---|---|---|
| 1951 | 199 | 12,749 | 100.00 |
| 1961 | 544 | 42,814 | 235.00 |
| 1971 | 1,124 | 1,15,961 | 825.00 |
| 1981 | 1,373 | 1,41,755 | 1,011.18 |

*Source* : Delhi Municipal Corporation Information, 1981.

urban areas and 55 per cent from rural areas. The basic amenities of water supply, latrines and electricity are throughout grossly inadequate and it is more acute in the 'bustees' of Old Delhi. The situation of slums aggravated by the presence of milch cattle and other animals within the slums. It is estimated that roughly about half of slum population is presently putting up with these nuisance.

Earlier people lived in squatter settlements of Delhi scattered on road-berms, slushy and sinking beds of the river and embankment of 'nallah' etc. Out of 1,400 haphazard clusters, 72 per cent have no water taps, 69 per cent have no lavaratory, 68.0 per cent on storm water drain and 63 per cent have no street lighting. It was estimated that about 2,000 acres of developed land, the market value of which exceeds to Rs. 200.00 million had been distributed amongst the urban poor and that is the biggest socialist measure taken in the city. Table 10.5 gives an idea of house allotment in resettlement colonies of Delhi to rehabilitate slum dwellers of Delhi.

These resettlement colonies of Delhi were planned with open spaces and proper landscapic provisions. Such resettlement schemes are generally launched with the promise of making the city clean, but filth and squalor is the predominant feature of the most of the areas occupied by these colonies, specially during the rainy season. Shortage of toilet facilities, and lack of regular cleaning is a very common feature of these localities. Due to the way in which narrow blocks have been laid out in several places, it has become difficult to provide proper drainage. Due to hurry in emergency for setting these colonies, no proper site selection was made which resulted that a few localities have to face flood problems in the drainage area of rivers Yamuna and Hildon.

TABLE 10.5 : **House allotted in resettlement colonies**

| Sl. No. | Name of the resettlement colony | Number of houses |
|---|---|---|
| 1. | Mangolpuri | 1,128 |
| 2. | Sultanpuri | 500 |
| 3. | Jwalapuri | 50 |
| 4. | Nangloi Phase III | 92 |
| 5. | Khyala | 34 |
| 6. | Haiderpur | 140 |
| 7. | Shakurpur | 120 |
| 8. | Trilokpuri | 800 |
| 9. | Kalyanpur | 150 |
| 10. | Khichhripur | 50 |
| 11. | Jahangirpuri | 1,006 |
| 12. | New Seemapuri | 324 |
| 13. | Nand Nagari | 340 |
| 14. | Dakshinpuri | 594 |
| | Total | 5,328 |

*Source* : Allotments of Flats, '*The Hindustan Times*', February 23, 1977.

The number of persons living in slums is on the increase every day in Madras. This is a general phenomenon during the past 30 year. According to a survey conducted by the Corporation in 1953-54, there were 306 slums in the city containing 57,436 families, having a population of 265,000. It has increased to 548 slum localities with 59,573 huts, containing a population of 412,168 in 1961 which was about 23.8 per cent of the total population of the city. According to present estimates, 31.87 per cent of the city's population live in slums of different types. It was also noticed that in Madras the slum areas have a birth rate, 43 per cent higher than the non-slum areas ; only to be checked by a parallel (about 50%) excess in death rate. The infant mortality rate is also higher in case of slums to the extent of 70 per cent. Most of the slums (20%) of the city are located on the seashore and banks of water courses, the Buckingham canal accounting for one-half of such slums. New Washermanpet, Tondiarpet, Grace Garden, Singara Garden, Royapuram, Ammen Koil (North), Korukupet, Wedia Nagar, Choolai, Ammaiyar Nagar, Koraput etc. areas reveal worst condition of slums.

The slum dwellers are engaged in odd jobs involving heavy labour or petty occupations. Among women, general labourers occupy the first place, followed by domestic servants like cooks, maids and then by cleaners and sweepers. It has been observed that people's tendency to congregate in and around their place of work is assumed to be one reason for the development of slums and blighted areas. Apart from 41.5 per cent of the workers who have no fixed place of work, another 36 per cent of them live within a distance of about 3 kms. from their place of work. Thus, it means that about 77.0 per cent of workers living in slums, live in and around their usual place of work, which has encouraged them to live in slums. The acute shortage of dwellings is still causing growth of slums in the city. According to a recent study, housing demand in Madras is estimated at 0.7 million in the next two decades, half of them for low income group and daily wage earners. Though it is a tough task to provide suitable accommodation to such a huge number of slum dwellers but through proper planning measures the problem of further influx of population in cities may be stopped through certain remedial measures.

The slums of Ahmedabad city are not only confined to the industrial areas alone, they also exist in the midst of non-industrial areas. But the problems of industrial slums are acute due to aggregation of all the unhygienic features of an insanitary life which exacts its own toll. In the city 85.0 per cent of the housing deficit has influenced the low income people. About 0.415 million persons were living in slums in 1978 and another 0.30 million were headed into insanitary 'chawls' in Ahmedabad (Ahmedabad Municipal Corporation, 1980). Further, out of a total member of 18,652 households living in the slums as many as 16,971 (90.9%) inhabit in one room tenements, 1,396 (7.5%) in two room tenements and the rest in more than two rooms. In certain wards of the city, such as Jamalpur I, Kalupur, Shahpur II, Asarwa and Baheranpura more than 90 per cent of tenements live in one room. These localities possess all sorts of problems associated with essential human needs. There is no proper facility of drinking water and drainage system, which has caused health hazards and spread of diseases. Due to low income and unhygienic condition in their working environment as well as living conditions, most of the industrial workers were found to be deficient in working capcity. Slum dwellers living near the old textile industries are the worst ; hit by such environmental problems of unpotable water supply, foul air and insanitary situation.

Slums in Kanpur city are only located in a few pockets of the city, but in every part of the city. Even the newly developed colonies have slum dwellings. Most of the slum localities are located in the inner part of the city and near to the old industrial establishments. In Kanpur, there are three types of slums—'ahatas, newer slums and temporary hutments. A survey conducted by Kanpur Nagar Mahapalika in 1978 revealed that there are 835 'ahatas' inhabited by 2,59,552 persons. About 1,58,511 persons live in 'bustees' such as Kakadeo, Babupurwa, Ompurwa, Jajipurwa etc. Most of the temporary hutments are located in outskirts of the city along G.T. Road and Railway lines. The socio-economic conditions of slum dwellers have forced them to live in these slums. One cannot imagine to what low depths, the human life can sink in these slum localities. There are common latrines in most of the 'ahatas' of the city. The apathy of the local authorities in the removal of waste dumps and the stench from the open drains, stagnant pools of sludge attracting flies and rodents, are a source of health hazard in all the slum localities of the city. Water taps are only available in a few slum localities, such as Sisamau, Darashanpurwa, Babupurwa, Jajipurwa and Aryanagar. Other slum localities have to use well water which is contaminated. In addition to this, due to poor financial condition, an ordinary slum-dweller cannot send his children to schools, which leads to juvenile delinquency amongst children.

The slums of Poona city developed after 1937 owing to growth of the city mainly for industrial and commercial purposes. According to a survey conducted by Gokhale Institute, there were 1,000 families living in slums or squatter settlements of the city. The Corporation also conducted a survey of slum localities in 1951, which recorded that 6,304 families were living in such localities. It was further recorded that in 1958 there were nearly 9,000 families living in slums. According to a present survey, about 22,500 families in the city of Poona live in slums (1984). Though the Corporation has tried to provide sanitary facilities in slums but the continuing process of migration in the city, has caused further growth of slums and thus creating stress on civic amenities.

The problem of slum-growth is not only recognised in metropolitan cities but also in cities which are approaching towards the million mark and other towns of the country. To know the problems associated with housing and increasing growth of squatter settlements, a case study of Varanasi (India) has been made for next lower recent metropolitan cities. The major growth of slums in Varánasi has taken place within a span of 35 years. After the partition, thousands of refugees from Pakistan settled in the city. Further, after 1971 Indo-Pakistan War, many of the Bangladesh is came and settled near the Railway station and outskirts of the city. According to a report of V.D.A, about 21.5 per cent of the total population lives in slums and squatter settlements. Most of the slums have developed around the periphery of the main city in pockets such as Sunderpur, Konea, Lallapura, Reori Talab, Shivpurwa and Saraiya Fakira Tola. Among all, Saraiya Fakira Tola locality is the biggest one where the worst condition of human habitation is visible. In addition, there are old buildings in the main city which contain sometimes more than 10 families with common facilities of latrine and water supply. it is estimated that about 85 per cent of the slum population lacks sanitary and water supply facilities. Most of the slum dwellers are employed in weaving, biri-making and daily labourers such as rikshaw pullers, in contruction. To know the building material used, a survey was conducted in a slum locality named 'Reori Talab' (1984), and it was found that 65 per cent of the dwellings are built with mud walls and thatched roofs, 30 per cent are covered with plastic sheets and tin sheets and only 5 per cent have permanent roofs. There are 63,013 persons living in 42 slum localities of the city, which require immediate solution. Every year the city requires 7,000 to 10,000 houses to provide better housing facilities to the inhabitants. For this purpose, construction of new houses and improvement in slum localities is urgently needed so that the environment of the city may be presumed as it is one of the most important tourist centres of the country.

## Suggestions to improve slums

The above study reveals that the slums particularly in million cities of India are growing frequently every day because of continuous rural-urban migration. If the same rate of migration will continue it will be very difficult to cope with the situation. To check the further growth of slums, it is essential to adopt some effective measures to stop the growth of urban centres, especially of cities and metropolises. There are two methods by which problem of uncontrolled urban growth can be tackled. The direct method is to clear the areas of blight and slums and house the affected dwellers in a descent environment. This method is called as 'slum clearance scheme', which refers not only cleaning slums but also properly rehabilitating the dishoused population. Population living in temporary hutments and 'jhugis-jhoparis', 'bustees' can be removed and may be rehabilitated to some other site. But before this there is urgent need of educating slum dwellers to keep their environment by healthy practices. Government should prepare schemes after intensive study of slums and attitude of slum dwellers. It has been observed in few cities of India that after getting houses in new residential areas, the slum dwellers again settle in the previous locality in which they were living or they form another slums on some new sites. So the 'scheme' must adopt the regulated step of educating, uprooting, rehabilitating and watching the slum dwellers.

The indirect method is to control rural-urban migration, develop agricultural incomes, redistribution of land to peasants, increase employment opportunities in small towns and create new settlements with employment avenues. Control of migration may prevent further worsening of urban congestion, but it does not by itself solve the already existing chaotic situation. Thus, slum clearance schemes will be more beneficial and get the most important consideration in urban planning. In addition, the 'slum removal programme' is also a most exciting project which was taken up in many cities. This can be implemented by one of the following methods (Report, 1964)—Declaration of the area as a clearance area. Such areas are selected where the only effectual remedy for improving conditions, is demolition of all such buildings which are unhygienic for living due to disrepair or sanitary defects. This can be secured by—(a) ordering the demolition of the buildings in the area by the owners themselves, and (b) acquiring the entire area by the Corporation and then demolishing the buildings.

The basic conditions laid down for the application of the scheme to an area are—(a) the area must contain 50 or more dwellings for the poor classes out of which one-third should be overcrowded and not capable of being rendered fit for human living at reasonable expense, (b) the re-development of the area as a whole is essential to provide housing accommodation for the lowest class of people. The clearance and redevelopment of the area must necessarily follow the scheme prepared by the Corporation and sanctioned by the Government. The sucessful implementation of the scheme depends mainly on the pychology of slum dwellers prepared by the society.

Further it can be noted that the efforts made by the Government and the private agencies have resulted a very nominal achievement in getting the slums cleared and the people rehabilitated in better urban areas. The growing urban population, because of the rural migration, is the main cause to bringforth the dangerous cancer for urban life but unless and until people are educated to hate the slum environment, it appears difficult to get rid of this problem. For example, in India the recent attempts have resulted only to provide better living only to 6-8 million people out of 33 million, the total slum population of the country. It was proposed to cover about 40 per cent of the remaining slum population (10 million) by the end of the Sixth Plan. The main problem in implementing these plans is lack of reliable data. But according to mid-term appraisal of Sixth Plan (1983), as against the plan target of covering about 10 million slum dwellers, only 3.77 million are covered. There is likely to be a shortfall in the achievement of the targets. To provide low cost dwellings to slum dwellers is also not a easy task. According to World Bank Studies two-third of the urban house-holds of Kanpur, Madras, Ahmedabad, Calcutta and Delhi cannot afford a house of Rs. 5,000, the cheapest one that modern construction technology can built. This shows very poor condition of urban residents living in squatter slums. Though the Government is preparing plans to demolish slum in these cities and rehabilitate them elsewhere, but it will require long term duration and larger exchequer.

The governments are only able to play a very marginal role in providing a roof over the heads of the urban poor. Social housing schemes were intended to serve housing for economically weaker sections with households income below Rs. 350 p.m. and households in low income between Rs. 350-600 p.m. In this regard in India the Sixth Plan document had a confession to make. 'It now appears that houses constructed for a particular income category are largely occupied by families in the next higher income group since the instalments payment required for the houses are beyond the paying capacity of the income group for which the houses were constructed. This is the biggest problem the low income group is facing. For this, Corporation should be very strict for allotting such dwellers to slum dwellers only. There must be a National Policy for improvement and rehabilitation of slum dwellers in the country. This will help in solving the problem of urban ecosystem which is deteriorating day by day, as slums contribute much to be degrading environmental quality of the cities. Voluntary organization should come forward to help the concerned authorities for slum improvement in big cities.

It appears that cities will continue to increase rapidly even in the coming century. The challenges of our cities where millions have to face spatial tyrrany, the administrative and academic problems of our times in planning, development and management will remain. The goal of 'World Peace' to a large extent depends upon what happens to the cities of the developing countries specially of China and India. Unless the nature of the underlying forces that operate both at the national and international levels is clearly understood and a new relationship is built, the tragic elements in the horizon of our cities will continue to loom large. Our cities have slums primarily because 'the power to clean or prevent them has been taken away from us'. Urban planning, urban ideology, urban patterns, urban linkages and urban resources were all issues of fundamental importance—not only for the individual countries but also for the international community. Therefore, the slums being the greatest blot on administration, especially on Municipal administration, require integrated approach both on national and international levels for rural and urban together. The philosophy of separate urban identity is to be revised and rural development for social and economic facilities must be at par with the urban counterpart. Thus, the underlying deteriorating forces, whole over the world in general and in India in particular, will be checked to generate an healthy urban environment without slums and blights.

## REFERENCES

Badami, J.V. (1976), *Slums of Kanpur Poor Man's Struggle for Shelter*, Kanpur Development Authority, 53-66.

Benniger, Christopher (1970), Models of Habitat Mobility in Transitional Economic, *Ekistics*, 29.

Bergel, E.E. (1970), The nature of slums, in *Slums and Urbanization* (edited). Desai, A.R. *et al.*, Popular Prakashan, 39.

Burgess (1939), The growth of city, in *The City*, (edited), Robert E. Park, Chicago, University Press, 32.

Chatterjee, A.B. (1967), *Hawrah : A Study in Social Geography*, Kashipati Bharti Studies I, Calcutta, 74.

Charles, Abraham (1946), *The Future of Housing*, Harper and Row, New York.

Clinard, M.B. (1970), Theories of Slums, in *Slums and Urbanization* (edited), Pillai and Desai, Popular Prakashan, 51.

Desai, A.R. and Pillai, S.D. (eds) (1970), *Slums and Urbanization*, Popular Prakashan, Bombay, 1-4.

Dogra, Bharat (1985), Life in hovels under the Sky, *The Hindustan Times*, March 24th.

Dubey, K.K. (1976). Use and Misuse of land in 'KAVAL' Towns of Uttar Pradesh, *N.G.S.I.*, Department of Geography, B.H.U., 43-46.

Engles, Frederick, (1936), *The Condition of Working Class in England* (edited), Allen and Unwin, London.

Kumra, V.K. (1982), *Kanpur City : A Study in Environmental Pollution*, Tara Book Agency. Varanasi, 154-164.

Leo, F. Schnove (1965), On the spatial structure of cities in two Americas, in *The Study of Urbanization*, (edited), Philip M. Houser, John Wiley and Sons, New York, 366.

Lewis, Oscar (1958), The Culture of Vecindad in Mexico city, *UN/UNESCO Publication*, Proceeding of Seminar on Urbanization Problems in Latin America.

Malkani, H.C. (1957), *A Socio Economic Survey of Baroda City*, M.S. University of Baroda.

Mazumdar, D.N. (1960), *Social Contours of an Industrial City—Social Survey of Kanpur*, Asia Publishing House, Bombay.

Mehta, S. and Kulkarni, P. (1983), Location choice among slum dwellers, a human response to urban environment in *Environmental Management* (edited), L.R. Singh *et al.*, The Allahabad Geographical Society, 188.

Nambier. P.K. (1970), The slums of Madras city, in *Slums and Urbanization* (edited), Desai and Pillai, Popular Prakashan, Bombay, 176-178.

Pathak, C.R. (1983), Environmental Problems of Calcutta City, Published in *ISEMA* abstract, NATMO, Calcutta.

Ramchandran, P. (1970), The Slums : A Note in facts and solutions with special reference to Bombay in *Slums and Urbanization* (edited), Desai *et al.*, 161-170.

Reports, Different reports from the Municipal Corporation and Government

Agencies published timely.

Singh, H.H. (1972), *Kanpur : A Study in Urban Geography*, Indrashani Prakashan, Varanasi, 79-82.

Trivedi, R.K. (1970), Slum Areas of Ahmedabad City in *Slums and Urbanization* (edited), p. 211-215.

some of the cases in point. The geographer is competent to play his part in portrayal of the present situation of pollution level of Ganga or existence and quality of waste-lands in the country. These documents could be permanent sources of references for action plan and monitoring. His synthetic analysis of the facts may open up new avenues of thinking and approach. We wish very much that geographers in this country come forward to contribute towards the national efforts and establish their capabilities.

## REFERENCES

Burton, I., R.W. Kates and G.F. White (1978), *The Environment as Hazard*, Oxford University Press, New York.

Dutt, G.K. (1982), Mapping the Environment, *International Journal on Environmental Studies*, 19, 25-35.

Dutt, G.K. (ed.) (1980), *Atlas Cartography*, National Atlas and Thematic Mapping Organisation, Monograph 3, Calcutta (Preface).

Dutt, G.K. (1985) Thematic Cartography : Problems and prospects in India, Souvenir of the XVII All Rajasthan Geographers Conference, Central Arid Zone Research Institute, Jodhpur.

Dutt, G.K. and *et al.* (ed.), *Proceedings of the International Seminar on Environmental Maps and Atlases*, National Atlas and Thematic Mapping Organisation, Calcutta (forthcoming).

Dutt, G.K. and P. Nag (1985 A), *Life and Work of Professor S.P. Chatterjee*, National Atlas and Thematic Mapping Organisation, Occasional Monograph 1, Calcutta.

Dutt, G.K. and P. Nag (1985 B), *Scenario of National Atlas and Thematic Mapping Organisation in 2001 A.D.*, Department of Science and Technology, Calcutta.

Harvey, David (1970), *Explanation in Geography*, Edward Arnold, London, 115.

Kates, R.W. (1978), *Risk Assessment of Environmental Hazard*, John Willey, Chichester and New York.

Nag, P., (1983), Perception of the Environment : Issues and Challenges, *National Geographical Journal of India*, XXIX (Nos. 3 and 4), 217-31.

National Atlas and Thematic Mapping Organisation (NATMO) (1982), *National Atlas of India*, Department of Science and Technology, Calcutta.

White, G.F. (ed.) (1974), *Natural Hazards : Local, National, Global*, Oxford University Press, London.

**J. Singh**

ELEVEN

# POPULATION, POLLUTION AND THE URBAN ENVIRONMENT

DURING the last eighty years, the urban population in India living in towns and cities has rapidly increased from 25 million to 160 million. In comparison to many countries of the world, the level of urbanisation is far behind but even this population has started posing a problem to maintain 'minimum comfortable living conditions' in the urban areas. Though, nearly four-fifth of the total investments is cornered for the development of the urban habitat yet it has not been able to meet even the basic needs of its dwellers. About 30-60 per cent people are bound to live in clumsy houses and slums. Now urbanisation is posing a threat to environment. Urban environment is being degraded day by day with the onslaught of population growth and migration from rural to urban areas. In 1971, nearly 24 million people migrated from villages (Town and Country Planning Organisation, 1975). The large scale movement of countryside people has accentuated the grave situation of the cities. The crowded cities are facing more and more pressing problem with the arrival of new faces. In urban areas the finer values of human life and society is being polluted by the appearance of slums, improper solid waste disposal, water, air, noise pollution etc. Day by day the problem is worsening. Since most of the development expenditure is siphoned of to big cities particularly by our metropolitan centres without taking care of development. Owing to this reason, the urban environment of these metropolies are being adversely affected by various types of toxic elements. So it is a high time to evaluate and analyse the trend of urbanisation in various classes of towns and how it has created a threat to human survival in general, and dwellers of million cities in particular. Major portion of this ecological disaster is created by no one else than man himself, and that too, in course of his living. Further, the utter technological and industrial growth have added the agony of discomfort.

**Trend of urbanisation**

In most of the developing countries of the world, urbanisation is a recent phenomena. Particularly after the second world-war these countries have experienced not only tremendous urban growth but many of the cities have reached to the brink of 'urban explosion' which consequently caused 'urban chaos' over there. In India about 26 million were enumerated as urban residents in 1901 which reach to about 33 million in 1931. However, a real spurt was marked in 1951 when urban population touched the mark of 62 million. Further. the urban population became more than doubled in the past twenty years from 79 million in 1961 to 160 million in 1981.

The growth pattern of urban population has been varying in different inter-census periods. The percentage variation during 1901-31 was slow. Actually it was a period of loss, recovery and slow growth. After this urban population recorded a very high growth rate particularly in decades 1941-51 (41.42%), 1961-71 (38.23%) and 1971-81 (46.38%).

It is interesting to note that in 1901 urban-rural population ratio was 1 : 8.2 which became 1 : 3.3 in 1981. It was solely because of economic factors which 'push' ruralities to urban areas to get rid from the poverty of subsistence agriculture, and 'pull' them on account of high wages, opportunity of getting jobs and other facilities. But, less attention has been paid to such movement which is causing great concern to planners, administrators and environmentalists. It was estimated that by 2001 A.D. the population of urban areas may go well over 50 per cent of the total population.

**Disproportionate growth in various classes of towns**

The growth of urban population is found strikingly different in various sizes of towns. The proportion of urban population in

Class I towns increased from 22.02 per cent in 1901 to 60.37 in 1981, while medium sized towns (Class II and Class III towns) have maintained more or less the same proportion. In the case of small towns (20,000 persons) i.e. class IV, V and VI towns, the proportion of population has declined continuously as well as significantly from 49.63 per cent in 1901 to 13.64 per cent in 1981. Among these most adversely effected are class V (20.63% in 1901 to 3.61% in 1981) and class VI towns (6.38% in 1901 to 0.51% in 1981).

Excepting class V and VI towns, the urban centres of other categories have registered population increase. The population of class I towns has increased about seventeen fold, class II six times, class III five times and class IV two and half times. This reveals that concentration of population is increasing at a rapid pace in big cities as compared to medium and small sized towns. With the increase in population in bigger cities, the number of class I cities has rose up from 25 in 1901 to 216 in 1981. By the turn of the present century, Calcutta was the lone million-plus city in the country. In 1911, Bombay too attained this hallmark. By 1951 Delhi, Madras and Hyderabad even had a million residents. In the next 30 years, seven more cities—Ahmedabad (1961), Bangalore (1961), Kanpur (1971), Pune (1971), Nagpur, Lucknow and Jaipur—joined the million group. These twelve cities now account for more than a quarter (25.73%) of the nation's urban population whereas they shared 18.61 per cent in 1901. In this way they recorded more than eight fold increase in population during the period 1901-81.

From the foregoing descriptions it is apparent that bigger cities are attracting a large influx of migrants as compared to other categories of towns even to-day because the factors responsible for natural increase remain more or less the same for all the regions. Because of heavy influx into million cities for jobs, urban environment is being polluted in terms of slum development, air, water, noise pollutions and generation of more solid wastes. Still most of the development is taking place in these large urban centres which are already crowded. Subsequently, they are emitting pollutants in a big way. In Bombay above 3,000 industries are located. Several of them are chemical and pharmaceutical units. In addition, as many as 65 per cent of the country's pesticide units are located alone in Bombay. It is posing a serious environmental threat. It indicates that Bombay is sitting on living volcano. Air is not being polluted only by the industries located in its midst but with the growth of population, number of vehicles are also increasing which also create air pollution in large proportion. It is also causing concern for increasing noise and water pollution. A recent survey carried out together by various institutes of Calcutta has disclosed that eight out of every one thousand people in and around Calcutta have had their hearing ability impaired, not by any apparent physical or hereditary cause but by noise pollution. It also reveals that average traffic noise level of Calcutta is about 75 decibals which is beyond the acceptable limit. The situation has been grave and alarming for the last ten years. It is common notion that if there is more to use, there is also more to throw. Thus, the so called priviledged urban class discharge more solid waste which not only deteriorate the quality of the environment but is also injurious to the health of the passer-by as well as residents of that locality. Keeping in view the multifarious adverse impacts of these pollutants on the residents, the magnitude of these pollutants have been discussed here in sequel.

**Housing and slums**

With the continuous influx of migrants in addition to the natural increase, there has been an acute shortage of housing facilities which consequently give rise to slums. Table 11.1 shows an overall picture of housing shortage in urban India, which ultimately leads to slum development. The National Building Organisation (NBO) has reported the backlog in urban housing to the tune of 4.8 million in 1981 and is estimated to jump to

TABLE 11.1 : **Estimated housing demand, supply and shortage in Urban India**

*(in thousand)*

| Year | Demand | Supply | Shortage |
|---|---|---|---|
| 1951 | 12,500 | 10,305 | 2,195 |
| 1961 | 15,780 | 14,063 | 1,717 |
| 1971 | 21,640 | 18,413 | 3,227 |

*Source* : Bhargava and Dougre, (1977).

7.2 million by the end of 1991.

Because of acute housing shortage a considerable proportion of urban dwellers in metropolitan cities has been forced to live in slum and blighted areas. Studies have revealed that most of the slum population is composed of rural migrants, e.g. in Ahmedabad about 83 per cent and in Delhi nearly 93 per cent of the household heads in the slums are migrants from the countryside (Sinha, 1985). It is estimated that about 0.2 million people are migrating to Delhi every year and about 30,000 new 'jhuggis' are added to existing one. In Bombay too, between 1931 and 1881, there was a net migration of 1.06 million which gives a figure of 290 per day. If trend is not reversed, it is estimated that 70 per cent of Bombay's population will be living in slums at the turn of present century. The poverty of the countryside, push rural population to the cities but better jobs often become their beyond reach which compel to take shelter eitherin the slums or in pavements. Table 11.2 indicates that a vast majority of the population in the metropolitan centres lives in slums and single room hutments.

TABLE 11.2 : **Showing persons living in slums and single room dwellings in some million cities of India (1980)**

| City | Percentage of population living in single room | Population living in slums |
|---|---|---|
| Bombay | 77.6 | 4.50 |
| Calcutta | 64.8 | 3.60 |
| Madras | 48.4 | 0.91 |
| Hyderabad | 27.4 | N.A. |
| Pune | 51.9 | N.A. |
| Delhi | 58.0 | 1.50 |
| Kanpur | 61.9 | 0.53 |
| Lucknow | 50.8 | N.A. |
| Ahmedabad | N.A. | 0.41 |

*Source* : Respective Development Authorities.

In Bombay these slum dwellers live on about 3,200 ha of land, thus bringing the per ha population of the dwellers to over 1,000. Nearly 40 per cent slums have two to four people packed in one room, 35 per cent households have five to six people crammed into one living room and 1 per cent have 10 or more people living in a room (D'monte, 1985). About 0.1 million people live on the pavements in Bombay, whereas in Calcutta an estimated 80,000 people go to sleep on the streets every night. The situation is so appalling that in slum localities a quarter of the households do not even have access to community toilets, drinking water facilities, electricity and covered drainage system. Another interesting feature of Bombay slum dwellers is that 38 per cent earn a family income of between Rs. 350 and Rs. 600 per month (D'monte, 1985). The economic structure of slum dwellers in all the cities is more or less the same. A study has shown that in Kanpur about 45 per cent of slum dwellers have income of Rs. 125—Rs. 250 per month (Kumra, 1982). Invariably the low income capacity forces them to live in slums. It is a common sight to see slum localities in every corner of the city which in absence of basic civic amenities create the gamut of whole environment foul and polluted. Due to unhieginic conditions, infant mortality is found higher in slum areas than non-slum areas.

**Air pollution**

The air over our million cities is greatly polluted on account of exhausts of poorly maintained motor vehicles, varying types of industries, thermal power plants, burning of coal and wood and other fossil fuels. With the growth of population, number of vehicles are also increase which emnates gaseous and particulates in a large amount (Table 11.3).

With the growth of vehicles, city residents are faced with increasing accidents, congestion and pollution. There were 496 fatal road accidents in 1975 in Delhi which became more than double i.e. 1,179 in 1984 (Hindustan Times, 12 August 1985). A study on motor vehicle pollution has concluded that the air in major traffic functions of the capital is too poisionous to breathe, with levels of pollutants exceeding the safety limits (Hindustan Times 29th July, 1985). The fashionable Connaught Circles is worst affected. Today, motor vehicle of Delhi along with 70,000 heavy vehicle entering the city daily discharges about 170 tonnes of carbon monoxide, 70 tonnes of hydrocarbons, 8 tonnes of oxides of nitrogen, 2 tonnes of sulphur dioxide and huge quantities of particulate matter including lead (Hindustan Times Oct. 20, 1985).

TABLE 11.3 : **Growth of motor vechicles in five metropolitan cities**

| Cities | Year | No. of vehicles |
|---|---|---|
| Calcutta | 1971 | 100,486 |
| | 1978 | 1,520,075 |
| Bombay | 1971 | 177,415 |
| | 1978 | 265,414 |
| Delhi | 1971 | 204,078 |
| | 1978 | 449,053 |
| Madras | 1971 | 48,497 |
| | 1978 | 79,771 |
| Kanpur | 1971 | 35,746 |
| | 1978 | 47,714 |

*Sources* : (1) Planning Commission (1980).
(2) Kumra (1982).

The traffic hazards are also caused by reduced visibility due to smoke. From Table 11.4 it is evident that air pollution has

TABLE 11.4 : **A quality in some Metropolitan Cities of the country**

(*in micrograms per cu. m.*)

| City | Suspended particulate matter 1970 | 1979 | Sulphur dioxide ($SO_2$) 1970 | 1979 | Oxides of nitrogen 1979 | Standards |
|---|---|---|---|---|---|---|
| Ahmedabad | 307 | 243 | 11 | 71 | N.A. | 1. Maximum accepted level of suspended particulate : 150 micrograms/cu.m. (ISI standard) |
| Bombay | 241 | 275 | 47 | 83 | 23 | |
| Calcutta | 341 | 578 | 33 | 85 | 24 | |
| Delhi | 601 | 481 | 41 | 39 | 38 | |
| Hyderabad | 146 | 295 | 5 | 26 | 16 | |
| Jaipur | 446 | 379 | 4 | 17 | 10 | 2. Maximum permissible levels of $SO_2$ (ISI standard) : 60 micrograms/cu. m. |
| Kanpur | 544 | 344 | 12 | 25 | 9 | |
| Madras | 101 | 106 | 8 | 16 | 12 | |
| Nagpur | 262 | 386 | 8 | 12 | 22 | 3. Maximum acceptable of oxides of nitrogen : standard not established |

*Sources* : (1) N.E.E.R.I., Nagpur 1978-79 ; and
(2) P.H.E.N., Nagpur 1970.

reached serious levels in most of the million cities of the country. Because of increasing levels of air pollution, the highest percentage of persons affected by T.B. in the world are found in Kanpur, Bombay, Calcutta and Delhi. In Delhi industrial units emit 115 tonnes of sulphur dioxide and 1,288 tonnes of suspended particulate matter every month. These pollutants caused severe allergies, eye irritation and respiratory problems. Among all kinds of pollution, air pollution is one of the most serious one concerning the health and well being of the people. In almost all the big cities the levels of air pollution have exceeded the permissible limit.

**Noise pollution**

Though city's noise pollution has made a sudden jump yet being hazardous to health in terms impairement of hearing or total deafness increase in blood pressures and heart beats, giving rise to peptic ulcers, fatigue, psychic deficiency. In recent times noise is considered an important aspect of urban environmental pollution. The main sources of noise pollution in cities are the industries, automobiles, running trains and the loudspeakers, construction activities.

A study conducted by the ENT Department of Maulana Azad Medical College Delhi has revealed that the level of noise varies from 64 (Teen Murti Marg) to 92 dB (Daryaganj). Intensity of noise is higest at Chandani Chowk. In Kanpur, level of noise varies 40 dB at HBTI to 89 dB at Parade in Kanpur (Kumra, 1982). Bombay has been characterised as the noisiest city in India with an average level varying between 75 to 90 dB. Maximum intensity of noise was observed at C.T. Tank, Church Gate and V.T. Station. Besides these, noise level is also marked high all along the routes of local train which passes through the midst of city. Aircrafts create severe noise in the nearby residential localities. In this respect Bombay, Calcutta, Delhi and Madras are worst affected by incoming and outgoing aircrafts.

A recent study undertaken jointly by Bose Institute, Saha Institute and the Calcutta Medical College reveals the average traffic noise level of Calcutta is above 75 dB (A) which is beyond acceptable limit. According to international standards, the average noise level in residential areas should not be higher than 45 dB (A). During the day's busiest hours (4 to 6 P.M.), the maximum average sound level of 90 dB (A) was recorded at Moulali Crossing near Sealdah Station, the Ganesh Talkies crossing and near Esplanade, Manicktola crossing, Howrah bus terminus and Posta Market records sound level more than 80 dB (A). They have noticed that due to resonance column effect in the multistoried buildings on narrow streets or lanes, the noise level in top floors is the same or even higher than that in the ground floors.

Noise greatly affects the hearing capacity of industrial workers where machines operate day and night. According to Safety and Health Act, workers should not be exposed to noise level of more than 90 dB (A) over 8 hours per day. A study on noise level in textile mill like I alimali reveals that noise level varies from one section to another. Maximum noise level is recorded in weaving section (100-105 dB) while folding depths record should level from 69-75 dB. This indicates that in textile mills of India which has engaged about 8,00,000 workers, noise levels are above the threshold of danger in all the sections with potential to cause permanent hearing loss.

TABLE 11.5 : **Textile Machinery : Noise level in different sections**

| Machinery section | Noise level in dB (A) |
|---|---|
| 1. Card room | 87— 90 |
| 2. Draw frames and spinning frames | 89— 90 |
| 3. Winding machines | 85— 89 |
| 4. Warping machines | 85— 87 |
| 5. Large weaving shed | 100—105 |
| 6. Folding depths | 69— 75 |

*Source* : Purushottam (1978).

**Solid waste disposal**

People do not pay attention while disposing the garbage, ashes and other waste materials on roads, infront of houses and other public places, but it should be kept in mind that if it is not disposed-off properly, it will be hazardous for health and well-being of the residents. Solid wastes include garbage, rubbish, street sweepings, ashes, animal and agricultural wastes, dead animals and other industrial waste. With the increase in popula tion and industrial activities, the generation of solid waste also increase. According to the Rubbish Department of Kanpur, the amount of solid waste increased from 197,162 tonnes in 1966 to 251,137 tonnes in 1976. Solid waste production varies from city to city according to the population and then functional activities. In India, Lucknow produces maximum quantity of garbage (0.81 kg/capita/day), followed by Kanpur and Bangalore (Table 11.6). Calculation is based on population of 1971 census.

TABLE 11.6 : **Generation of solid waste in few million cities of India**

| City | Total solid waste disposal per day (in tonnes) | Per capita per day soild waste disposal |
|---|---|---|
| Calcutta | 2500 | 0.36 |
| Bombay | 3000 | 0.51 |
| Madras | 1200 | 0.38 |
| Delhi | 2100 | 0.58 |
| Kanpur | 850 | 0.67 |
| Lucknow | 700 | 0.87 |
| Bangalore | 1000 | 0.61 |

It is common practice in almost all the cities to dump the city refuse in lowlying areas which create apparently disorder in a big way by infusing into ground water which also fouls the air and give rise to mosquitoes. In our country the disposal of refuse from a community or individual is a most neglected aspect. Even the elite and priviledged urban class do not bother and throw their daily refuse haphazardly on the roads. Their cleanliness are confined to their houses only, and by throwing the wastes of their home they spoil the whole environment by which they themselves are not immuned. It has been observed that people from high income group generate more garbage than other in terms of paper, vegetable waste and covers, plastic pieces, glass and crockery pieces. Further, the population size of city is equally responsible for amount of total waste production in the cities (Table 11.7). Through different processes it polluted the land, water and air which ultimately give rise to incidence of various diseases with

TABLE 11.7 : **Showing percentage of type of waste according to the population of cities**

*(in lakhs)*

| Characteristics | Population | | | |
|---|---|---|---|---|
| | 2 | 2—5 | 5—20 | 20 |
| Paper | 3.09 | 4.74 | 3.80 | 7.07 |
| Plastics | 0.57 | 0.59 | 0.81 | 0.86 |
| Metals | 0.51 | 0.38 | 0.64 | 1.03 |
| Glass | 0.29 | 0.34 | 0.44 | 0.76 |
| Ash and fire earth | 46.60 | 39.97 | 41.81 | 31.74 |
| Total composite matter | 33.41 | 39.76 | 40.05 | 41.74 |
| Carbon | 12.56 | 12.51 | 11.95 | 15.92 |
| Nitrogen | 0.60 | 0.61 | 0.50 | 0.51 |
| Phosphorus as phosphate | 0.70 | 0.71 | 0.67 | 0.59 |
| Potassium | 0.70 | 0.73 | 0.72 | 0.69 |
| Density (kg/cu. m) | Varies from 350 to 550 | | | |

*Source* : N.E.E.R.I., Nagpur.

the transmission of biological vectors. In many places it remains for many days.

**Water pollution**

Water is an essential ingradient for the survival of human organism and for this purpose safe and adequate quantum of water is pre-requisite. But like air and noise it is also contaminated. On the one hand with the rapid growth of population and industries, the requirement of water is increasing ; while on the other hand, water is being polluted in the same proportion by its consumer itself.

Two types of problem related to water supply. There is a large section which do not have such facilities, while other one facing the shortage as well contaminated water with their own game. India's sixth plan document claims that nearly 27 per cent urban population is devoid of even drinking water facilities. Out of 1,027 towns which lacking drinking water supply systems, 902 are class IV, V and VI. The Central Public Health and Environmental Engineering Organisation (CPHEEO) of the Union Ministry of Works and Housing recommends 70-100 litres for class V and VI towns ; 100-125 litres per capita for class III and IV and 125-200 litres for class I and II. Data in Table 11.8 shows

TABLE 11.8 : **Availability of per capita per day water in selected million cities of India, 1978**

| Cities | Per capita supply (litres) daily |
|---|---|
| Bombay | 192 |
| Delhi | 197 |
| Calcutta | 71 |
| Poona | 249 |
| Ahmedabad | 115 |
| Bangalore | 15 |
| Hyderabad | 96 |
| Madras | 55 |
| Kanpur | 132 |

*Sources* : (1) Central Board for Prevention and Control Water Pollution.
(2) Hussain (1975).

that five out of nine 'million cities' in 1978 were below the recommended estimate. Among this the worst affected cities are Calcutta, Hyderabad and Madras. Besides, owing to the inadequate supply of water, per capita supply has also decreased in many cities with the growth of population. Among the city dwellers the hardest hit sections are people living in slums. The slum dwellers obtain their water requirement from public standpipers. It is common to see a long queue of women and children with their buckets in morning and evening. Even standing for hours to get water, the trickling water pipe could not meet their requirement. In absence of sewer system, they discharge waste water into open drains and sometime the stagnant pool of water becomes ideal places for mosquito breeding.

In most of the cities the same water is unfit for consumption from the bacteriological point of view. According to the C.P.H.E.E.O. study sixty-one per cent of the water supply of class I cities comes from surface sources, 9 per cent from ground water and the remaining 30 per cent from a combination of these sources and these sources are contaminated with surface runoff from the dumps of urban solid wastes and industrial wastes, various toxic chemicals, proper and untreated disposal of industrial and municipal water as well as sewage. This has created problem of water pollution. Besides, all these, the cremation of dead body on river bank also create pollution. This contaminated water is used by residents for bathing, drinking, industrial and other purposes causing harmful effects on living organisms.

Discharge of untreated sewer water into water bodies is the main source of water pollution in cities. The characteristics of sewage effluent in respect to Bio-chemical Oxygen Demand (B.O.D.) tells the gravity of pollution level in sewage water. From the Table 11.9 it is evinced that industrial cities have higher B.O.D. contribution in its sewer as compared to other cities on account of disposal of industrial waste water into sewers.

TABLE 11.9 : **Bio-chemical oxygen demand contribution in Sewage water of million cities of India, 1976**

| City | BOD contribution (gm/capita/day) |
|---|---|
| Madras | 27.2—38.5 |
| Bombay | 31.7—63.5 |
| Kanpur | 31.7—39.0 |
| Ahmedabad | 27.2—45.3 |
| Calcutta | 38.0—44.0 |
| Delhi | 30.5—38.3 |

*Source* : NEERI, Nagpur.

The high B.O.D. load of sewage water ultimately joins the water bodies which in turn deteriorates the quality of water. The main sources of water are either rivers or coastal sea water and all these are greatly polluted by various polluting agents. Table 11.10 shows source of pollution and effluent discharge per day. It would be fair to accept that almost every river and coastal water near major cities is getting polluted every day by innumerous forms of pollutants.

Delhi consumes about 1.20 million kilo litres of water every day and pours back 0.96 million kilolitres of toxic water into River Yamuna. The River Yamuna is polluted mainly due to continuous discharge of waste water and sludge from 13 storm water drains, out of which Najafgarh chain is the largest. The Central Board for the Prevention and Control of Water Pollution has categorically stated that quality of water in the Yamuna was of category 'E' which was fit for industrial use only, and not fit for bathing purposes, leave alone drinking. According to the findings of the

TABLE 11.10 : **Status of major Indian rivers**

| River | Cities | Polluting agents | Effluent flow per day |
|---|---|---|---|
| Ganga (Hooghly) | Calcutta | Chemicals, tar, paper, metallic salts, jute, sulphur, human, animal and domestic wastes | above 43 million litres |
| Bombay Coastal Water | Bombay | Oils and greases, industrial and community wastes, chemicals, fertilizer, acids, power plants and nuclear plant, wastes, viscose rayon wastes | — |
| Ganga | Kanpur | Tanneries, textile, Jute defence plants and other industrial wastes, human, animal and domestic wastes, dyes, starch, oils and greases | |
| Yamuna | Delhi | untreated human wastes, industrial effluents | Above 24 million litres |

*Source* : Halder (1984).

report, one of the main reasons of the rivers pollution is the increasing growth rate of urban population and increase in the number of 'jhuggi clusters' slum areas and the non-conforming uses of urban and rural villages. Besides these, there are 17 large industries in Delhi which also cause pollution in the Yamuna.

In Kanpur, River Ganga and Pandu are the sources of water supply as well as absorber of the city's sewage. So about 40 per cent of water supply is contaminated. The gravity of the situation can be gauged that about 40 mgd of waste water is discharged into river Ganga through sewers only (Kumra, 1982). Besides, a number of outfalls carrying effluents of textiles, tanneries, plastic and urban industries directly join the River Ganga. In Calcutta too the situation is critical. There are about 350 outfalls for discharge of Greater Calcutta waste water by paper mills, jute mills, textile mills, chemical factories, tanneries, engineering factories etc. Total volume of liquid wastes discharged through the various outfalls is 252 mgd, of which 77 mgd is contributed by various industries. Because of the increasing salt percentage (250 ppm in 1935-40 and 380 ppm in 1980), the Ganga water is today fast becoming unpotable (Halder, 1984). Besides water pollution count (WPC) has increased from 5-10 per cent in 1960 to

20-25 per cent in 1970. Service latrines is also a major source of pollution city's water. A report prepared by Central Inland Fisheries Institute, Calcutta says that the fishes in the Hooghly is highly contaminated with heavy metals and can lead to disastrous consequences if eaten. Fish forms one of the major diets of Calcutta's people. Elements like zinc cadmium, lead and even mercury are thrown into the rivers by the industries with effluents. According to N.E.E.R.I. the heavy metal intake has resulted into increase in bone diseases. Arthritis is also found to be a common disease here.

In Madras and Bombay, the sewage water is directly discharged into ocean through many open chains. In Madras the total quantity of sewage discharge is estimated as 30 mgd while in Bombay about 445 million litres of untreated sewage is discharged every day in the Arabian Sea. If the water of the River Ganga which has highest self purifying capacity is not fit for consumption then the water of other rivers like Yamuna, Gomati and Sabarmati must have crossed the permissible limit. These all are explicitly enough to indicate the saga of awful situation regarding water pollution. As a result, water is becoming undrinkable, fishes and other aquatic life specimens are diminishing, diseases like diarrhoea, indigestion, jaundice, hepatitis, stomach and liver troubles, high blood pressure on account of arising salinity are the other outcomes of grave concern.

**Conclusion**

In all probability it can be said that the environment of our metropolitan centres are far from satisfactory in either way, and day by day with the establishment of newer industries and with continuous influx of rural migrants, the quality of our environment is becoming alarming. The day is not far when the fumes of factories and other pollutants will reach the limit from where return is not possible. So it is high time that every precaution and prevention should be taken so that urbanisation which has become a national process will may not become problem to all of us. The National Environmental Engineering Research Institute has confirmed that the levels of sulphur dioxide and other particulate matter in big cities have exceeded the permissible limit as prescribed by WHO. Hence the industries whose wastes are physiologically hazardous, should be immediately closed down. Electrostatic precipitator should be used to minimise the air pollution especially by power plants. In vehicular traffic such devices should be developed which could reduce the smoke emission levels. Before discharging sewer water into rivers, it should be treated properly, and effluents of industries should not be poured into either ocean or rivers. Now time has come that environmental considerations should have priority whenever an industry is to be located anywhere. Development should be in harmony with the environment. In recent years every city development authority has launched a massive programme of plantation in city areas, it is a welcome step. This measure will not only beautify the city but it will also reduce the air and noise pollution. To a large extent the present day urbanisation problem can be solved if villagers and urban dwellers find a confortable living in their own places. Besides all the steps to improve the environment, until people do not perceive the problems in their own interest, all efforts will be in vain. Therefore, people's co-operation and keen involvement in cleaning their own environment is must and to achieve this goal awareness about the causes and consequences of the problems should be inculculated in them through mass-media, training camp and other sources. If the pollution control measures are not adopted in metropolitan and other big cities now, then pollution levels may cross the point of redemption. This it is imperative for every citizen to create and emit least pollution either in terms of need or greed.

## REFERENCES

Bhargava, G. and P.N. Dougre (1977), Housing for urban poor, *The Economic Times*, 14th July.

Dayal, D'monte (1985), "How the Other Half Lives ....." *Hindustan Times*, Sunday Magazine, 20 October.

Haldar, A. (1984), *Pollution Rules Over Calcutta, Calcutta*, .92.

Hussain, S.K. (1975), *Quality of Water, Water Supply and Sanitary Engineering*, Oxford and I.B.H. Publishing Co., 79.

Kumra, V.K. (1982), *Kanpur : A Study in Environmental Pollution*, Tara Book Agency, Varanasi, 163.

Planning Commission (1980), *Report of the National Transport Policy Committee*, New Delhi.

Purushottam, S. (1978), The noise nuisance, *Science Today*, The Times of India Publications, February, 35.

Sinha, V.C. (1985), *Resolving India's Urbanisation Problems : The Gandhian Way*, Khadi Gramodyog, XXXI (9), 368.

Town and Country Planning Organisation (1985), *Towards a Human Settlement Policy*, New Delhi, 55.

PRADIP KUMAR ADHIKARY

TWELVE

# THE POLLUTION LOAD IN THE GANGA WATER

OUR Prime Minister has launched very seriously a scheme and he is working with firm determination for restoring the Ganga water to its pristine purity. In spite of the fact that the sacred river has a big place both in our mythology and literatures and it reflects the culture and hèritage of India from ancient times, the Ganga has been turned into a most polluted river for we have meted out a very selfish treatment to it for centuries.

## The Ganga Basin

In this context while pondering for a discussion on the pollutional load of the Ganga water, we should at least be aware of the general physical background of the Ganga basin. Since human activities both agricultural and industrial, are the principal monitors of pollution load in the Ganga, a first hand knowledge of the physical setting of the vast basin is a pre-condition for any such analysis. The Ganga basin is bounded by the Himalayas on the north and by the Vindhayas on the south and after traversing so vast an area it reaches to the Bay of Bengal. From the source at Gangotri, the river after traversing nearly 250 km in the Himalayas descends into the plain at Rishikesh. Further down from here is Hardwar, where thousands of people take a dip in the sacred 'Har-ki-Pauri' : River Ganga receives half a dozen of tributaries in the upper region which helps to transport all soluble as well as solid matters to the river. The river ceases to be known as Ganga hundred km downstream of *Rajmahal hills* and bifurcastes into Bhagirathi and the lower portion of which beyond Kalna (West Bengal) is known as Hugli and Padma which form the boundary between India and Bangladesh. The total length of the river course is about 2,525 km and the total area of the basin in India is about 861,404 km² and that nearly 60 per cent of the water flowing through the river comes from the catchment areas of the northern plains which is primarily an agricultural region. Human activities in this basin are dependent on Ganga water for its heterogeneous uses. Both agricultural and industrial practices in this basin give birth to major pollution agents i.e., chemical pollutants, sewer wastes and agricultural pollutants. Here, excessive use of chemical fertilisers and pesticides cause tremendous pollution of river water after being washed away by rain, irrigation water permeating through the subsoil down to even the ground water level. In this connection, it may be noted that a quantity of 1.15 m tonnes of only chemical fertilisers and 15,000 quintals of pesticides are used within 80 km stretch on either sides of the Ganga channel cn an average in a year. (Table 12.1).

TABLE 12.1 : **Total chemical fertilisers and pesticides used in different States**

*(in tonnes)*

| | Fertilisers | | | Pesticides | | |
|---|---|---|---|---|---|---|
| | Nitro-genous (N) | Phos-phorous (P) | Potas-sium (K) | Organo-Phos-phorous | Organo-Chlorine | Others |
| West Bengal | 82,194 | 9,265 | 16,767 | 2,02,224 | 1,94,339 | 3,52,284 |
| Bihar | 1,23,192 | 16,228 | 8,745 | 13,676 | 73,491 | 1,18,707 |
| Uttar Pradesh | 5,38,508 | 55,335 | 35,422 | 35,422 | 6,29,092 | 2,14,259 |
| Madhya Pradesh | 4,547 | 3,242 | 275 | 42 | 2,472 | 803 |

*Source* : The Anti Pollution Board of West Bengal, Bihar and U.P.

**Pollution of water**

Pollution in the environmental elements is an inevitable outcome of human craving for betterment of living standards through the increasing efforts for heavy industrialisation and urbanisation. People use Ganga water for heterogeneous purposes, like drinking, irrigation, industry, bathing, performing religious rites, nurturing fisheries and also use the river as repository for dumping industrial effluents and sewage wastes of urban centres. Two other uses of the river water are, the generation of hydro-electricity and water transport which also deplete dissolved oxygen in water. Most of these activities in the basin area advertently or inadvertently vitiate the purity of water of the river. Besides, the reckless felling of trees in the basin as well as excessive use of fertilisers and pesticides all have vital roles in the matters of pollution of the Ganga water.

A survey conducted by WHO in the year 1973 showed that only a half of the urban population have a drinking water supply in their houses and virtually 12 per cent of the rural population have access to safe drinking water in this basin. This factor together with the inadequate waste treatment facilities and the generally low level of personal hygiene of rural India leads to very high level of enteric diseases. The presence of faecal material in the Ganga water poses the most immediate hazard to human health since faecal material from either human or animal sources contain pathogenic micro-organisms. Here, diseases transmitted by faecal material are often caused by viruses, bacteria, protozoa or metazoa. It is calculated that bacteria constitute over 30 per cent of the total wet volume of faeces and counts of $10^{10}$ anaerobes, and $10^{8}$ aerobes per gram wet weight of faeces are reported under Indian condition. It is found in a study that chlorinated unfiltered water from Hugli river distributed to hydrants in Calcutta contained pathogenic vibrios and salmonellae in at least some of the samples, and protozoa and viruses are known to be more resistant to disinfection than bacteria (Table 12.2). The Table 12.3 will show the survival time of bacteria in different samples of water of the Ganga in and around Patna.

Strangely enough, acute shortage of water in Agra in times of drought force authorities to supply water from a partially treated sewage 190 km (WHO, 1973) away from New Delhi. In such a situation the distinction between water treatment and waste water reclamation becomes very slight. A large number of industrial units, towns and cities along the entire stretch of the river Yamuna and Ganga are mainly responsible for the pollution of Ganga. Disposal of untreated effluents of the industrial units and

TABLE 12.2 : **Rural Organic Pollution Load (B.O.D.) n the Ganga Basin (1977)**

| | | Rural B.O.D. load generated From : (kg/per day) | | | Percentage of total load | | Total B.O.D. per rural inhabitant (gram/day) |
|---|---|---|---|---|---|---|---|
| | | Total load | Human origin | Cattle origin | Human origin | Cattle origin | |
| 1. | Haryana | 100,138 | 49,551 | 50,587 | 49.48 | 50.52 | 30 |
| 2. | Uttar Pradesh | 2,017,991 | 1,262,683 | 755,308 | 62.57 | 37.43 | 24 |
| 3. | Bihar | 1,340,167 | 963,175 | 376,992 | 71.87 | 28.13 | 21 |
| 4. | West Bengal | 478,101 | 354,290 | 123,811 | 74.10 | 25.90 | 20 |
| 5. | Rajasthan | 410,393 | 155.837 | 254,556 | 37.97 | 62.03 | 40 |
| 6. | Madhya Pradesh | 527,488 | 242,924 | 284,564 | 46.05 | 53.95 | 33 |
| 7. | Himachal Pradesh | 18,966 | 5,967 | 12,999 | 31.46 | 68.54 | 48 |
| 8. | Delhi | 11,505 | 8,415 | 3,090 | 73.14 | 26.86 | 21 |
| | Total Ganga Basin | 4,904,749 | 3,042,842 | 1,861,907 | 62.04 | 37.4 | 24 |

*Source* : State Pollution Control Boards of U.P., Bihar and West Bengal.

TABLE 12.3 : **Survival time of bacteria in different water samples in and around Patna**

*(Number of days required to kill bacteria)*

| Sources of water samples | Escherichia Coli-form. | Faecal Streptococcus. | Vibrio-cholera. | Coli-form count. (MPN/ 100 ml) | Organic Pollution (B.O.D in Mg/ Litre). |
|---|---|---|---|---|---|
| Ganga at Patna | 7.5 | 7.5 | 1.0 | 90,442 | 2.3 |
| Ganga at midstream | 7.0 | 6.0 | 1.5 | 11,000 | 0.5 |
| Gandak at Hazipur | 15.0 | 10.0 | 1.5 | 1,700 | 1.1 |
| Son at Koelwar | 13.0 | 16.0 | 5.0 | 3,273 | 1.5 |
| Dugwell at Patna | 30.0 | 16.0 | 7.0 | 2,400 | 1.2 |
| Tubewell at Patna | 10.0 | 16.0 | 6.0 | 5 | 0.0 |
| Double distilled water | 12.0 | 10.0 | 3.0 | 0 | 0.0 |

*Source* : Survey report of the pollution control boards of the state of West Bengal, Bihar and Uttar Pradesh.

the sewage wastes from the towns and cities contaminate water of the river. The contamination of Ganga water may be classified into organic and inorganic and contents which are generally found in the form of suspended, colloidal and dissolved matters. Simultaneously, as mentioned earlier, innumerable living organisms are also present in bacterial and aquatic form in polluted Ganga water. Discharge of untreated sewage water into the river is the main source of water pollution in cities which in turn use water for drinking from the river with minimum purification. The Table 12.4 shows the extent of organic pollution which *permeate through the environment* and ultimately reach the river Ganga for its entire length within the basin.

The pollution level of the river Ganga at any particular place at a specific point can be understood by reading the BOD (Biological Oxygen Demand) of sewage waste water which is poured into the river through out falls and open drains. The BOD test gives the amount of molecular oxygen required by bacteria to reduce the carbonaceous material. The purity of water is indicated by its ability to maintain a certain amount of dissolved oxygen. It is this dissolved oxygen that is used by fish and aquatic plants. Table 12.5 shows the Organic Pollution load (B.O.D.) generated from urban and rural sectors in the Ganga basin.

**TABLE 12.4 : Origin and destination of urban Organic Pollution (B.O.D.) discharge**

*(Figures in parenthesis indicate per cent of the total BOD discharge)*

| States | Origin of B.O.D. discharge (kg/day) | | Total B.O.D. discharge (kg/day) | Destination of B.O.D. discharge | | B.O.D. content in waste-water (mg/l) |
|---|---|---|---|---|---|---|
| | Domestic | Industrial | | Surface water | Land | |
| Haryana | 19,335 | 11,665 | 31,000 | 31,000 | 0 | 545 |
| | (62.4) | (37.6) | (100) | (100) | 0 | — |
| Uttar Pradesh | 3,71,626 | 7,18,274 | 10,90,000 | 2,75,000 | 8,15,000 | 737 |
| | (34.1) | (65.9) | — | — | — | — |
| Bihar | 56,172 | 50,828 | 1,07,000 | 90,000 | 8,000 | 304 |
| | (52.5) | (47.5) | — | (92.5) | (7.5) | — |
| West Bengal | 2,66,076 | 1,12,924 | 3,79,000 | 3,79,000 | 0 | 436 |
| | (70.2) | (29.8) | — | (100) | 0 | — |
| Rajasthan | 40,712 | 11,288 | 52,000 | 51,000 | 1,000 | 258 |
| | — | (78.3) | (21.7) | — | (0.9) | — |
| Madhya Pradesh | 43,591 | 11,409 | 55,000 | 30,000 | 25,000 | 166 |
| | (79.3) | (20.7) | — | (54.5) | (46.5) | — |
| Himachal Pradesh | 1,900 | 0 | 1,900 | 1,900 | 0 | 350 |
| | (100) | (0) | — | (100) | 0 | — |
| Delhi | 2,42,565 | 7,435 | 2,50,000 | 1,72,000 | 77,500 | 261 |
| Total Ganga Basin | 10,27,830 | 9,38,970 | 19,65,900 | 10,39,400 | 9,26,500 | — |
| | (52.3) | (47.7) | — | (52.9) | (47.1) | — |

*Source* : Concerned State Pollution Boards.

TABLE 12.5 : **Organic Pollution Load (B.O.D.) generated from urban and rural sectors in the Ganga Basin (1977)**

| | B.O.D. load kg/day | | | Per cent of total Basin load | Rate of B.O.D. load generated per day | |
|---|---|---|---|---|---|---|
| | Urban Sector | Rural Sector | Total load | | Per person (in grams) | Per km$^2$ (in kg) |
| Haryana | 77,107 | 100,138 | 177,245 | 2.39 | 41 | 8.3 |
| Uttar Pradesh | 1,329,674 | 2,017,991 | 3,347,665 | 45.18 | 34 | 11.4 |
| Bihar | 174,862 | 1,340,167 | 1,515,029 | 20.45 | 21 | 11.0 |
| West Bengal | 442,636 | 478,101 | 920,737 | 12.43 | 30 | 18.2 |
| Rajasthan | 75,054 | 410,393 | 487,447 | 6.55 | 39 | 4.7 |
| Madhya Pradesh | 143,303 | 527,488 | 670,791 | 9.05 | 34 | 3.3 |
| Himachal Pradesh | 4,138 | 18,966 | 23,104 | 0.31 | 47 | 4.0 |
| Delhi | 257,952 | 11,505 | 269,457 | 3.64 | 45 | 181.5 |
| Total Ganga Basin | 2,504,726 | 4,904,749 | 7,469,475 | 100.00 | 31 | 9.1 |

*Source* : State Water Pollution Control Board of Uttar Pradesh, Bihar and West Bengal.

THE DO-BOD PROFILE OF THE GANGA WATER-I.

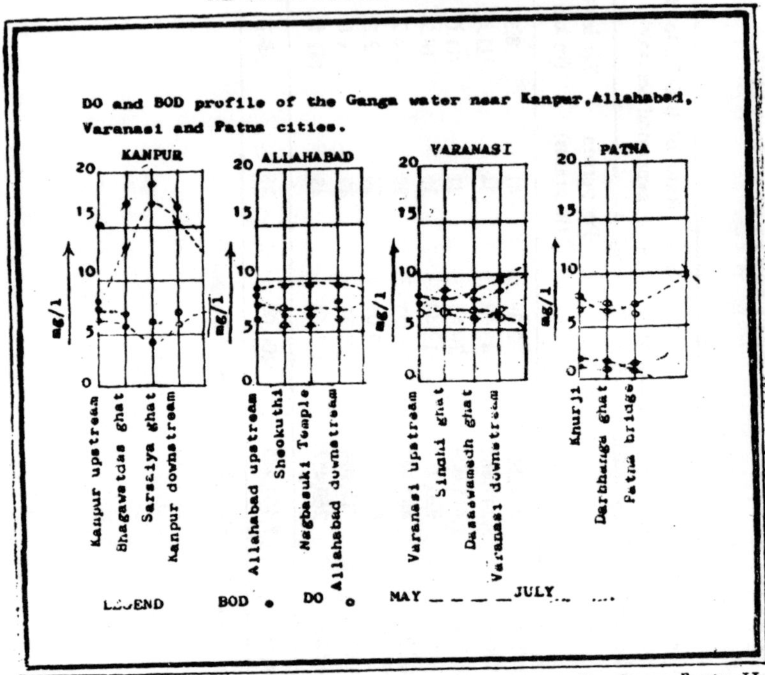


Source: The Ganga Basin II.

FIG. 12.1 : *The DO-BOD Profile of the Ganga Water—I*

If the effluent from a treatment plant has a higher content of organic pollutants, the effluent will have a higher B.O.D. In other words, more oxygen is taken out from the water to break down the sewage. And this B.O.D. contribution as per the survey conducted by N.E.E.R.I. on Calcutta, Kanpur and Delhi were on average 41, 35 and 34 respectively. This fact further showed that B.O.D. contribution per capita per day in sewage water of industrial cities are higher than that of other cities because of disposal of industrial wastes into sewers. And it reveals that if sewage water with high B.O.D. load is pumped out to the river, whole water of the river for considerable distance is polluted. This has been the case for Ganga water pollution. Tables 12.5 and 12.6 will show organic pollution load (B.O.D.) generated from urban and rural sectors in the Ganga Basin (1977) and the Ganga water quality

TABLE 12.6 : **Ganga water quality and the B.O.D. generation in the catchment areas of twelve selected stations. (May, 1981 and 1982 average)**

| Monitoring Stations | Ganga water quality | | | | | | Mean flow in Ganga $m^3$/second | B.O.D. generation in catchment area kg/day | B.O.D. rise if waste water added to stream flow mg/l |
|---|---|---|---|---|---|---|---|---|---|
| | Catchment area | BOD mg/l | DO mg/l | COD mg/l | Total Coliform MPN/100 ml | BOD/COD ratio | | | |
| Hardwar | 22,685 | 2.0 | 5.7 | 8.6 | 2,400 | 0.2 | 381 | 100,430 | 3.05 |
| Garmukteswar | 9,798 | 5.1 | 6.8 | 11.4 | 1,430 | 0.4 | 170 | 195,599 | 13.32 |
| Kannauj | 29,364 | 7.3 | 6.2 | 13.4 | N.A. | 0.5 | 52 | 343,815 | 76.53 |
| Kanpur | 4,274 | 15.3 | 6.5 | 28.4 | N.A. | 0.5 | 86 | 171,440 | 23.07 |
| Allahabad | 12,065 | 9.3 | 7.4 | 22.0 | 31,250 | 0.4 | 133 | 105,923 | 9.22 |
| Varanasi | 11,799 | 9.3 | 6.4 | 20.0 | 12,075 | 0.5 | 285 | 111,963 | 4.53 |
| Patna | 33,265 | 1.3 | 7.4 | 16.0 | 28,315 | 0.1 | 1,108 | 354,926 | 3.37 |
| Munger | 16,187 | 2.2 | 6.4 | 22.3 | 24,200 | 0 1 | — | 179,386 | — |
| Farakka | 22,895 | 0.03 | 8.1 | 16.2 | 125,500 | 0.001 | — | 157,495 | — |
| Nawadwip | 6,048 | 0.6 | 7.5 | 14.4 | 58,500 | 0.04 | 1,013 | 58,194 | 0-66 |
| Dakshineswar | 3,947 | 2.0 | 6.0 | 19.8 | 240,000 | 0.1 | — | 198,542 | — |
| Dimond Harbour | 1,207 | 8.0 | 7.1 | 185.4 | 8,250 | 0.04 | — | 74,286 | — |

*Source* : State water pollution control board of Uttar Pradesh, Bihar and West Bengal.

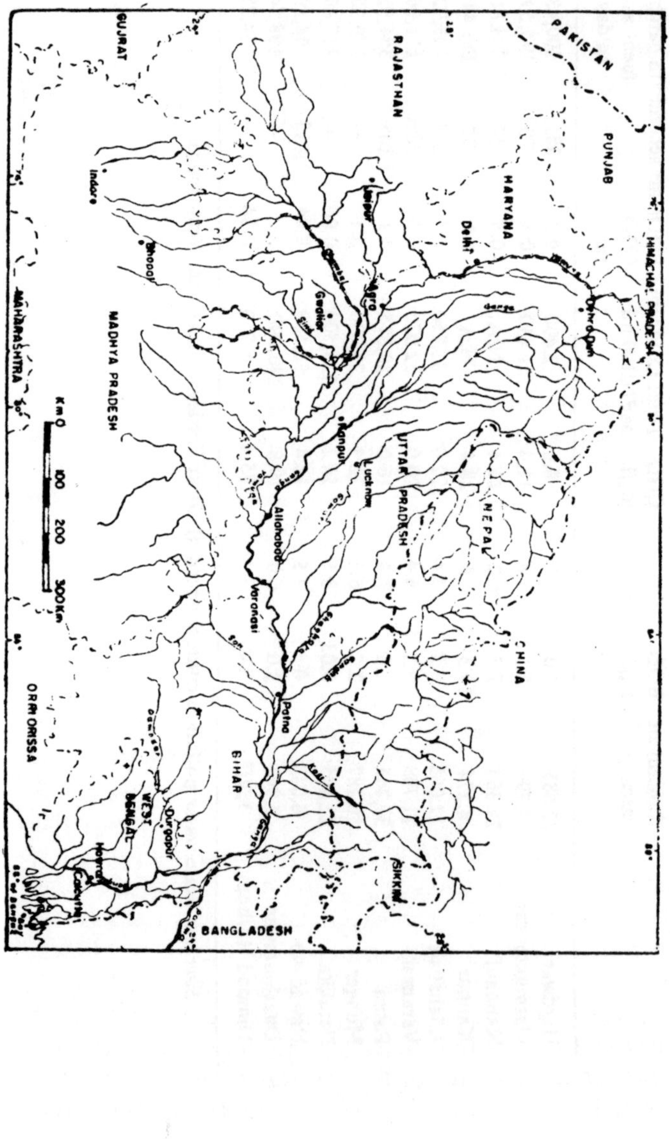


FIG. 12.2 : *The Ganga Basin*

and B.O.D. generation in the catchment areas of twelve selected stations. As for example Delhi consumes about 1.2 million kilo litres of water every day and in turn it pour back 0.96 million kilo litres of toxic waste water into the river Yamuna. And when chemically analysed, this contaminated water is found to contain 250 thousand kilograms of dissolved solids, 125 thousand kg of suspended matter and 125 thousand kg of B.O.D. Industrial effluents account for only 23 thousand kilolitres of waste water mainly from printing and electro-plating units. The remaining 0.94 million kilolitres are discharged into Delhi sewers. The real assault on the Yamuna starts a little downstreams of Wazirabad Waterworks. Sewer pipes discharge about 120 million gallons of waste water and filth daily into the river without any biological treatment. One hundred million gallons of waste water of west and north Delhi is discharged by the dreadful Nazafgarh drain. Thus the transformation of a healthy river into a virulent sewer is almost complete. The matter of utmost concern is that only 46.4 per cent of the effluent water and sullage flowing back into the river Yamuna through seventeen open drains is only being treated before discharge which ultimately reach Ganga at Allahabad. Sewage plants are located only at Keshopur, Coronation Pillar and Okhla. Thus the total quantity of untreated waste water getting into the river Yamuna is 515 thousand kilo litres a day. Further it is also remarkable that whatever the treatment plants do is of primary nature. Hindon river for example also is a most polluted river of our country and joins Yamuna at Okhla.

After Delhi, Kanpur city is also the most ill-famed for its pollution hazards particularly concerning air and water. More than 40 per cent of the water supply of the city is contaminated. The Ganga and Pandu rivers receive the city's sewage and lethal amount of toxic effluents and at the same time are sources of drinking water supply. About forty million gallons of waste water is discharged daily into the river Ganga through sewers only. The analysis of sewage waste water from the city showed that concentration of B.O.D. was 270 milligrams per litre. In addition, a number of outfalls carrying effluents of textile, tanning, plastic and rubber industries directly join the river. The physio-chemical characteristics of B.O.D. was found to be 33.4 mg per litre which is above the prescribed standards. The mean value of oxygen demand (B.O.D.) was observed as 3.05 mg per litre to 5.5 mg per litre at Bithorghat where no outfall joins the river. The self-purifying capacity of the river Ganga is higher than the river Yamuna at Delhi.

The D.O.--B.O.D. profile-I, shows the value of D.O. reaching a critical low when B.O.D. reaches the peak, both D.O. and B.O.D. indicate higher concentrations in the monsoon than those during the pre-monsoon period. The D.O.--B.O.D. profile in Allahabad although flatened is comparable to Kanpur. At Varanasi, the trend of rising B.O.D. from upstream to down streams is similar to that of dry month. But the D O. sag is more evident in the monsoon than in the pre-monsoon month, where the sag shows an increasing trend without sign of recovery. Near Patna the pre-monsoon D.O. sag characteristics show an almost reverse trend with similar trends of B.O.D in water.

At Calcutta, the Bhagirathi-Hugli river basin is badly affected by the innumerable industries of the 'Hugli-Industrial-Belt'. This industrial belt extends north-south to over 90 km stretch of the Hugli estuary. Hundreds of industries which have been identified as the polluting industries have come up on either sides of the river Hugli. Discharge of industrial effluents from the polluting industries like chemical, textile, leather, paper, drug manufacturing, dye and paint, automobile and rubber industries and distilleries etc. vitiate the whole water of the river. Also, the untreated domestic waste water along with the industrial effluents cause much harm to human and acquatic life. In this connection, effects of pollution on the dwindling supply of popular hilsa fish from the estuarine river may be mentioned. The total volume of liquid waste discharge into the river through various outfalls is 252 million gallons per day, of which 77 million gallons per day is contributed by various industries. The daily average pollutional load contributed by Calcutta in terms of B.O.D. total solids, suspended matters and dissolved matters is normally high. To add to this, is the burnt-oil and coal discharged by thousands of ships calling at Calcutta Port every year. Moreover, sewage wastes from Kidderpore and Garden Reach areas flowing into the river is another instance of deterioration of the quality of the river water.

TABLE 12.7 : **Physical quality of the Ganga water at selected stations for the month of May and July, 1981-82**

| Monitoring Stations | Conductivity | | pH value | | JTU* turbidity | | Temperature | | Stream velocity (metres per sec) | |
|---|---|---|---|---|---|---|---|---|---|---|
| | May | July | May | July | May | July | May | July | May | July |
| Hardwar | 661 | 490 | 8 | 7 | 22 | 35 | 26 | 13 | 0.60 | 0.35 |
| Garmukteswar | 771 | 809 | 8 | 7 | 32 | 45 | 28 | 18 | — | — |
| Kannauj | 357 | 286 | 7 | 8 | 8 | 4 | 32 | — | — | — |
| Kanpur | 399 | 286 | 8 | 8 | 12 | 1 | 29 | 15 | 0.50 | 0.59 |
| Allahabad | 439 | 228 | 8 | 8 | 39 | 471 | 32 | 21 | — | 0.90 |
| Varanasi | 332 | 255 | 9 | 8 | 44 | 480 | 32 | 20 | — | 0.40 |
| Patna | 472 | 238 | 8 | 8 | 70 | 1700 | 30 | 20 | — | — |
| Munger | 378 | 357 | 8 | 8 | 80 | 1400 | 29 | 18 | — | — |
| Farakka | 325 | 211 | 8 | 7 | 90 | 663 | 31 | 20 | 0.35 | 0.44 |
| Nawadwip | 327 | 277 | 8 | 7 | 195 | 155 | 31 | 21 | 0.56 | 0.42 |
| Dakshineswar | 430 | 224 | 8 | 7 | 158 | 515 | 31 | 20 | 0.27 | 0.27 |
| Diamond-Harbour | 2060 | 223 | 8 | 7 | 560 | 215 | 31 | 22 | 0.09 | 0.22 |

*JTU—Jackson Turbidity Unit.
*Source* : State Water Pollution Control Boards of Uttar Pradesh, Bihar and West Bengal.

TABLE 12.8 : Seasonal variations of the mean concentrations of the indicators of the Organic Pollution in the Ganga basin for the year 1980-82 as studied

| Monitoring stations | BOD | | | Total Kejldahl Nitrogen (mg/l) | | | DO | | | Coliform MPN/100 | | |
|---|---|---|---|---|---|---|---|---|---|---|---|---|
| | PM | W | S | PM | W | S | PM | W | S | PM | W | S |
| Diamond Harbour | 1.30 | 1.47 | 15.58 | 0.41 | 0.47 | 2.20 | 6.73 | 7.47 | 6.20 | 31333 | 8833 | 22600 |
| Farakka | 0.47 | 0.54 | 0.72 | — | — | — | 7.80 | 8.70 | 8.84 | 9500 | 52050 | 23138 |
| Varanasi | 6.00 | 5.20 | 6.49 | 5.65 | 5.66 | 4.02 | 8.20 | 11.14 | 7.03 | 31885 | 25930 | 14010 |
| Allahabad | 7.10 | 5.40 | 6.95 | 4.12 | 4.60 | 3.80 | 7.90 | 10 72 | 7.08 | 38100 | 15440 | 17570 |
| Kanpur | 15.55 | 14.15 | 16.39 | 0.33 | 1.54 | 43.00 | 6.50 | 8.02 | 4.35 | — | — | — |
| Hardwar | 13.20 | 3.17 | 2.50 | 0.01 | 0.28 | 1.19 | 5.25 | 6.80 | 4.87 | 43 | 2400 | 2400 |
| Bhagalpur | 1.57 | 1.60 | 3.60 | NA | NA | NA | NA | NA | NA | NA | NA | NA |

*Source* : States Pollution Control Boards of Uttar Pradesh, Bihar and West Bengal.
PM—Post Monsoon
W—Winter
S—Summer

**Measures to check pollution**

It is observed that contaminated water supply is the common feature of most of the cities in India. Slum areas are the most hard hit in big cities. Contaminated water supply to the Trans-Yamuna colonies in Delhi, is the root cause of all diseases like scabies, malaria, typhoid, disentry and jaundice. In Delhi about 2,800 cases of hepatitis and jaundice were reported in 50s due to water intake from a canal, virtually an open sewer for water supply. In Calcutta the current outbreak of hepatitis is also due to the intake of polluted water. Local authorities have instructed the Public Health Engineering Department for examination of water samples from all afflicted localities. However, the report revealed a different picture as regards the quality of potable water supply. For this contradictory report, some circles speculate that the Ganga pollution might have caused this sudden outbreak of water-borne diseases and that they suggest boiling and proper chlorinisation of drinking water at the household levels.

To get rid of these recurrent incidence of epidemic and to ensure our own survival, we must check the undue aggressive human needs upon the Ganga water. Some control measures must be enforced soon, lest it should be very difficult to cope with the problem of steadily increasing pollution level of the river Ganga. Hence, to protect Ganga from being polluted by the discharge of sewage waste of towns and cities and industrial effluents, treatment plants should be installed at suitable locations and that industrial wastes and city's sullage may be discharged into the river only after treatment. In Delhi, pollution control board has been fighting a protracted battle with big industries to make them install treatment plants in their respective units. The board brought two big industrial houses of the city to the court. Except a few, most of the industries by now have installed or installing pollution control measures as a result of strict action by the board. In West Bengal too several industries have no facilities at their factories for treatment of effluents which flow defectively into the river Hugli while others have submitted proposals for setting up such plants or improved facilities. According to the pollution control board, the effects of pollution from the inadequately treated effluents are not immediately felt because they are released either directly into the river or through sewers. They do however, result in an increase of level of pollution which could have long term effects on human health. In the downstream, Calcutta depends on the Hugly river for 80 per cent of its drinking water supply, therefore, legal steps should be taken in all the cities to control water pollution along the entire course of the river Ganga. This law should equally be enforced to the cities like Delhi, Kanpur, Calcutta and the Hugly riverside industrial belt. Legislation should be brought in to incorporate suitable treatment facilities for effluents in the designs of industries at the time of granting licence to a new industry particularly in Ganga basin. Steps should also have to be taken to curb the pollution of the groundwater through the sub-surface percolation of municipal or industrial waste water through soak-pits, septic tanks, wells and mineshafts. Human and animal excreta are also potential sources of pollution. A rational policy on canal irrigation must also be implemented as regards the tapping of water from the river Ganga by different states in the upper reaches of the river, so that a normal flow of the river is maintained throughout the year. Sufficient drainage arrangements for water must be made in order to check the piling up of salts and alkali in the agricultural lands. Table 12.9 shows the salt loads in irrigation water in the Ganga basin. Generally the total quantity of salts added to the Ganga basin every year by irrigation is 17 million tonnes. The groundwater irrigation contributes another 24.74 million tonnes of salts every year. Thus, the salinity level of Ganga water is raised by 88.5 milligrams a litre every year.

The farmers of the entire Ganga basin should be provided with necessary guidance as regards the optimal use of chemical fertilisers and pesticides so that the harmful effects of the concentrated doses of residual chemicals of excess fertilisers and pesticides which get access into the Ganga water, could be averted. Construction of underground and covered surface drains might be the good answer to the problem of Ganga pollution and that sewage wastes treatment measures should be initiated in all the outfall points along the river at the earliest.

### Quality control methods

Only the treatment of wastes is not the way of achieving the Ganga water standard and should by the last resort in a develop-

TABLE : 12.9 : **Irrigation Water and its salt loads in the Ganga Basin (1977)**

| States | Irrigation Water (million cubic metres per annum) | | | Salt load in Irrigated Water (million kg per annum) | | | Salt concentration in Irrigation Water (in milligrams per litre) |
|---|---|---|---|---|---|---|---|
| | Surface Water | Ground Water | Total Supply | In Surface Water | In Ground Water | Total Salts | |
| Haryana | 4,371 | 4,335 | 8,706 | 874 | 2,168 | 3,042 | 349 |
| Uttar Pradesh | 44,551 | 30,456 | 74,907 | 8,911 | 15,178 | 24,089 | 322 |
| Bihar | 15,601 | 6,024 | 21,625 | 3,120 | 3,012 | 6,132 | 384 |
| West Bengal | 8,521 | 2,746 | 11,267 | 1,704 | 1,373 | 3,077 | 273 |
| Rajasthan | 7,010 | 3,455 | 10,465 | 1,402 | 1,728 | 3,130 | 199 |
| Madhya Pradesh | 4,607 | 2,291 | 6,898 | 921 | 1,145 | 2,066 | 300 |
| Himanchal Pradesh | 95 | 40 | 135 | 19 | 20 | 39 | 289 |
| Delhi | 244 | 237 | 481 | 49 | 118 | 167 | 347 |
| Ganga basin | 85,000 | 49,484 | 1,34,484 | 17,000 | 24,742 | 41,742 | 310 |

*Source* : States Pollution Control Board of Uttar Pradesh, Bihar and West Bengal.

ing country like that of ours, because it is normally the most costly method. Though a big amount of Rs. 250 crores have been allotted by the Union Government during the seventh five year plan for controlling pollution in the Ganga water and restoring it to the level of pristine purity, the sum is inadequate for the entire project throughout its course in the context of the vastness of the deal. At the present market price a single waste water treatment plant or installation of an effluent treatment plant at any industrial unit would involve an expenditure of 25 to 30 lakhs of rupees. In the mean time, states of Uttar Pradesh, Bihar and West Bengal have submitted plans for pollution control measures to the Union Government involving several crores of rupees. And Uttar Pradesh and Bihar have got their plans approved by the project authority.

Basically, the approach to the pollution problem must be considered in its totality. The first method is to be achieved by augmenting the flow of Ganga for increased waste dilution during lean season with the control of tapping excessive water for irrigation and industrial purpose in the upper reaches, which could be substituted by harnessing groundwater resources wherever possible. In addition, this assimilation capacity may be further increased by redistributing the effluents with regulated discharge and by its transfer to different locations and by underground disposal. The second method of Ganga pollution control by reducing waste generation can be achieved by changing production process with appropriate technology, and by improving the sanitation and good house-keeping in the rural areas of the Ganga basin. Lastely, pollution control could be a success with resorting to the methods of reducing wastes after generation, i.e. by recovery, recycling and effluent reuse. For example, slag from steel plants may be profitably reused in cement production and bagassee from sugar factory may be reused in spirit manufacturing and paper board manufacturing. Similarly, waste water from the cities of Delhi, Kanpur and Calcutta may be directed to the agricultural fields instead of the river Ganga and Yamuna for irrigation and fertilisation. Waste treatment is not always fit in the context of our country since it is expensive. The industrialists of our country have not yet considered the external damage and social costs caused by the wastes discharged in the river. The situation with regard to municipal waste water treatment in many towns and cities along the river Ganga and its tributaries is more complicated by the fact that in most cases conventional sewerage systems are not found to be existing. Without waste water collection and transportation to central plants, the cost advantages of large scale treatment plants in cities like Delhi, Kanpur and Calcutta can not be achieved. However, it may be necessary to arrive at a solution to this problem because the capital cost of installing underground sewers in large cities is usually prohibitive. But, in view of the gravity of the situation, it is as important for municipal waste water to receive treatment as it is for industrial wastes to arrest any further pollution of the Holi Ganga.

## Policy Alternatives

Ideally, the Ganga Authority and other pollution control boards must provide incentives for maintaining the purity of Ganga water at the lowest cost to the society. State level pollution control boards should be created for maintaining the quality of Ganga water throughout the course of the river. Industrial houses using water and discharging wastes should have representations in these boards instead of only being instructed for the direct control of discharges, in order to achieve the expected standard of the river water. Besides, a number of other concessions such as subsidies, tax relief and grants for capital etc. should also be given to encourage them (industries) to help in maintaining water quality of the river Ganga in the greater interest of the nation.

## REFERENCES

Government of India (1984), *The Ganga Basin*, Central Board for the Prevention and Control of Water Pollution, Department of Environment.

Karan, P.P. (1980), Changes in the environmental pollution in the Calcutta-Hooghlyside industrial strip of India, *Environmental Studies*, 15 (3 and 4).

Rao, K.L. (1979), *India's Water Wealth*, Orient Longman.

# SECTION III

# REGIONAL ISSUES

S. K. SINGH

# THIRTEEN

# SOCIAL ASPECTS OF SOIL CONSERVATION AND AFFORESTATION IN KUMAON REGION

THE Kumaon region (78°—45′ to 81°—5′ E and 28°—44′ to 30°—49′ N) covering an area 21,033 km², with a population of 2,383,163 persons (1981) is situated in the Central Himalaya. In the north, the water divide separate it from Tibbet, while towards the east the river Kali all along marks its border with Nepal. In south, west and north district boundaries of Bijnor, Moradabad, Rampur, Bareilly and Pilibhit form the demarcation line (Fig. 1).

The region has a unique geographic location between the plateau of Tibbet in the north and the Ganga plains to the south. This study attempts to present some of the approaches to resource managements and eco-development processes. Damage to mountain ecosystems does not stop at the actual site of damage; but it affects the downstream areas. It has been observed that the effects of deforestation in terms of soil erosion, landslides and loss of agricultural productivity in upstream areas of the Kumaon hill are more severe, while its after-effects cause unprecedented floods in the Ganga plain.

There are two types of erosion in the hill areas, viz. natural erosion and man-made. Natural erosion is caused by heavy rainfall in intensive bursts from June to September falling on steep slopes which are geomorphologically unstable and fragile in nature like all the Himalayan region. Man-made erosion is caused by the pressure of the people on the land. The pressure is due to rapid growth of population which makes an ever increasing demands of forest products. It ultimately leads to their destruc-

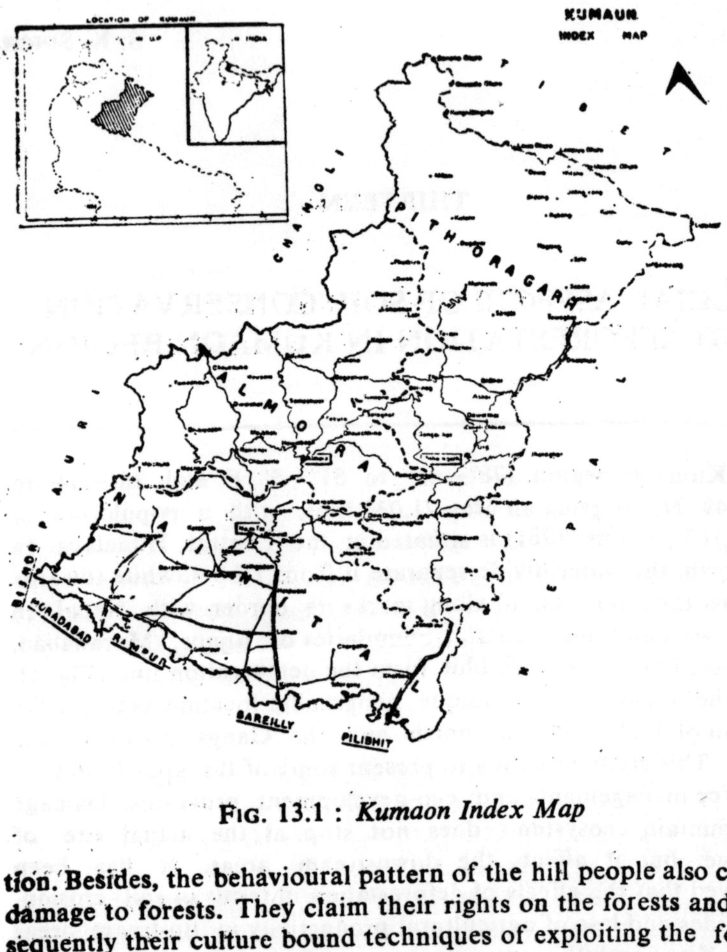


FIG. 13.1 : *Kumaon Index Map*

tion. Besides, the behavioural pattern of the hill people also causes damage to forests. They claim their rights on the forests and consequently their culture bound techniques of exploiting the forests and destruct the existing stock of forests. It is man rather than the natural forces which has upset the natural balance in the hills. People's behavioural pattern needs to be changed. The rapidly deteriorating ecosystem in the hills can be saved by changing, modifying and adopting people's behaviour. What factors determine the behaviour of the hill people ? What are the factors responsible for causing ecological imbalances ? What are the socio-economic compulsions of the hill peoples ? Do the soil conservation and afforestation programmes of the state and central governments reflect the priorities of the hill people ? If not what sort of incentives or motivations are being given or needed ? Since soil conservation and afforestation programmes involve community decisions and people's participation, whether that sort of institutions or organizations exist or are needed to ensure people's participation ?

Based on a study of about thirty social forestry projects under the auspices of the Ramganga Soil Conservation Division, Ranikhet: Civil Soyam Forest Division, Almora; and Trimukhi Van Kheti ; and also some soil conservation works in farmers fields in Bhikiasan and Hawalbagh blocks of Almora district, an attempt has been made to analyse the social aspects of these programmes.

**Factors determining the hill peoples**

*Behavioural patterns*

Forests are linked with the daily life of the people of their region. Since time immemorial the hill people got all forest produce freely. They are used to cultivate forest lands and fetch fuel and fodder from the forests without spending any amount. The forests were considered to be a common property of all the inhabitants who had complete freedom of grazing their cattle and gathering fuel and fodder. The British administration in Kumaon commenced in 1815, and through several processes of land settlements, village boundaries were demarcated and the rights of grazing of land, cut fire wood and timber were recognized. Land settlement of 1823 (Sambat 1880 Vikrami) called *Assisala Bandobast* is considered very important because of the fact that for the first time village boundaries and rights of the people in the forests in their neighbourhood were recognized. In accordance with this arrangement, they developed the ideas that every compartment of the environment can be utilized by them as and when needed.

Similarly, under the *Kumaon Nayabad Act*, people were given the right to extend cultivation in contiguity of their cultivated land. Though the right was repealed in 1955 yet clandestine efforts to extend agriculture on marginal and sub-marginal lands continues. The hill people are poor because of the small size of their holdings and unfavourable topography. A typical feature of land distribution in the hills is that a very high proportion of land belongs to the government or to the community. This should have enabled the government to regulate and monitor landuse on a scientific basis but the actual situation is just the opposite. Both the government land (reserved forests) and community land (civil soyam/panchayat forests) are in a poor shape due to over commercialization, monetary motives and owing to over exploitation. Table 13.1 gives an idea about the distribution of the land under different categories in the Kumaon region. It is seen that 47.5 per cent of the land is directly under government's control (reserved forests) and 29.4 per cent (items 2 and 4) are under civil soyam or panchayat forests. Thus, 76.9 per cent of the land is either under government control or under community which is again subjected to government control. It shows that our policy makers or administrators have allowed the lands under their control to be over exploited which have proved injurious to the community and society at large. The land under private control for agricultural and horticulture purposes amounts to 22.3 per cent only.

TABLE 13.1 : **Distribution of land under different categories in the Kumaon region (1982)**

| Land Category | Area in km² | Per cent |
|---|---|---|
| Forest (reserved) | 10,605 | 47.5 |
| Civil soyam forest and panchayat forests | 3,763 | 19.2 |
| Net area sown | 3,284 | 17.5 |
| Barren and uncultivable waste | 2,005 | 10.2 |
| Under tree crops and orchards | 827 | 3.4 |
| Old and current fallow | 254 | 0.9 |
| Area under non-agriculture use | 168 | 0.8 |
| Cultivable wasteland | 127 | 0.5 |
| | 21,033 | 100.0 |

*Small size of farm and related environment*

Almost all land holdings in the hill region are uneconomic Table 13.2 shows that 68.3 per cent of the holdings are less than one ha, and 19.8 per cent of the holdings are between 1 to 2 ha. Only 10 per cent of the land is irrigated. As mentioned earlier, most of the holdings belong to subsistance level. In the hill areas it has been observed that 96.6 per cent of the holdings are less than

TABLE 13.2 : **Distribution of land by size-group of holdings in Kumaon region**

| Size of holding (ha) | No. of holdings | Percentage | Area of holding (ha) | Percentage |
|---|---|---|---|---|
| <0.5 | 1,16,541 | 39.1 | 30,741 | 9.6 |
| 0.5—1.0 | 86,891 | 29.2 | 62,088 | 19.5 |
| 1.0—2.0 | 59,218 | 19.8 | 81,578 | 25.6 |
| >2.0 | 36,605 | 11.9 | 1,49,703 | 45.3 |
| Total | 2,98,255 | 100.0 | 3,24,110 | 100.0 |

*Source* : Agricultural Census, 1980-81.

one ha and account for 88 per cent of the cultivated land, and only 7.7 per cent of the land is irrigated. The land holdings are highly fragmented and scattered. The average size of holding worked out to 0.38 ha. The income composition indicates that animal husbandry is the major component of the farm income though it varies from place to place. The large number of animals kept in the farm households cause a lot of environmental problems.

### *Why hill people keep more animals* ?

The hill farm families have to keep a large number of animals. Except the buffalo, all animals are sent to civil soyam or panchayat forests for grazing. This leads to over-grazing and the soil is left without any cover and the entire system is impoverished. It is futile to blame the hill people for keeping a large number of animals without looking into the reasons, which are as below :

(a) Animal husbandry forms an important component of the income of the farm families. The milk produced from the animals is low, so more animals are kept to meet the present need.

(b) The maintenance cost of animals is very low due to long hours of grazing. Besides, female labour brings a head load of grass or fodder leaves from the forests almost whole the year round.

(c) The farmers by and large do not use chemical fertilizers and in lieu use cow dung as manure to maintain the soil fertility.

(d) The tragedy of common property (every body's property is no body's property) apply to the civil soyam or panchayat pastures. No body makes the initiative of restricting his own cattle from grazing the common lands. In the absence of any government or voluntary organizations it is difficult to control grazing or encourage stall feeding.

The hill people are too poor to invest for the development of their land in order to use it round the year. The benefits of reducing floods, siltation of dams and soil erosion from the view point of downstream beneficiaries or from the national point of view is irrelevant to the hill people. The hill people act with disregard for future because day-to-day survival is tied with forests. It is difficult for them to comprehend about the future. The government actions are regarded as unwarranted restrictions on their traditional rights. In this way the hill people's rationality, the national rationality or collective rationality are at tangent to each other. Therefore, herculean efforts have to be made in this regard through education, extension, motivation and incentive.

*Lack of enthusiasm for soil conservation*

The hill farmers have been constructing terraces to save their lands from erosion hazards particularly in those fields which are irrigated. There are two motivating forces: firstly, farmers would like their scarce resources like irrigated land to be preserved ; and secondly, they would like to conserve the water and not allow soil nutrients to flow down. The conservation work undertaken by the government's departments entails construction of terraces by farmers on 50 per cent subsidy and there is no emphasis on : (a) water harnessing or harvesting techniques, (b) increasing land productivity, (c) changing cropping pattern, and (d) increasing cropping intensity. It is seen that in many places there are small springs with some irrigation potentials while due to the lack of scientific techniques these sources become waste and cause water-logging. Because of the reason, farmers are able to take only one crop in such lands. The irrigation department has its own norms regarding potential water resource and their approach suits only to large projects which are virtually non-existent in the hills.

Soil conservation measures like terracing and levelling without increasing land productivity is uneconomical to farmers. There is a great need to explore possibilities of those water harnessing or harvesting techniques which are simple in design and can be easily managed by the farmers at the local level as hydrants or ponds with polythene lining. The requirement is of an integrated approach combining terracing, water harvesting and agricultural development at the farm level. Where terracing in uneconomical, change in cropping pattern may be adopted as the solution to erosion hazards.

*Social aspects of afforestation*

The forest department administers the social forestry projects under two schemes, viz. the soil conservation division and the civil soyam forestry divisions ; whereas the planning department administers the social forestry projects through development blocks under *Trimukhi Van Kheti*. Case studies of thirty social forestry projects show that in most of the cases *Trimukhi Van Kheti* have been a flop. It highlights that such system is not always conducive to the success of the programme. It was also found that planning of the project was done hastily on the recommendation of the *Gram Sabha* to afforest their civil soyam land. There is no discussion with the village community on the subjects : what species have to be planted, where, and why? There is no discussion on the extent to which their need for grass, leaf, fodder and fuel and timber could be met from the project and from alternative sources. and what role they have to play in making the project a success. Almost all the people think that these projects are of forest department or block development projects and hence the concerned departments should ensure their success. Representatives of the *Gaon Sabha* or panchayat forest hope that the concerned department would provide a *chowkidar* for three or four years or till the maturity of plants. Some consider that the *chowkidar* should be from the same village so that he may look after the interests of his villagers enthusiastically and keep strict vigilance on the trespassers of other villages. On the other hand, some people feel that the *chowkidar* should be from some other village so that he could be serious on the treespassers of their own village. This indicates their apathy towards the project.

It has also been found that a lot of time is taken by the revenue authorities in identifying the land earmarked for the project, and in some cases other villages claim that they had also rights on that land. It appears to be a serious problem as tempering of walls, pilferage of grass and fuel is done in several cases not only by people in the same village but by other villagers as well who claim the area as their pastured land and do their utmost to wreck the project.

The choice of trees in these projects is a bone of contention between the forest department and the people. Though this choice depends on elevation, soil type, rainfall, temperature, soil moisture and wind condition, whereas the forest department claims that both the conifers and broad leaved varieties are planted considering all above facts. But the real situation inclines towards mono-cultural practices. The people's complaint is justified on the ground that in the projects studied on an average there are more than 60 per cent of the area was allotted to Chir (pine). There is a strong feeling that the social forestry projects reflect only the monetary aspect or commercial objectives of the forest department. Foresters neither care the needs of the people nor priority of the area. It seems forest department considers that pine plantation is the only remedy for degrading soils. Besides, they feel that it can withstand the grazing and fire and can also survive under current practices of the villagers. It also enables them to fulfil the assigned targets. The people clamour that they need varieties of fodder trees which would give them returns after ten years. Pine would give fuel after twenty five years, resin from seventy years and timber after hundred years. People cannot perceive the future to that exent.

**Modifications in laws to encourage private forestry**

Under the *Kumaon Zamindari Abolition Act of 1956*, private lands under forests were taken by the government which shows that government is keen to have forestry as a state monopoly. This may not be very conducive in order to develop social forestry in the hills as government monopolies could not activise peoples participation. Under the protection of *Trees Act of 1975* people are not allowed to cut trees even on their own private lands. Instead of restrictive measures, there is a need of scientific management of the land and the trees. Restrain and impediments on the use of private lands will damage the cause of social forestry. This has caused lot of distruction towards the forest management. The people have a suspicion that the forest department will simply seize the trees as they are matured. When farmers are advised to plant fodder trees on steep slopes of their farms rather than cereals which give poor returns, they retort.

**Ideal institution for developing the lands**

It is thought that *Van Panchayat* is the model for all social forestry programmes and efforts should be made to convert all civil soyam forests as *Van Panchayat* forests. Since many *Van Panchayat* forests are in a good shape it is some time concluded that they are the right answer. However, it may be said that only those *Van Panchayts* are successful which are called *Lath Van Panchayats* (panchayats using a stick). Here force is used to set right the recalcitrant members of the community. In this way discipline can be enforced. It has been observed that wherever *Van Panchayats* are in a good shape, nearby reserved forests are found in bad shape. It may be mentioned that civil soyam forestry programmes are successful only where the *Gram Sabhapati* is strong, dedicated and has a good reputation for honesty and integrity. Unless good people come forward to hold these positions, social forestry programmes cannot attain success. Hence, the conclusion that *Van Panchayat* is the solution to all the problems of civil soyam land does not seem to be realistic. However, there are some plus points in favour of *Van Panchayats*. People are familiar with *Van Panchayats* and they have a historical background and some success stories. Therefore, a revised *Van Panchayat Act* wherein *panchayats* are made autonomous units and forest departments are directly made responsible for technical advice, input supply system, education and training, may be thought of as solution.

**Suggestions**

On the basis of the above analysis following suggestions have been made :

(a) There are a large number of animals in the hill region, so steps should be taken to reduce cattle population.

Appropriate legislation or incentives should be made to check bovine population and to encourage stall feeding.

(b) Forest Officers working on civil soyam or panchayat forest divisions should be made conversant with community organisation, rural institutions. rural society, agricultural needs of the people for forestry products.

(c) The conflict between what the people want and what the forest department gives need to be resolved by giving a close and second look to existing programmes. Research efforts are needed to make to evolve hybrid species which are suitable for the hill areas. It is felt that technologies which are more productive, ecologically sustainable and low cost are lacking, hence a technical support system on a continuing basis needs should be developed by the forest department.

(d) A *khasra* and *khatauni* of the civil soyam or panchayat lands in which villages have rights should be provided to all the *Gram Sabhas* of the hill districts by the revenue authorities and they may be encouraged to ponder over the scientific management of these lands. In this way delay by revenue authorities for the identification and demarcation of land could be avoided. This will also speed up civil soyam forestry programmes.

(e) The revised *Van Panchayat Act of 1976* needs further revision for making *Van Panchayats* as autonomous units so that it can be free in preparing plan for production, distribution and marketing of the *Van Panchayat* forest products. Since a single *Van Panchayat* may not form a viable management unit so several *Van Panchyats* may join together to form a viable unit. The technical support, input supply, research, training and extension support may be provided by the forest department itself.

(f) The children and young people who are the custodians of the future of the hills could not remain ignorant of the problems of hill land and its conservation, it is imperative that forestry or ecology be taught as a course in the schools from the primary stage. The high level schools should have a farm representing the silvi-agricultural systèm in the region.

(g) The forest department should introduce an extension education programme in social forestry. Extension education material as manuals, advertisements, booklets, audiovisual aids, slogans, and films which are within the perceptual capacity of the people 'particularly women' should be prepared and disseminated. Women extension specialists may be appointed to educate the women for preservation of the forests. Demonstration of social forestry may be laid by forests department at strategic points.

(h) The existing rural institutions as *Gaon Sabha*, block development and *Van Panchayats* need activisation for a deeper involvement in the soil conservation and afforestation projects. The local schools are most suitable for making a centre of demonstration of the silvi-agricultural system for the area. The *National Commission on Agriculture, 1976* (*Forestry*) recommended that for the protection of forests, Tree Protection Brigades, *Van Raksha Dals*, Society of Tree Lovers, Honorary Tree Warders may be tried. A proliferation of institutions may not be helpful. Voluntary agencies and a task-force of young educated men and women of the local area trained in forestry and agriculture may be helpful in filling the communication gap between the forest department and the people by working as a catalyst and involving the people.

## Conclusions

Soil conservation and afforestation in the hill areas are long felt programmes for making livable environment. Besides, it will provide high returns in long term. Though these programmes are incompatible from the viewpoint of farmers who want quick returns, but in long run the society can enjoy sustainable benefits. These programmes require community participation with dedication to community. For this, a continuous educational and extension programme should be corporated so that they can feel that their schemes are in their own interest. For protection and conservation of these two important resources, some strict laws should be enforced, so that their future can be safeguarded.

## REFERENCES

Ashis, Madhava (1980), A suggested core strategy for development of the U.P. Hills. Paper presented to the Planning Commission New Delhi. 1980.

Central Soil and Water Conservation Research and Training Institute, (C.S.W.C.R.T.I.) (1977), *Operational Research Project on Watershed Management, Fakot (Tehri-Garhwal)*, Dehra Dun.

Food and Agricultural Organisation of the United Nations, (1978), *Forestry for Local Community Development*, 1978, Rome.

Gupta, P.N. (1979), *Afforestation, Integrated Watershed Management, Torrent Control and Land Use Development Project for U.P. Himalayas-Siwaliks*, U.P. Forest Department.

Jackson, M.G. (1981), Livestock in the economy of the Himalayan Foothills, lecture delivered to the staff of the Indo-German Dhauladhar Forest Project, Himachal Pradesh, (mimeo).

Ministry of Agriculture and Irrigation (1976), *Report of the National Commission on Agriculture*, Part IX Forestry. Government of India, New Delhi.

Seckler, David (1981), *Management of Social Forestry Programme : A Prospectus*, Draft Report, The Ford Foundation, New Delhi (mimeo).

Shah, S.L. (1978), 'Rapporteur' report on farming systems in hill areas. *Indian Journal of Agri. Economics*. XXXIII (4).

Srivastava, B.P. and Pant, M.M. (1979), Social forestry on a cost-benefit analysis framework, *The Indian Forester*, 105 (1), 2-35.

Vivekanand Laboratory for Hill Agriculture (1979), *Programme of Integrated Land and Water Resource Appraisal, Planning and Management in the Hills of U.P.*, Report, Indian Council for Agricultural Research (mimeo).

V.K. Asthana

FOURTEEN

# ANALYSIS OF PHYSICAL ENVIRONMENT

ENVIRONMENTAL consciousness emerged as a consequence of rapid industrialization and callous attitude in the disposal of industrial waste. Towards the end of 1960s a crusade was lanuched in America for "clean air, clean water". The movement later developed as a world-wide campaign, patronized by United Nations. The movement is popularly known as "environmental crusade". In the past there had been a variable degree of emphasis on environmental studies in geography. In the beginning of the twentieth century, there was a distinct attitude amongst geographers to disassociate themselves from environmental determinism. Even after second world war, when studies on land forms revived, it concentrated itself to the aspect of mapping and searching empirical parameters of descriptive nature. An integrated approach of viewing the landscape as an outcome of the interaction between man and land has been lacking. This functional approach is the need of time, where the two components of the environment, i.e., man and the nature could be seen as interdependent to each other. Nonetheless, the spatial context inherent in geography should not be forgotten.

Geography, according to Schaefer (1953), is a branch of knowledge with unique integrating function, synthesizing more specialized fields in space as history does in time. Eyre (1964) also agrees with Schaefer when he proclaims, "geography either stands or falls as an integrating discipline". This new emphasis in geography is a deviation from the old Kantian geography which tried to isolate geography from other major disciplines of science.

In order to introduce environmental consciousness in geography, it is imperative that geography follows ecological principles of research as well, which is more integrating and is developing parallel to other major sciences. The latest application in geographical research is the ecosystem approach, which according to Tansley (1935), "the whole complex of organisms—both animals and plants—naturally living together as a sociological unit". The two important components of the ecosystem is therefore the biome and the habitat. In a geographical perspective the component of biome is rather anthropocentric. We can, for our convenience, consider the ecosystem as a complex relationship of man and environment. Stoddart (1972) has outlined four main properties of an ecosystem for geographers. They are :

(a) *Monistic* : which brings man, plant, animal and environment in a single framework.
(b) *Structural* : Studying the complex in more orderly, rational and comprehensible way (geography of land forms, settlement patterns, and communication net works).
(c) *Functional* : Quantification of the interaction and interchanges between component parts.
(d) *Open-system* : Ecosystem is an open system tending towards a steady state and obeying the laws of open system thermodynamics (idea of vegetation climax, soil maturity, grade system of river etc.).

Bouldig (1956) has further cautioned that "the potential value of a system clearly depends on the correct selection of components when the system is structured and this normally presupposes considerable experience with the problems or data involved". The scope of the present study is to confine the component of physical environment and its analysis. It becomes obvious, at the outset, to recognise the structure of the physical environment.

The complex of physical environment includes factors like (a) geology, (b) landforms, (c) climate, (d) hydrology, (e) soil, and (f) vegetation. All these factors are intricately interrelated and a wholistic study of all the factors may be very difficult to summarise. Nonetheless, the author would attempt to refer in general terms about these factors and concentrate on the factors of landforms and hydrology.

**Identifying the integrated unit of physical environment**

Identifying a physical unit is a very difficult proposition, as the factors of a physical unit are very complex and intricate. However, a drainage basin can be considered as the lowest unit of observation because of the character of its integration. Doorknamp and Cook (1974) observe :

> "River network occurs within drainage basins (catchments), which frequently have finite boundaries (divides or watersheds). There is a natural input into a drainage basin from precipitation, and a natural output by way of water and sediment discharged alongwith the river channel".

In the light of the above argument it is rather self-conclusive that a drainage basin possesses the essentials of a general system and thus can be considered as the smallest integrated unit of physical environment. Asthana (1968) has done detailed investigation in the aspect of landform studies in Almora. This study is based on his work as well.

**Structure of the drainage basin**

The structure of the drainage basin usually is discussed in three sub-heads : (a) linear aspect, (b) areal aspect, and (c) relief aspect.

*Linear aspect*

Linear aspect of a drainage basin is studied in terms of orders of streams in the drainage basin and the associated attributes of number and average length of streams with the respective orders. Most widely used system of stream order was adapted by Strahler (1964) which was initially proposed by Horton (1945). This system proposes that the finger tip channels developing at the periphery of the drainage basin is order ($u_1$), and when the two first order streams unite they make a stream order ($u_2$). Progressively, the order of stream changes in simple arithmatic order when the stream of the same order joins together. Its order remains unchanged when a lower order stream joins with a higher order stream like ($u_1$) joining ($u_2$) will not effect the order of ($u_2$).

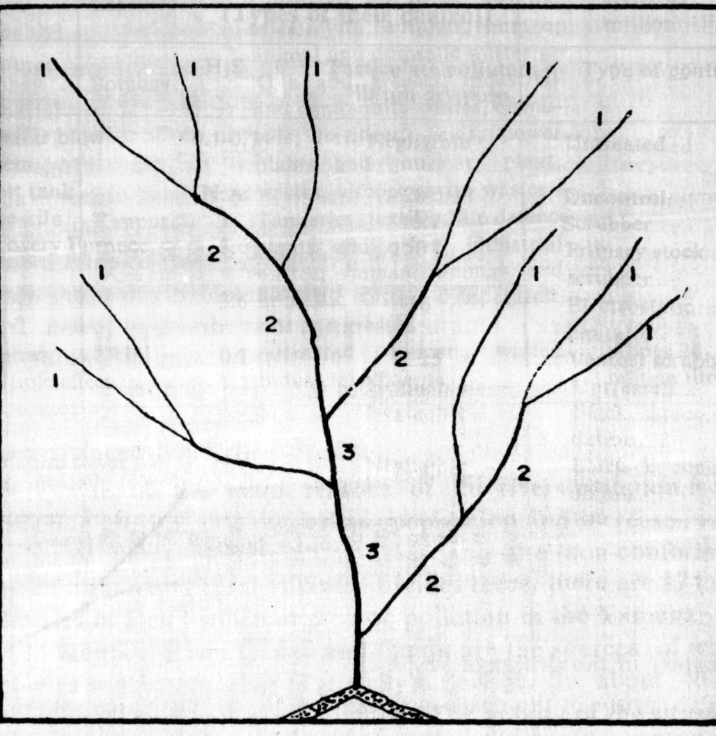

FIG. 14.1 : *Odering system of the streams*

Within the proposed system of orders in a drainage basin. variety of properties can be measured. Some of these are given below in a tabular form (Table 14.1).

Horton (1945) observed following relations between the order (u) and number (N) and mean length of order u, $\bar{L}u$. These are referred as laws of Horton.

*First law of Horton* : "The number of streams of different orders in a given drainage basin tend closely to approximate

TABLE 14.1 : **Parameters of drainage basins**

| Sl. No. | Variable | Symbol | Dimension | Explanation |
|---|---|---|---|---|
| 1. | *Linear* Steam Order | u | O | |
| 2. | Number of Streams order u | Nu | O | |
| 3. | Total No. of Streams in a basin | (ΣN)u | O | |
| 4. | Bifurcation Ratio | Rb= Nu/Nu+1 | O | |
| 5. | Total Length of Streams of order u | Lu | Length | |
| 6. | Mean Length of Streams of order u | $\bar{L}$u= Lu/Nu | Length | |
| 7. | Drainage Density | Du= (Σlu)/Au | L-1 | |
| 8. | Area of a basin | Au | $L^2$ | |
| 9. | Total basin Relief | H | L | Difference between highest and the lowest pts of a drainage basin. |
| 10. | Stream Channel slope | θc | Degree | |
| 11. | Valley side slope | θg | Degree | |
| 12. | Elongation Ratio | Re | O | Maximum Length of drainage basin (Lb)/Diameter of a circle with same area of drainage basin. |
| 13. | Hypsometric Integral | Ihyp. | % | *Refer Strahler |

*Source* : Adapted from Cooke & Door Knamp (1974).

an inverse geometric series in which the first term is unity and the ratio is bifurcation ratio".

*Second Law of Horton* : "The average length of each of the different orders in a drainage basin tend closely to approximate a direct geometric series in which first term is average length of streams of the first order".

The two laws mentioned above help in understanding the nature of drainage basin and specially the bifurcation-ratio (Rb) which is an important parameter for understanding the shape of the drainage basin as illustrated in Fig 14.2. Chorley (1969) has pointed out, that, "the bifurcation ratio, for a given density of

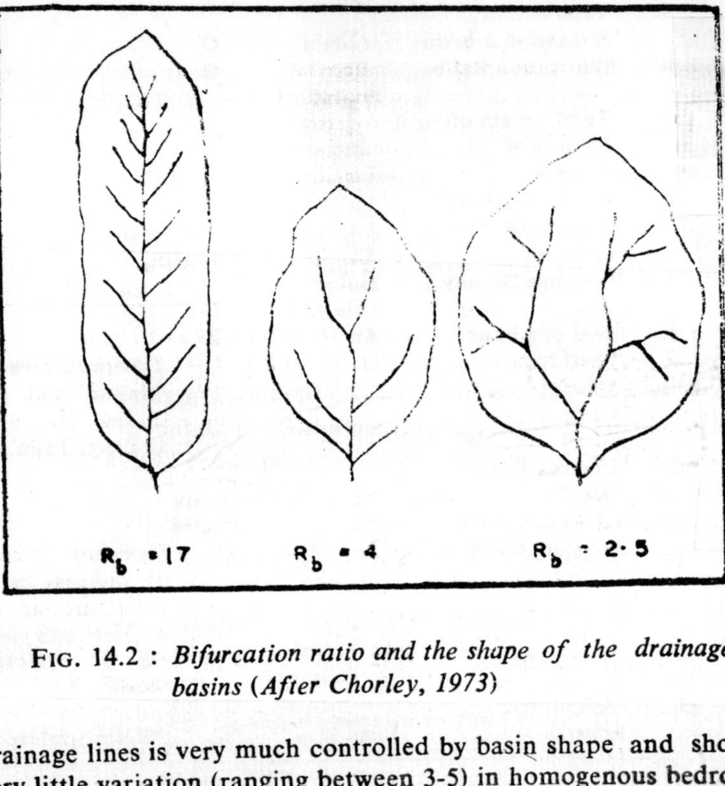


FIG. 14.2 : *Bifurcation ratio and the shape of the drainage basins (After Chorley, 1973)*

drainage lines is very much controlled by basin shape and shows very little variation (ranging between 3-5) in homogenous bedrock from one area to another".

It is apparent from the above statement and Fig 14.2 that the initial formation of a drainage basin records a higher bifurcation ratio and develops an oblong shape, while with the gradual lapse of time, bifurcation ratio decreases and the basin becomes circular to elongate. Normally the range of bifurcation ratio is confined in between 3-5, except in abnormal situations.

Asthana (1975) has analyzed about 11 drainage basins in Almora region whose properties are shown in Table 14.2.

A glance of this table will reveal some interesting conclusions confirming the laws of Horton, as stated above. Bifurcation Ratio (Rb), which was calculated by fitting the data of order (u) and (Nu) on a semilogrithmic graph and drawing a regression line by least square method. In a straight line equation (Log y=bx+a) the value of bifurcation ratio (Rb) is $\log^{-1} b$. It can be broadly seen that the drainage basins of Lower order say 3 or less than 3 have higher Rb, i.e., 4-7 while those of higher order, i.e., 4 or 5 have lesser Rb, i.e., less than four. On the other hand, the Rl (length ratio) generally increases as the order of the basin increases.

*Areal aspect*

There are various parameters to measure the areal aspect of a drainage basin, which cannot be included within the limited scope of this work. However, a brief discussion on some generally used parameters can be undertaken. Areal aspects generally include the properties of a drainage basin, like area (Au), drainage texture (Du), drainage frequency (Fu) etc. By implication shape and form characteristics of a drainage basin can also be included in the areal aspect : they are elongation ratio (Re), circularity index (Rc) etc. All these parameters have been discussed in detail by Cooke and Doorknamp (1974).

Some of the parameters of the areal aspect of a drainage basin has been analysed by Asthana. Some salient indices of the referred analysis has been presented in Table 14.2. It can be seen that the basins of smaller areal coverage have coarse drainage texture (Du), while the basins of larger areal coverage have fine texture. It is, however, an agreed norm that the maximum integration in drainage basin is attained at a later stage in a larger basin. Otherwise, also, fine drainage texture broadly corresponds to a mature topography, where the granularity in the basin is very small. In other terms length of streams per unit of area is large.

*Relief aspect*

Relief aspect includes the height and slope parmeters of a drainage basin. Few of these have been included in Table 14.1. On an average, in a mountainous relief the channel slope (c) increases as the relief (H) increases (see Table 14.2). Elongation

**TABLE 14.2 : Parameters of drainage basins around Almora and its environs**

LINEAR

| Sr. No. | Name of the Drainage Basin | Order (u)/Vs Number (N) | | | | | | Order/Vs Ar. Length (km) | | | | | |
|---|---|---|---|---|---|---|---|---|---|---|---|---|---|
| | | $u_1$ | $u_2$ | $u_3$ | $u_4$ | $u_5$ | Rb | $u_1$ | $u_2$ | $u_3$ | $u_4$ | $u_5$ | Rl |
| (1) | (2) | (3) | (4) | (5) | (6) | (7) | (8) | (9) | (10) | (11) | (12) | (13) | (14) |
| 1. | Rengad | 34 | 6 | 1 | — | — | 5.76 | .35 | .6 | 2.2 | — | — | 2.48 |
| 2. | N. Gadhera (Trib) | 18 | 2 | 1 | — | — | 4.19 | .42 | .74 | 1.5 | — | — | 1.87 |
| 3. | N. Gadhera (Trib) | 17 | 2 | 1 | — | — | 4.08 | .61 | 1.0 | .7 | — | — | 1.06 |
| 4. | Mahadeogad | 64 | 11 | 1 | — | — | 7.87 | .42 | .6 | 4.5 | — | — | 3.24 |
| 5. | Benigad | 40 | 4 | 1 | — | — | 6.24 | .40 | .77 | 1.7 | — | — | 2.05 |
| 6. | Jam Gadhera | 115 | 21 | 5 | 1 | — | 4.74 | .45 | .74 | 3.5 | 2.8 | — | 2.76 |
| 7. | Kanari-Gadhera | 53 | 8 | 3 | 1 | — | 3.58 | .42 | .75 | 1.7 | 2.0 | — | 2.5 |
| 8. | Trib of Gagas | 42 | 8 | 3 | 1 | — | 3.35 | .51 | .5 | 1.6 | 1.6 | — | 1.7 |
| 9. | Nanakosi | 210 | 27 | 8 | 2 | 1 | 3.74 | .44 | .74 | 1.51 | 1.25 | 4.5 | 3.89 |
| 10. | Kaligad | 329 | 43 | 12 | 3 | 1 | 1.4 | .41 | .64 | 1.2 | 2.5 | 7.5 | 2.03 |
| 11. | Kuchgad | 260 | 56 | 14 | 4 | 1 | 3.9 | .46 | .65 | .91 | 2.5 | 7.8 | 5.2 |

| | | AREAL | | | | | RELIEF | | | |
|---|---|---|---|---|---|---|---|---|---|---|
| Sr. No. | Name of the Drainage Basin | Area (Au) Sq km | D. Tex. (Du) km² | Relief (H) (m) | El. Ratio (Re) | Max. Length (Lm) km | Channel Slope (θc) | Side Slope (θg) | Hy. Int. (%) | Stage |
| (1) | (2) | (15) | (16) | (17) | (18) | (19) | (20) | (21) | (22) | (23) |
| 1. | Rengad | 10.42 | 4.42 | 600 | .214 | 3.5 | 5°42 | 11°20 | 30.4 | Mon. |
| 2. | N. Gadhera (Trib) | 6.78 | 4.0 | 1100 | .54 | 4.5 | 15°48 | 27° | 37.5 | Mon. |
| 3. | N. Gadhera (Trib) | 4.99 | 6.64 | 971 | .60 | 3.4 | 12°12 | 28°24 | 50.2 | Mat. |
| 4. | Mahadeogad | 17.08 | 5.95 | 73767 | .85 | 4.57 | 43°6 | 18° | 42.7 | Mat. |
| 5. | Benigad | 7.76 | 6.92 | 945 | .99 | 3.0 | 10°18 | 27° | 48.3 | Mat. |
| 6. | Jam Gadhera | 35.61 | 5.7 | 1162 | .77 | 2.25 | 7°0 | 27° | 44.7 | Mat. |
| 7. | Kanari-Gadhera | 15.31 | 5.95 | 992.33 | .83 | 2.6 | 7°18 | 27° | 38.4 | Mon. |
| 8. | Trib of Gagas | 12.31 | 5.25 | 967.33 | 1.01 | 12.5 | 9°24 | 28° | 50.2 | Mat. |
| 9. | Nanakosi | 55.04 | 8.95 | 80.33 | .68 | 7.4 | 30°20 | 51°30 | 41.0 | Mat. |
| 10. | Kaligad | 88.50 | 5.6 | 1129.33 | .70 | 9.5 | 35°15 | 58° | 58.5 | Young |
| 11. | Kuchgad | 91.70 | 5.2 | 1109.33 | .54 | 14.84 | 27°26 | 48° | 41.5 | Mat. |

ratio (Re) also has a direct correspondence with increasing relief (H), of a drainage basin. There is an interesting feature to note with regard to the channel slope (θc) and the valley side slope (θg). It can be seen from Table 14.2 that higher the channel slope still higher is the valley side slope, e.g., if the channel slope ranges between 5°—12°, the valley side slope ranges between 11°—27° (except Mahadeogad). On the other hand, when the channel slope becomes 25°, the side slope rises to 45° and more. There is almost a double range of increase within channel slope (θc) and side slopes (θg). Another important attribute of a drainage basin is revealed by elongation ratio (Re). It is a ratio between the maximum length of the drainage basin and the diameter of a circle with the same area as that of the drainage basin. When Re is about 1 the basin is generally circular while it is elongated when it is more than 1 and flattened in case it has a Re less than 1. We can see from Table 14.2 that some basins are more or less circular, e.g , basin nos. 4, 5, 8, while all others have lower values of Re indicating thereby that they have a flattened shape. The two situations arise respectively due to young relief and river captures. In the young relief larger number of lower order streams develop, while the higher order streams have smaller length and number, consequently they develop a circular drainage basin. On the other hand, when a fully developed basin is truncated by river capture the length of main channel sharply decreases, while the tributaries are well developed. This stiuation gives rise to flattened shape of the basin.

Strahler (1964) developed a very popular index to analyse the relief aspect of the drainge basins. It is known as Hypsometric Integral (Ihy). The graph is drawn between two relative variables of drainage, area (a/A) and height (h/H) ; where a is the cumulative areal component of contours starting from the confluence point, A is the total area of the drainage basin ; and h is the cumulative relief component of the successive contours and H is the total relief of the drainage basin. The curve drawn is a hypsometric curve, whose area can be calculated in percentage, as shown in the Table 14.2. It can be seen that almost all the drainage basins have either youthful stage of development or early mature stage of development.

Hack and Goodlet (1960) have divided some functional units of a drainage basin which are very useful parameters to describe the functional aspect of the basin. These units are :

(a) Nose : driest and convex area, usually on the periphery of the drainage basin. It functions as shed-off area of runoff to the drainage basin.

(b) Side slope : receives water from the nose, has a straight profile, wettier than nose.

(c) Hollow : concave portion of the valley, marks the convergence area of tributaries, water component of the valley increases in this unit.

(d) Foot slope : gentler lower side of the valley side slopes, covered with regolith and talus material.

(e) Channel way of Valley Floor : river course plus adjacent flood plain.

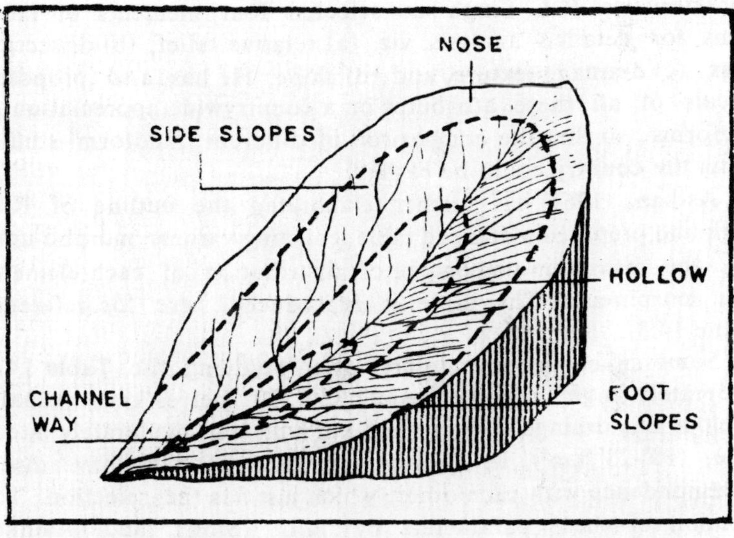


FIG. 14.3 : *Functional subdivision of a drainage basin (Afer Hack & Goodlet, 1960)*

It had already been higlighted that the drainage basin should be adapted as the smallest integrated unit for the analysis of a physical environment. We have, therefore, described in brief various parameters of analysing a drainage basin.

The next stage is develop a scheme of synthesizing the information provided by the aforesaid analyses. The *synthesis* is an exercise of integrating various parameters together with a view to delineate areal units of uniform characteristics. These units are referred as "morphounits". R.L. Singh and Asthana (1967) have tried to evolve a scheme of recognising such morphounits. The scheme envisages four stages of evaluating the attributes of landforms and finally recognising morphounits. The four stages of studying landform were initially suggested by Weaver (1965), i.e. : (a) formulating the basic criteria, (b) changing the scale of observation, (c) dismantling landforms, and (d) reassembling landforms. The first two stages refer to the aspect of fixing objective of study and deciding the unit of area under observation. The third stage refers to the analysis part and the fourth is the aspect of synthesis and recognising homogeneous units of physical characteristics. R.L. Singh has selected four elements of landforms for detailed analysis, viz. (a) relative relief, (b) dissection index, (c) drainage texture, and (d) slope. He has also proposed a scale of all these attributes on a countrywide appreciation of landforms ; so that the comparison of different landform studies across the country could be easier.

Asthana (1967) has further elaborated the outline of R.L. Singh and prepared a detailed table, enlisting various morpho-units and the maximum percentage of distribution of each elements in a morphounit. The table is reproduced here for reference (Table 14.3).

Some salient features emerge after studying the Table 14.3. Moderately high relative relief (50-150 metres/km$^2$) usually develops fine drainge texture (73.5 km/km$^2$) and moderately steep slope, (10°-25°/km$^2$). In essence, all these elements have direct correspondence with each other, which justifies the selection. The selection of these parameters by R.L. Singh and Asthana (1967) in Table 14.4 will be discussed at length in subsequent pages. The morphological units, thus recognised are shown in Fig. 14.4. Each unit has a code consisting of capital letters R.D.T.S, which respectively stands for relative relief (R), dissection index (D), drainage texture (T), and slope (S). All these letters carry small letters at their bases which are explained in the following Table 14.3.

**Functional relationship**

The area around Almora was analysed in detail separately

TABLE 14.3 : **General scale of landform elements**

| Landform Elements | Code | Associated range of value |
|---|---|---|
| I. *Relative Relief* | R | |
| Extremely low | Rel | 0-10 metre/sq km |
| Moderately low | Rml | 10-20 m/sq km |
| Low | Rl | 20-40 m/sq km |
| Moderate | Rm | 40-80 m/sq km |
| Moderately High | Rmh | 80-160 m/sq km |
| High | Rh | above 160 m/sq km |
| II. *Dissection Index* | D | |
| Low | De | 0-0.1 |
| Moderate | Dm | 0.1-0.3 |
| High | Dh | 0.3 and above |
| III. *Drainage Texture* | T | |
| Very coarse | Tvc | 1.25 km/km$^2$ |
| Coarse | Tc | 1.25-2.5 km/km$^2$ |
| Moderate | Tm | 2.5-3.75 km/km$^2$ |
| Fine | Tf | 3.75-5.0 km/km$^2$ |
| Very fine | Tvf | >5.0 km/km$^2$ |
| IV. *Slope* | S | |
| Level | Sl | <2° |
| Gentle | Sg | 2-5° |
| Moderate | Sm | 5-10° |
| Mod. Steep | Sms | 10-25° |
| Steep | Ss | 25-40° |
| Very steep | Svs | 40-60° |
| Escarpment | Se | >60° |

*Source* : Adapted from Singh, R.L. (1967).

considering its landform elements mentioned earlier. Major distribution of different elements was carefully studied and specific boundaries were delineated by the method of superimposition. The process can now be replaced by detailed computer programming and by the cluster analysis method (Mather, 1976). Characteristic distribution of each element with respect to different morphological units is tabulated Table 14.4, and the delimitation of the boundaries of morphounits is shown in Fig 14.4.

Table 14.4 reveals that there are two first order units, i.e., valleys and divides, five second order units based on major tracts of valleys and divides and seventeen third order units having specific distribution of morphometric attributes. It is evident that the valleys generally have an altitudinal range of

TABLE 14.4 : **Identification of morphometric units**

| Morphological-units | Morphometric | |
|---|---|---|
| | 1. Altitude | 2. R. Relief |
| I. *Himachal valleys* | | |
| A. Kosi-Sual valley | | |
| (i) Kosi valley | 12—1500 m | Mod. High (60) |
| (ii) Sual valley | 12—1500 m | High (60) |
| (iii) Kuchgad valley | 12—1500 m | High (60) |
| (iv) Kali Rao valley | 12—1500 m | Mod. High (60) |
| (v) Kaligad valley | 12—1500 m | Mod. High (70) |
| (vi) That-gad valley | 13—1600 m | High (75) |
| (vii) Left upper tract of Nanakasi | 12—1500 m | Mod. High (35) |
| B. *Gagas velley* | | |
| (viii) Gagas valley | 12—1500 m | Mod. High (80) |
| (ix) Jam Gadhera | 12—1500 m | Mod. High (75) |
| (x) Naurar Gadhera | 15—1800 m | Mod. High (60) |
| (xi) Upper tract of Jam Gadhera | 12—1500 m | High (60) |
| II. *Dividing ranges* | | |
| C. Kosi-Gagas divide | | |
| (xii) Nothern divide | | |
| (a) Aira Dev range | 15—1800 m | High (70) |
| (xiii) Central divide | | |
| (a) Ranikhet ridge | 15—1900 m | High (40) |
| (a1) Siahi Devi | 15—2100 m | High (70) |
| ($a^2$) Southern extension | 12—1500 m | High (50) |
| (b) Central tangle | 15—1800 m | High (80) |
| (xiv) Southern divide | | |
| (a) Upper | 15—1900 m | High (80) |
| (b) Lower | 15—1900 m | High (60) |
| D. (xv) Kasi-Sual divide | 13—1800 m | High (70) |
| (xvi) Sual-Kali Rao | 15—1900 m | High (60) |
| E. (xvii) Kosi-Saryu divide | 16—1900 m | High (50) |

*Note* : Figures in brackets indicate percentage coverage.

| Attributes | | | Code formulation |
|---|---|---|---|
| 3. Diss. Index | 4. Drainage Texture | 5. Average Slope | |
| Med. (60) | Fine (60) | Mod. Steep (80) | I Ai (Rmh Dm Tf Sms) |
| Med. (90) | Coarse (60) | Mod. Steep (70) | I Aii (Rh Dm Tc Sms) |
| High (60) | Med. (60) | Steep (80) | I Aiii (Rh Dh Tm Ss) |
| Med. (90) | Coarse (60) | Mod. Steep (70) | I Aiv (Rmh, Dm Tc Sms) |
| Med. (100) | Fine (45) | Mod. Steep (100) | I Av (Rh Dm Tf Sms) |
| Med. (70) | Fine (70) | Steep (70) | I Avi (Rh Dm Tf Ss) |
| Med. (100) | V. Fine (60) | Mod. Steep (70) | I Avii (Rmh Dmi Tvf Sms) |
| Med. (90) | Fine (45) | Steep (40) | I Bviii (Rmh Dm Tf Ss) |
| Med. (95) | Fine (40) | Mod. Steep (80) | I Bix (Rmh Dm Tf Sms) |
| Med. (80) | Coarse (70) | Steep (70) | I Bx (Rmh Dm Tc Ss) |
| Med. (90) | Fine (55) | Mod. Steep (80) | I Bxi (Rh Dm Tf Sms) |
| Med. (100) | Fine (60) | Steep (70) | II C xii a (Rh Dm Tf Ss) |
| Med. (80) | Fine (40) | Mod. Steep (60) | II C xiii a (Rh Dm Tf Sms) |
| Med. (90) | Fine (60) | Steep (65) | II C xiii a1 (Rh Dm Tf Ss) |
| Med. (100) | Fine (50) | Steep (100) | II C xiii a2 (Rh Dm Tf Ss) |
| Med. (60) | Fine (40) | Mod. Steep (70) | II C xiii b (Rh Dm Tf Sms) |
| Med. (80) | Fine (40) | Steep (60) | II C xiv a (Rh Dm Tf Ss) |
| Med. (100) | Coarse (50) | Steep (90) | II C xiv b (Rh Dm Tc Ss) |
| Med. (80) | Fine (60) | Mod. Steep (35) | II D xv (Rh Dm Tf Sms) |
| Med. (80) | Fine (50) | Mod. Steep (65) | II D xvi (Rh Dm Tf Sms) |
| High (70) | Fine (70) | Steep (100) | II E xvii (Rh Dh Tf Ss) |

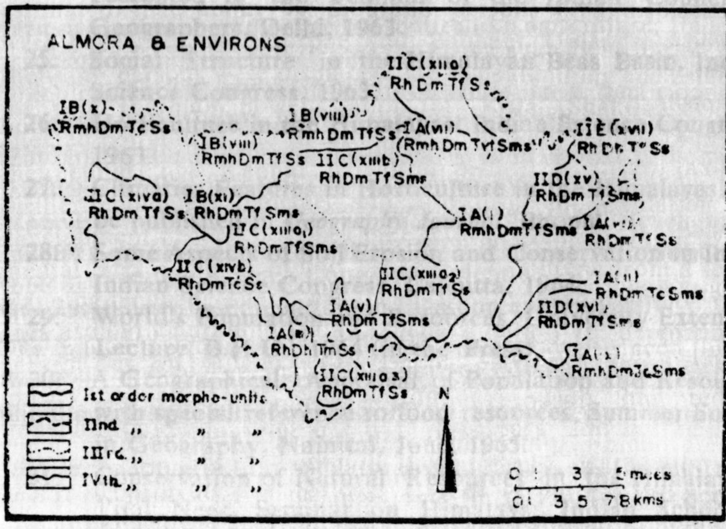


FIG. 14.4 : *Morphological units of landforms*

1200-1500 metres and have high to moderately high relative relief, high and medium dissection index, coarse to fine texture and moderately steep to steep slope. On the other hand, the divides have higher altitude ranging between 1500-1900 meteres having high relative relief, fine texture, medium dissection index and moderately steep to steep slope. The difference lies primarily in the texture and slope condition of the two first order units ; valleys have generally coarser values of texture as compared to divides. This is because of the larger frequency of lower order streams in the divide areas compared to fewer higher order streams in the valley areas. Similar differences can be seen in the subsequent lower order units. The exercise of recognising morphological units is very fruitful in describing the topographic characteristics of the area, and delineating broadly homogenous tract of landforms. These landform regions can now be seen in the functional relationship of certain cultural aspects, like landuse, accessbility and settlement distribution.

Functional relationship of morphounits and physical attributes of an environment is an exercise towards the wholistic approach of viewing the landscape. The attributes of the physical environment should not limit itself to mere description but enable a reader to understand the utility of each attribute in terms of human activities. There can be various parameters to indicate the relationship. In the present endeavour, however, only one parameter of human environment is taken, i.e., settlements. Asthana (1976) has tried to establish the relationship of physical attributes and the density of settlements. He has found out some adhesive parameters and non-adhesive parameters. The terms adhesive and non-adhesive refer to the favourable or non-favourable sites of settlements.

The adhesive values of the parameters have been selected by drawing ogive curves between percentage settlement density and physical attributes and later selecting the values of interquartile area on the axis of physical attributes. Following Table 14.5 indicates the corresponding adhesive and non-adhesive values of each physical attributes :

TABLE 14.5 : **Physical attributes**

| | R. Relief (oo m) | D Index | Texture (Km-1) | Slope (Degrees) | Altitude (ooom) |
|---|---|---|---|---|---|
| **Adhesive values** | 3.5-4.0 | 0.16-0.21 | 2-3.5 | 18-32 | 1.5-2 |
| **Non-Adhesive values** | 3.5 and 4.0 | <0.16 and >0.21 | <2 and >3.5 | <18 and >32 | <1.5 and >2 |

*Source* : Adapted from Asthana (1968).

When all these values were superimposed, three categories emerged viz., maximum coincidence of adhesive values, maximum coincidence of non-adhesive values and partial coincidence of the two values. The three areas were designated as, (a) areas suitable for settlements, (b) areas not suitable for settlements, and (c) areas partially suitable for settlements. Fig. 14.5 shows such areas, which go in conformity with the settlement density observable in the area. This, therefore, proves that such exercises can be very fruitful in delineating potential areas of human activities and also *deterrent* area for the same.

## Conclusion

The whole *treatise* on the analysis of physical environment needs

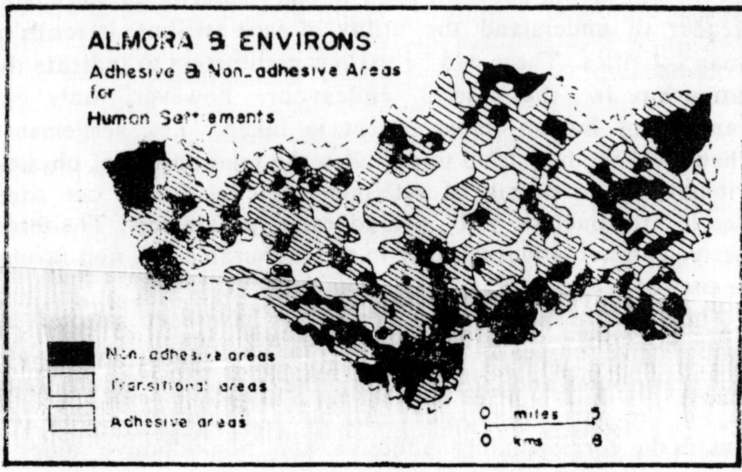


FIG. 14.5 : *Functional relationship of physical factors and human settlements*

to be understood in wholistic perspective and more meaningful purpose be highlighted in such a study. The integrating approach of not only the two main compartments of geography viz. ; physical and human, should be kept in view, but efforts should be made to co-operate with other sciences, as well. The study of environment is no more a restricted domain of geographers. It will, therefore, be a murderous act if we continue to keep ourselves excluded from other disciplines and falsely create a hallow around us about the uniqueness of the subject. Nonetheless, the expertise and skill of a geographer in analysing and describing the spatial pattern and its organisation will remain its distinctive feature, even in times to come.

## REFERENCES

Asthana, V.K. (1967), Morphometric evaluation of landforms in Almora and its environs, *National Geographical Journal of India*, xiii (1), 37-54.

Asthana, V.K. (1968), *Landforms and settlement in Almora and Environs*, Ph.D. thesis (unpublished) Banaras Hindu University, Varanasi, 1-213.

Asthana, V.K. (1975), The drainage characteristics of basins around Almora and its environs, *National Geographers*, X, 87-103.

Asthana, V.K. (1976), Study of rural settlements in Almora and environs, in Singh, R.L. *et al.* (eds.), *Geographical Dimension of Rural Settlements* ; National Geographical Society of India, Varanasi, 131-9.

Bouldig, K. (1956), General system theory : the skeleton of science, *General Systems*, I, 11-7.

Chorley, R.J. (1969), The drainage basin as the fundamental geomorphic unit, in Chorley R.J. (ed.), *Introduction to Physical Hydrology*, Metheun, 40.

Cooke, R.U. and Doornkamp, J.C., (1974), *Geomophology in environmental management* ; Clarendon Press, Oxford, 8.

Eyre, S.R. (1963), Determinism and the ecological approach to Geography, *Geography*, 49, 376.

Hack, J.T. and J.C. Goodlet (1960), Geomorphology and forest ecology of a mountain region in the central appalechians, *U.S. Geological Survey, Professional Paper*, 347, 66.

Horton, R.E. (1945), Erosional developments of streams and their drainage basins ; Hydrophysical approach to quantitative morphology, *Bulletin of Geological Society of America*, 56, 275-370.

Makhan, O.S. (1967), The brown forest soils and the grey-brown podzolic soils of the Himalayan forests ; Unpublished monograph, Department of Geography, Banaras Hindu University, Varanasi.

Mather, P.M. (1976), *Computational Methods of Multivariate Analysis in Physical Geography* ; John Wiley, 1-502.

Peltier, Louis (1950), The geographical cycle in periglacial regions as it is related to the climatic geomorphology, *Annals of the Association of American Geographers*, 40, 214-36.

Schaefer, F.K. (1953), Exceptionalism in Geography : A methodological examination, *Annals of the Association of American Geographers*, 43, 226-9.

Singh, R.L. (1974), *Morphometric Analysis of Terrain*, The National Geographical Society of India, Bull. No. 22, 1-24.

Stoddart, D.R. (1972), Geography and the ecological approach : The ecosystem as a geographical principle and method in Davies, K.D. Wayul (ed.), *The Geographical Revolution in Geography*, University of London Press, 302-6.

Strahler, A.N. (1964), Quantitative geomorphology of drainage basins and channel networks ; in Chow, V.T. (ed.), *Handbook of Applied Hydrology*, McGraw Hill, New York, 4-11.

Tansley, A.G. (1935), The use and abuse of vegetational concepts and terms ; *Ecology*, 16, 206.

Weaver, Glen D. (1965), What is landform ? *Professional Geographer*, XVII (i), 11-4.

Haushila Prasad

FIFTEEN

# PHYSICAL ENVIRONMENT AND SETTLEMENTS IN MUSSOORIE REGION

The study region occupies a south-western portion of Uttar Pradesh (U.P.) Himalayas comprising parts of Dehra Dun and Tehri-Garhwal districts. It lies in between the Ganga and the Yamuna drainage systems, just north and north-east of the well-known Dun Valley and Siwalik Range. This area has been particularly selected in order to find out how far morphometric components are responsible for identifying the difierent categories of physical environment. Obviously physiography plays a vital role in shaping the environment. The genesis of the physical complexities which form environment, ranges from hazardous to favourable for human activities. Settlements can be considered as one of the most prominent compromise of all the physical and human factors in an area. Owing to this reason, settlements have been studied in relation to different morphometric attributes for identifying the types of environments.

The present region has been delineated on the basis of physical environmental factors such as water partings in the north and east, Bandal nadi in the south-east, Yamuna river in the south-west, and 3000 feet contour and Baldi nadi in the south. However, the consideration of environmental factors for the delineation of the region or its parts has to be modified and adjusted in order to coincide with the revenue village units. This has been considered necessary because in the present study the settlements have been thought to be the human activities interacting with the physical environment. The data and records on settlements are available on the basis of revenue villages. Hence, a basic frame-work comprising of environmental factors and the administrative units have been developed for the present study.

Kotidhar, Nagdhar, Gorphadhar and Dhancholidhar ranges form the northern boundary of this region and also make the water divide of Aglar river—an important tributary of the Yamuna river, lying in the northern part of this study region with Bhadrigad and Nogungad. In the east, Singhudhar range makes the water divide of Aglar and Bandal nadi system.

Inclusion of the parts of Dun Valley lying just to the south has been avoided as it forms a separate and distinct natural sub-unit of the U.P. Himalayas. Thus the region 'Mussoorie' is a distinct physical environmental unit centred around Mussoorie Town in the Kumaun Himalayas.

The study region embraces an area of 592.57 $km^2$ and total population of 56,784 (1981). The height in the region varies from 520 M (at the confluence of Bin nadi and the Yamuna river) to 3022 M (at Nag Tilla located at Nagdhar range). In conformity with the general trend of the Himalayan environment, the main ranges have asymmetrical slopes—the steeper towards southern sides and gentler on the northern sides. The physical environment of the area is rugged due to complex action of denudational processes.

### Methodological development

The study of man-land relationship is quite accepted anology since the time of Soucer and Huntington, whereas the locational association of man and his activities can be vividly studied in the concerned area in relation to human settlements as well. The study of human settlements must be understood as a compromise of all the possible physical and cultural factors of the environment of the concerned area. This study should also be considered as an example of how the human activities have adjusted with the physical environment.

The International Geographical Union Commission on National Settlement System has taken into account a much broader view of Settlements. Accordingly, settlements are now treated as 'system' and 'sub-system' rather than isolated entities. The United Nations Conference on Human Settlements

(Stockhom, 1972) was aimed at drawing global attention for the betterment of settlements. Similarly, other international bodies also have partially or wholly expressed their concern for settlements. Settlements are treated as human occupance units in environment. The human activities cannot be limited to settlements alone, as socio-economic environmental elements are also associated with these activities. The relationships developed between various elements have influenced and modified the environment through settlements. On the other hand, the social, economic and circulatory organization of space has been superimposed on the physical environment.

Several scholars have acknowledged the concept of settlements in a much wider perspective. Doxiadis (1968) has considered five elements of settlements which are (a) nature, (b) man, (c) society, (d) shell and (e) networks. Doxiadis, while studying ekestics, the science of human settlements, considered the settlements in four dimensional conception emphasizing equally on human and physical factors. According to him, the human settlement consists of (a) the content or man alone and in society and (b) the container or the physical settlement, which consist of natural and man-made or artificial elements. Garner (1967), considers settlements as a complex set of man-man relationship ; while Baker (1969) has given more importance to the elements of the landscape and studied settlements which includes the various phenomena and processes resulting from complex man-land relationship. Garner also has considered the following six premises related to settlements location and spatial behaviour.

(a) The spatial distribution of human activities reflect an ordered adjustment to the factor of distance.

(b) Locational decisions are taken in general, so as to minimize to functional effects of distance.

(c) All locations are endowed with a degree of accessibility, but some locations are more accessible than others.

(d) There is a tendency for human activities to agglomerate to take advantage of scale economies.

(e) The organization of human activity is essential hierarchical in character.

(f) Human occupance is focal of character.

Losch's (1954) model of economic landscape and Christaller's (1966) theory of Central Place are the relevant theories regarding the organization of space through settlements, while Philbrick (1957) has considered the arrangement of functions according to the hierarchy of settlements.

In India, Chatterjee (1956), Ahmed (1960), R.L. Singh (1976), K.N. Singh (1969), H.H. Singh (1962), O.P. Singh (1971), Nag (1979) and so many other geographers have given their ideas regarding settlement pattern, their location and organization of settlements. They have also analysed and discussed the effects of various physical, social and cultural factors on the settlements in the different parts of the country.

The modern science of geography is specifically based on the study of environment in relation to man. During the early decade of 20th century persons like Gilbert (1909), Russell (1949), King (1967) and other scholars were studying physical environment then discovering the fundamental principle of landforms development. During the later half of the 20th century geo-scientists like Kesseli (1954) and Hammond (1964) etc. have advocated for the geographic approach to the study of physical environment. They emphasized on the outer features of the landscape rather than their genetic aspects. Devis (1924), Wooldridge (1958) and others have considered landforms as the best means to understand the evolution of physical environment all over. The whole study was then confined only around the origin and evolution of landforms which show the interaction between erosional and weathering process operating on rocks of varying structure and lithology under varied environmental conditions.

During the early decade of the twentieth century the studies were mostly of qualitative nature. But some geo-scientists like Wentworth (1930), Horton (1945), Raisz (1946), Strahler (1950), Dov Nir (1957), Thornbury (1961), Dury (1963), Leopold (1964), Weaver (1965), Clark (1967), Zakrzewska (1967) have given more emphasis on quantitative analysis of the characteristics of the physical environment. They have also developed new trends in this field.

Zakrzewska has given the idea of geography of landforms which "encompasses an empirical descriptive landform analysis followed by an explanation and an analysis of landform in relation to other physical and cultural phenomena". Dury has mentioned that "the systematic study of landforms has barely reached a full scientific status". In order to study of landforms

Weaver has made the following suggestions : (a) formulating the basic criteria, (b) changing the scale of observations, (c) dismantling landforms and (d) re-assembling landforms. Strahler's concepts concerning 'relief' and 'drainage' analysis, and Horton's postulations regarding the "drainage basin parameters" are well received. Dov Nir, Raisz and Wentworth have worked on "slope" and these also are considered as significant development in this field.

Available literature shows that in most of the workers have delineated the regional landform units into physiographic and natural regions ; while a few have demarcated on the basis of morphometry. In this regard the contributions of Chorley (1957, 1962), Schumm (1963) and King (1967) have been significant.

During the last fortyfive years, there has been a great advancement in geo-environmental studies. But, little attention has been paid in this regard by geographers in India. Chatterjee (1946) and Chhibber (1948) paved the way for geomorphological investigations in our country. Chatterjee put forth his ideas while investigating the physiographic evolution of Chotanagpur, while Chhibber studied the classification and origin of plateau and contributed towards the scientific enquiry of the drainage and landforms. A pioneer and systematic study in this regard in India has been done by R.P. Singh (1960), Mukherjee (1961) and Sinha (1966) who have also worked systematically in this field ; while R.L. Singh (1974) paved a new path for geoanalytical studies in India by studying the physical environment of Rajmahal Highlands, the neighbourhood of Almora and Mirzapur Upland. Ahmed has also done good work in the field of geomorphology.

The following scholars have worked on similar lines in the Banaras Hindu University. They are Asthana (1967), Kharkwal (1969), Agrawal (1970), Gupta (1977), K.N. Singh (1978), S.B. Singh (1978), Yadava (1978) and H. Prasad (1980).

## Morphometric basis of environment and settlements

Here attempts have been made to explore the process of adjustment of human and cultural forces over the physical environment in the form of settlements and associated spatial imprints. Efforts have been made by several geographers to study similar problems in other parts of the lesser Himalayas and also in the Vindhyan region. The most popular approach was to examine such inter-relationship in the 'original grid' frame work. The density and distribution of settlements were explained in term of morphometric characteristics of the grid. This approach, however, has its own limitations. The combined and inter-acting morphometric characteristics of grid or morpho-units were left to the readers to draw their own conclusions. Later Asthana (1968) tried to develop a responding study by incorporating more sophisticated quantitative techniques which, however, is also not free of criticism.

The present study is on different lines whereas the quantitative and spatial corrections were blended to find out a reasonable approach for identifying different types of environment based on physical attributes.

## Methodology

In order to study the consequences of physical factors on different components of settlements, a fresh methodology has been adopted here. Representative morphometric values of different attributes were correlated which contributed to a viable frame work in order to investigate the complex relationship between settlements and physical environments, which is based on the landforms of the region. After a detailed investigation, the study area has been divided into different environmental regions according to the nature of settlements and their association with the physical environment. Two broad categories of areal units, *i.e.*, (a) revenue village and (b) forest settlements were identified. The settlements in villages and reserved forests were counted on the basis of District Census Handbook (1981) of Dehra Dun and Tehri-Garhwal and concerned topographical survey sheets at a scale of 1 : 50,000. The inter-relationships of different physical and other factors which were experimented spatially and quantitatively were found useful for better understanding between human factors and physical environment. For this purpose, the correlation between different morphometric attributes and settlements were also calculated. It is evident from the present study that there is a positive correlation (=0.341) between 'absolute relief' and 'relative relief'. It is also noticed that the degrees of slope too is positively related to the 'relative relief'. The other correlations between different morphometric attributes and settlements are indicated in Table 15.1.

TABLE 1.51 : **Correlation between different morphometric attributes and settlements**

| | AR | RR | DI | Sl | DD | DF | St |
|---|---|---|---|---|---|---|---|
| AR | — | 0.193 | 0.442 | 0.576 | —0.164 | —0.021 | —0.053 |
| RR | 0.193 | — | 0.396 | 0.591 | 0.015 | 0.062 | —0.227 |
| DI | 0.442 | 0.396 | — | 0.092 | 0.213 | 0.141 | —0.112 |
| Sl | 0.576 | 0.591 | 0.092 | — | 0.132 | 0.061 | —0.242 |
| DD | —0.164 | 0.015 | 0.213 | 0.132 | — | 0.554 | —0.093 |
| DF | —0.021 | 0.062 | 0.141 | 0.061 | 0·554 | — | —0.081 |
| St | —0.053 | —0.227 | —0.112 | —0.242 | —0.093 | —0.081 | — |

AR=Absolute Relief ; RR=Relative Relief ; DI=Dissection Index ; Sl=Slope ; DD=Drainage Density ; DF=Drainage Frequency ; and St=Settlement.

The above values show that the organisation, distribution and pattern of settlements are very much affected by the morphometric attributes and physical environment in the Mussoorie Region. A detailed study of settlements and their association with different attributes is made here.

**Altitudinal zones and settlements**

Quantitatively absolute relief and settlements have negative relationship ($r$=0.052). This indicates that settlements are poorly associated with absolute relief. Further, it can be explained that with the increase of height, there is decrease in the number of settlements (Table 15.2).

Table shows a well-marked concentration of settlements between 1200 M—2400 M height groups. This range covers 86.48 per cent of the settlements and 78.23 per cent of the study area. Further, to a smaller extent within this range marks the more favourable conditions for settlements in height group of 1650—2100 M. This category covers 41 per cent of settlements in the area under study. The favourable areas for settlements are distributed almost all over the study region in patches with exceptions in northern area, eastern parts of central ridge and deep gorges of rivers in the area. The settlements are also not noticed on the peaks and highly slopy regions.

TABLE 15.2 : **Absolute relief and settlements**

| Absolute relief (M) | No. of settlements | Percentage | Density of settlements (settlements/km²) |
|---|---|---|---|
| Below 900 | 13 | 0.77 | 3.23 |
| 900-1050 | 78 | 4.63 | 8.73 |
| 1050-1200 | 98 | 5.81 | 2.54 |
| 1200-1350 | 133 | 7.88 | 4.99 |
| 1350-1500 | 196 | 11.62 | 4.66 |
| 1500-1650 | 142 | 8.42 | 1.99 |
| 1650-1800 | 224 | 13.29 | 4.17 |
| 1800-1950 | 205 | 12.16 | 3.33 |
| 1950-2100 | 263 | 15.60 | 3.01 |
| 2100-2250 | 131 | 7.78 | 2.10 |
| 2250-2400 | 164 | 9.73 | 2.82 |
| 2400-2550 | 29 | 1.72 | 0.76 |
| 2550 and above | 10 | 0.59 | 0.49 |

**Relative relief and settlements**

The analysis of seven categories of relative relief in relation to settlements, brings forth the conclusion that the existence of more favourable conditions are found below relative relief of 550 M. This group of relative relief covers 90.11 and 76.25 per cent of settlements and area under study respectively. The highest concentration of settlements is between 35.0 M and 450 M where 736 out of 1,686 settlements occupy 26.92 per cent to give a density of as high as 4.78 settlements per km² (Table 15.3). This favourable area settlements in relation to relative relief are also similar and mostly common to the areas favourable for settlements in terms of absolute relief.

TABLE 15.3 : **Relative relief and settlements**

| Relative relief (M) | No. of settlements | Percentage | Density of settlements (settlement/km²) |
|---|---|---|---|
| Below 250 | 56 | 3.32 | 2.11 |
| 250-350 | 169 | 10.02 | 2.26 |
| 350-450 | 763 | 45.25 | 4.78 |
| 450-550 | 532 | 31.52 | 3.78 |
| 550-650 | 155 | 9.19 | 1.51 |
| 650-750 | 11 | 0.65 | 0.48 |
| Above 750 | 1 | 0.05 | 0.25 |

However, the outlying parts of the study area which were not favourable for settlements regarding absolute relief are found in some cases to be marginally favourable in case of relative relief The eastern parts of Central Ridge is comparatively not suitable for settlement, while comparing absolute and relative relief the composite area attractive for settlement is reduced; as considerable area from western parts of the Central Ridge to the Yamuna river is found unfavourable regarding relative relief. There are a number of sizeable pockets having similar characteristics. The correlation between relative relief and settlements is negative ($r = -0.227$).

**Dissection index and settlements**

The relationship of dissection index with absolute and relative relief is found to be directly proportional. The area favourable for settlements regarding the dissection index should normally be on lower range of both morphometric attributes, *i.e.*, absolute and relative relief. The area having density around three settlements per km² may be brought to be more convenient or rather a norm for settlement in the region. Such density of settlements is observed in the range of 0.2–0.4 of the dissection index, which covers 75.25 per cent of the area and 84.05 per cent of the settlements (Table 15.4).

TABLE 15.4 : **Dissection index and settlements**

| Dissection index category | No. of settlements | Percentage | Settlement density (settlement/km²) |
|---|---|---|---|
| Below 0.2 | 219 | 12.98 | 1.85 |
| 0.2-0.3 | 1054 | 62.50 | 3.30 |
| 0.3-0.4 | 364 | 21.57 | 2.88 |
| 0.4 and above | 50 | 2.95 | 1.75 |

Comparing the areas favourable for settlements regarding dissection index with those favourable in terms of absolute and relative relief, we do get a composite picture and observe how both the types of relief interact in space to contribute to a suitable environment. Fringes of the study region and parts of the ridges are again found to be unsuitable. Dissection index and settlements are negatively correlated ($r = -0.110$) in this region.

## Slope and settlements

The study of slope in relation to settlements makes a special case for investigation. Relatively, the moderately steep slope (30° – 35°) covers 38.32 per cent of the area and 49.42 per cent of total settlements to give density as high as 3.70 settlements per km². The areas with less slope (below 15°) are generally considered to be more suitable for settlements, have a uniformly low density in this region except of course the areas below 15 degrees which have a very high density (9.53 settlements/km²). This may be treated as an exceptional case as the concerned area covers only 0.62 per cent of the total area of the study region (Table 15.5).

TABLE 15.5 : **Slope and settlements**

| Slope category (degree) | No. of settlements | Percentage | Settlement density (settlements/km²) |
|---|---|---|---|
| Below 15 | 35 | 2.07 | 9.53 |
| 15-20 | 51 | 3.05 | 2.53 |
| 20-25 | 101 | 5.99 | 2.71 |
| 25-30 | 213 | 12.63 | 2.34 |
| 30-35 | 838 | 49.70 | 3.70 |
| 35-40 | 445 | 26.39 | 2.24 |
| Above 40 | 3 | 0.17 | 0.22 |

It is apparent from the analysis that the favourable area for settlements covers the central part of the study area and some patches in river valleys. The correlation value ($r = -0.242$) shows the negative arrangement between slope and settlements.

## Drainage density and settlements

The settlement density, around 3 settlements per km² considered previously as a norm for this region, falls in the range of 2.4 km/km² of drainage density. The associated area (70.44%) contains 82.23 per cent of the settlements in the study area (Table 15.6). The distributional pattern of settlements does not make a zone or a region.

The drainage density of 2–4 km/km² category makes patches instead of regular and continuous zone. Thus the area considered to favour the settlement process are limited and much scattered. The correlation between settlements and drainage density is also

TABLE 15.6 : **Drainage density and settlements**

| Drainage density (km/km$^2$) | No. of settlements | Percentage | Settlement density (settlements/km$^2$) |
|---|---|---|---|
| Below 1 | 9 | 0.53 | 1.15 |
| 1-2 | 65 | 3.86 | 1.14 |
| 2-3 | 744 | 44.13 | 3.48 |
| 3-4 | 597 | 35.42 | 2.98 |
| 4-5 | 236 | 13.99 | 3.37 |
| 5-6 | 29 | 1.72 | 2.05 |
| Above 6 | 6 | 0.35 | 6.33 |

negative, *i.e.*, $r = -0.093$.

### Drainage frequency and settlements

The corresponding value of drainage frequency are 6—12 drainage lines per km$^2$, which is favourable to the settlements in this study region. This drainage frequency (6—12) category covers 72.63 and 78.77 per cent of area and settlements of the study region respectively (Table 15.7).

TABLE 15.7 : **Drainage frequency and settlements**

| Drainage frequency | No. of settlements | Percentage | Settlement density (settlements/km$^2$) |
|---|---|---|---|
| Below 6 | 145 | 8.60 | 1.26 |
| 6-9 | 753 | 44.67 | 3.13 |
| 9-12 | 575 | 34.10 | 3.02 |
| 12-15 | 171 | 10.14 | 4.46 |
| Above 15 | 42 | 2.49 | 4.78 |

The drainage frequency also gives a similar picture of the settlement distribution pattern likewise the drainage density. One of a few pockets has been seen as reduced in size. The correlation value ($r = -0.081$) shows the negative relationship between drainage frequency and settlements in the present region.

### Environmental zones

The composite picture shows the interaction of different morphometric attributes in relation to settlement, it would have been better in the absence of special physical factors which have checked proper development of settlements. Notwithstanding, the settlements cannot be studied in isolation owing to the associated cultural components. Roads, social amenities, lust for urbanism, level of technology and the like have impressed the settlement pattern to a large extent. The large tract of forest also does not permit to assess the expected settlement pattern. Further, the question of settlement is not restricted to the favourable morphometric attributes, but also to the relative significance of a settlement site in the scale of preference of the inhabitants.

Keeping in view the above problems, the present attempt is to make a synthesis of the different morphometric attributes favourable for settlements. Table 15.8 is aimed at bringing out the related range of morphometric attributes generalized for various settlements densities.

As indicated earlier, the density of above 3 settlements per km$^2$ may be taken as an indicator of adequately settled area. Such an area is characterized by 'absolute relief' varying from 1050—2100 metres 'relative relief' from 350—550 m; dissection index from 0 2—0.4; slope from 15—35 degrees; drainage density from 3—6km/km$^2$ and drainage frequency from 6—15 streams per km. The low density areas have mostly high absolute and relative relief low and very high dissection index, very steep slope and low drainage density and drainage frequency.

If the above model is tested in relation to the map of distribution of settlements (Fig. 15.1), we will find that the areas with favourably morphometric attributes are not even fully settled. There is scope for further settlements even in the high density region.

On the basis of detailed environmental analysis, this region has been divided into five environmental zones. These zones are showing the nature of adjustment with the human and cultural environments of the study region. The different environmental zones are as follows :

### Severe hazardous environmental zone

This environmental zone occupies the Nagdhar and Gorphadhar ranges with upper catchment of Paligad and Gochugad which is located in the northern and north-eastern part of the study area. This hazardous environment possesses very steep slope, high altitude, fine drainage texture with dense reserved forest. This

TABLE 15.8 : Density of settlements and associated morphometric ranges

| Settlement density (settlements/ $km^2$) | Absolute relief (M) | Relative relief (M) | Dissection index | Slope (degree) | Drainage density (km/km ) | Drainage frequency (No. of line/ km) | Environmental zones |
|---|---|---|---|---|---|---|---|
| Below 1 | 2250 and above | Above 650 | 0.1-0.2 | Above 40 | 4-5 | 9-12 | Severe hazardous environmental zone |
| 1-2 | 1550-1650 | 550-650 | Below 0.2 and above 0.4 | 35-40 | Below 2 | Below 6 | Moderate hazardous environmental zone |
| 2-3 | 1050-1200<br>2100-2400 | Below 250<br>250-350 | 0.3-0.4 | 15-30 and 35-40 | 3-4 and 5-6 | 6-9 | Less hazardous environmental zone |
| 3-4 | Below 900<br>1800-2100 | 450-550 | 0.2-0.3 | 30-35 | 2-3 and 4-5 | 6-12 | Moderately favourable environmental zone |
| Above 4 | 900-1050<br>1200-1500<br>1650-1800 | 350-450 | — | Below 15 | 6 and above | Above 12 | Favourable environmental zone |

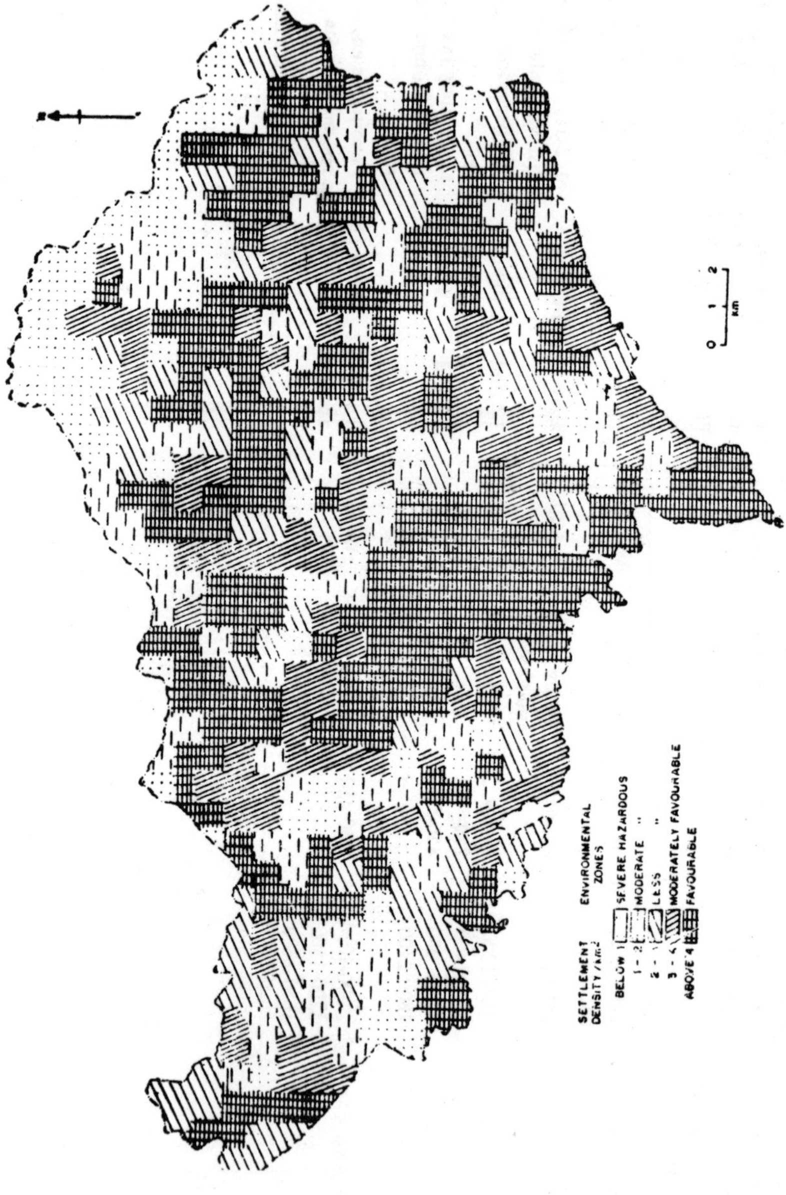


FIG. 15.1 : *Physical environmental zones*

zone also enjoys with high mountain peaks and ranges, *i.e.*, Nagtibba (3022 m), Singhudhar, Gorphadhar and Koridhar with rugged topography and highly dissected terrain. Some small pockets of this zone are also visualized in the scattered form on the Central Ridge which are at Top Tibba, Deo-ka-Tibba, Tarudhar and Dhanaulti in the east, Landour and Banong hill in the centre and Karasudhar in the western part of the region (Fig. 15.2). These main ranges have asymmetrical slope—steeper towards southern sides and gentler in the northern sides. This environmental zone has practically no settlement which shows the severeness of the region with an areal coverage of 67-91 km$^2$ (Table 15.8).

**Moderate hazardous environmental zone**

This hazardous zone is well visualized all over the study region and covers 213.56 km$^2$ area. The settlement density of this zone is 1—2 settlements/km$^2$ (Table 15.8). It is extended all over the study area. In the northern part it possesses Koridhar, upper and middle catchment of Kyarigad, central part of Paligad and in the east, Tharudhar and Gochugad valley fall under this category (Fig. 15.2). This environmental zone is well marked with the Yamuna Valley, Bharond Valley, lower Bin, upper parts of Kotinadi and Saurna river, Kimiari Khala and Karasudhar in the west and south-western region. It is also found in the upper parts of Phedugigad and Diuligad in the south-eastern part of the study region.

**Less hazardous environmental zone**

This hazardous zone possesses settlement density of 2-3 houses per km$^2$ with an areal coverage of 159.53 km$^2$. This environmental zone covers Bin and Koti valleys in west, lower catchments of Arnigad, Bandal and Baldi nadi in south, Patalgad in centre, middle and lower parts of Paligad and surrounding areas of Kyari village in north (Fig. 15.2). Some patches of this zone are also found in scattered forms in the study region.

**Moderately favourable environmental zone**

This environmental zone contains 3-4 houses per km$^2$ as settlement density and only 36.44 km$^2$ area lies under this region. This environmental zone is found mostly in the central part of valley

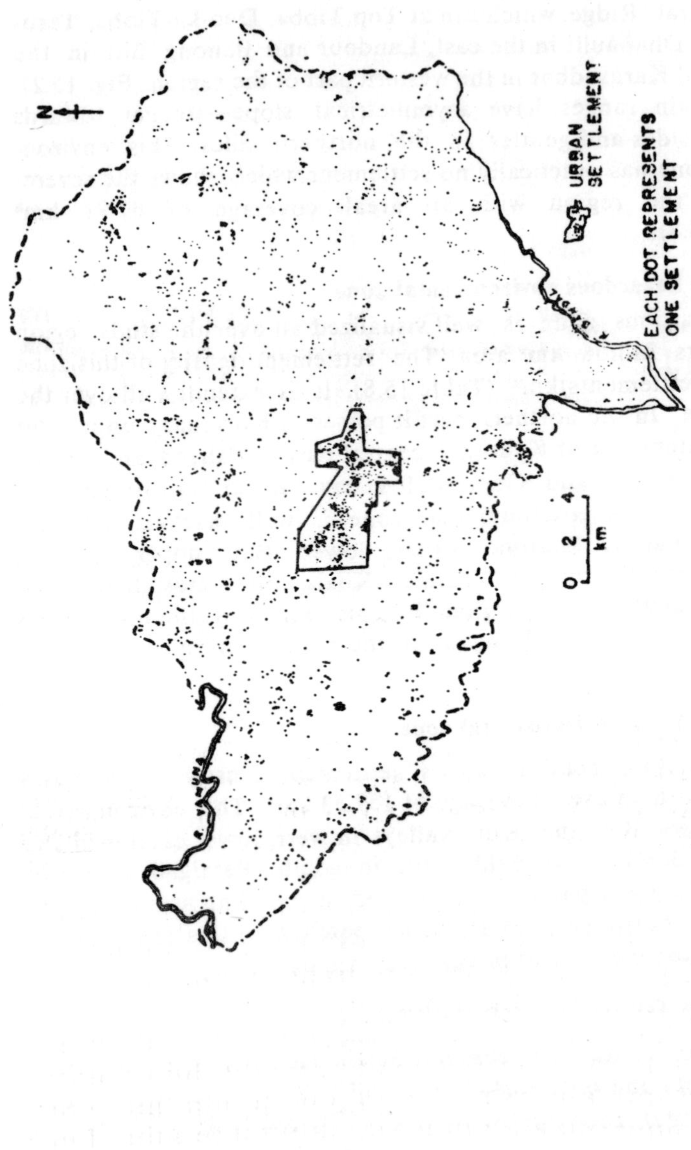


FIG. 15.2 : *Distribution of settlements*

region of Arnigad, Bandal nadi, Kiarkuli nadi and lower part of Paligad (Fig. 15.2).

**Favourable environmental zone**

This zone is most suitable for the location of settlements in the study area. This category also includes Mussoorie-Landaur urban complex. Settlements density is found above 4 houses per km$^2$, while only 45.27 km$^2$ area comes under this category. This environmental zone occupies Central Ridge with top of the hills which are more or less flat and terraced with gentler slope in some localities. Some small patches of this environment is also visualised at Pali and Kyari villages in the north, Jaunpur and Khaskudauri village in the north-western part of the study region. Lower Bandal and Baldi valley in south also come under this category with lower Saurna and Arnigad in the south-west and west of the study area. In the central part of study region, Mussoorie and Landour towns are mainly responsible to raising the settlements density. These hill towns attract the people towards the Central Ridge which is famous for a well known tourist centre in the U.P. Himalayas. This area is well connected by roads, cart-tracks and footpaths with the other parts of the study region. The flat hill tops and terraces with gentle to moderate slope are also other factors which play an important and vital role for high settlements density in this region.

**Conclusion**

The variation in the morphometric attributes of the Mussoorie region has been enormous, which has catered a settlement density ranging from below 1 to above 4 per km$^2$. It is particularly interesting to note that favourable conditions at a high absolute relief becomes the focus for human settlements and the best example for this case is Mussoorie and Landour towns. Here the slope is gentler, dissection index is moderate, relative relief is moderate to moderately high, drainage density is moderate to moderately fine and drainage frequency is low to medium.

The detailed study of the morphometric attributes with settlement density indicates that the relationships are not always linear. However, different environmental zones can be identified with the help of morphometric attributes and settlement densities. This study will also be helpful in identifying the zones which can possibly sustain a higher density of settlements. Such an exercise will in any case become necessary with ever increasing population in U.P. Himalayas. A desired man-environment relationship can be projected on the basis of the study done here.

## REFERENCES


Agrawal, M. (1972), Morphometric analysis of some drainage basins of Jabalpur Plateau, *National Geographical Journal of India*, Varanasi, vol. 18 (3 and 4), 148-68.

Asthana, V.K. (1967), Morphometric evolution of landforms in Almora and its environs, *National Geographical Journal of India*, XIII (1), 37-44.

Baker, R.H. (1969), A study of rural settlements in Cooke and Johnson (eds.), *Trends in Geography*.

Chatterjee, S.C. (1946), Physiographic evaluation of Chota Nagpur, *Geographical Review of India*, VIII, 37-49.

Chhibber, H.L. (1948), The drainage pattern observed in India and adjacent countries, *National Geographical Journal of India*, Varanasi, 2 (1), 2-5.

Clarks, J.L. (1967), Morphometry from maps in *Essay in Geomorphology*, by G.H. Dury (ed.) Heinemann, London, 235-74.

Chorley, R.J. (1957), Climate and morphometry, *Journal of Geology*, 65, 628-38.

Chorley, R.J. (1958), Aspects of the morphometry of a polycyclic basin, *Journal of Geography*, 124, 370-3.

Chorley, R.J. and Morgan, M.A. (1962), Comparison of morphometric features in Unaka Mountains, Tennessee and North Cordina, and Dortmoor, England, *Bulletin of Geological Society of America*, 73, 17-34.

Christaller, W. (1966), *Central Places in Southern Germany*, Englewood Cliffs, New Jersey (Translated by C.W. Basnin).

Davis, W.M. (1924), The process of geography in United States, *Annals of the Association of American Geographers*, 14, 159-215.

Davis, W.M. (1924), *The Geographical Essays*, Ginn & Company, New York.

Dov. Nir (1957), Landform analysis of Mt Karnel, *Geographical Review*, 47, 564-9.

Dury, G.H. (1963), *The Face of the Earth*, Pelicon Book, England.

Dury, G.H. (1966), *Essay in Geomorphology*, Heinemann Educational Books Ltd., London.

Doxiadis, C.A. (1968), *Ekestics : An Introduction to the Science of Human Settlements*, Oxford University Press, London.

Garner, B.J. (1967), Models of urban geography and settlements location, in Chorley, R.J. and Petter Hagget (eds.), *Socio-Economic Models in Geography*, Methuen, London, 304-55.

Gilbert, G.K. (1909), Convexity of hill-tops, *Journal of Geology*, 17, 344-50.

Gupta, B.P. (1977), *Champawat : A study in landforms and settlements*, Ph.D. thesis, submitted to the Department of Geography, Banaras Hindu

University. Varanasi.

Hammond, E.H. (1964). Analysis of properties of landform geography : An application to broad scale landform mapping. *Annals of the Association of American Geographers*. 54, 11-19.

Horton, R.E. (1945), Erosional development of streams and their drainage basins : Hydro-physical approach to quantitative morphology, *Bulletin of Geological Society of America*. 56, 275-370.

Kesselli, J.E. (1954). A geomorphology suited to the needs of geographers, *Annals of the Association of American Geographers*. 44. 220-21.

Kharkwal, S.C. (1968). *Landforms and settlements of Naini Tal and its environs*. Ph.D. Thesis submitted to Banaras Hindu University, Varanasi.

King. C.A.M. (1967). *Techniques in Geomorphology*, Edward Arnold (pub.) Ltd., London, 51-60.

King, L.C. (1967), *The Morphology of the Earth*, Oliver and Boyd, Edinburgh, 275-85.

Leopold, L.B., M.G. Wolman, and J.B. Miller (1964), *Fluvial process in Geomorphology*, W.H. Freeman & Co., London.

Lösh, A. (1954). *The Economics of Location*, New Haven.

Mukerjee, A.B. (1961), Morphogenetic nature of the Bhur, *Indian Geographical Journal*. 36, 2-13.

Nag, P. (1979), Alternative approach for the identification of randomness in point pattern, *Geographical Review of India*, 39, 193-4.

Philbrick, A.K. (1957). Principles of areal functional organization in human geography. *Economic Geography*, 33, 299-336.

Prasad, Haushila. (1980). *Landforms and settlements of Mussoorie and its environs*. Ph.D. Thesis, Department of Geography, Banaras Hindu University, Varanasi.

Raisz, E. (1946). Landform, landscape, land-use and land type maps, *Journal of Geographers*, 45, 85-90.

Russell, R.J. (1949), Geographical geomorphology, *Annals of the Association of American Geographers*, 39, 1-11.

Schumm, S A. (1963). Sinuosity of alluvial rivers of the Great Plains, *Bulletin of Geological Society of America*, 74, 1089-99.

Schumm, S.A. and Mosley, M.P. (1973), *Slope Morphology*, Dowden, Hutchinson and Ross, Inc. Stroud-burg, Pennsylvania, 13-60.

Singh, H.H. (1968), *Kanpur city : A study in Urban Settlement*, Varanasi.

Singh, K.N. (1968), *Southern Upland of Eastern U.P. : A study in landform and settlements, distribution*, Varanasi.

Singh, O.P. (1971), Toward determining hierarchy of service centres : A methodology for central place studies, *National Geographical Journal of India*, xvii, (1), 165-77.

Singh, O.P., S.K. Singh and H. Prasad (1978), Central places in Rewa-Panna Plateau, *National Geographer*, 13 (1), 67-74.

Singh, R.L. (1971), *India : A Regional Geography*, National Geographical Society of India, Varanasi, India.

Singh, R.L. (1976) (ed.), *Rural Settlement in Mansoon Asia*, National Geographical Society of India, Varanasi.

Singh, R.L. (1974), Morphometric analysis of terrain, *The National Geographi-*

*cal Society of India*, 22, 1-24.

Singh, R.P. (1960), Geomorphological evolution of the Highlands of Raj Mahal, *National Geographical Journal of India*, 6 (1), 1-13.

Singh, S.B. (1978), *Landforms and settlements in Simla and its environs*, unpublished Ph.D. Thesis, Department of Geography, Banaras Hindu University, Varanasi.

Sinha, N.K.P. (1966), *Geomorphic evolution of the Northern Rupununi Basin Guyana*, Ph.D. Thesis, submitted McGill, University, Montreal, Canada.

Strahler, A.N. (1950), Davis's concept of slope development viewed in the light of recent quantitative investigations, *Annals of the Association of American Geographers*, 40, 209-13.

Thornbury, W.D. (1961), *Principles of Geomorphology*, John Wiley and Sons, New York.

Weaver, G.D. (1965), What is landforms ? *The Professional Geographer*, 17 (1), 15-8.

Wentworth, C.K. (1930), A simplified method of determining the average slope of land surface, *American Journal of Science*, Series 5, 20, 184-94.

Wooldridge, S.W. and D.L. Linton (1955), *Structure, Surface, and Drainage in South East England*, Philip, London.

Wooldridge, S.W. (1958), The trend of Geomorphology, *Transactions of the Institute of British Geographers*, 25, 29-35.

Yadva, S.R.S. (1978), *Hazaribagh plateau : A study in landforms and settlements* unpublished Ph.D. Thesis, Department of Geography, Banaras Hindu University, Varanasi.

Zakrzewaska, B. (1967), Trends and models in landform geography, Review Article, *Annals of the Association of American Geographers*, 57, 128-65.

S.C. MUKHOPADHYAY

SIXTEEN

# LANDSCAPE SPHERE, ENVIRONMENT AND APPLIED GEOMORPHOLOGY
## The Tista Basin

THE present chapter deals with the studies on landscape sphere and environment—its conceptual framework with special reference to the applied aspects of the Tista Himalayan terrain (The Tista basin)—a case study. In the first part a discussion has been made in some details on the concept of landscape sphere and environment mentioning the topic of general resurgence of interest in such aspects of the environmental sciences relevant to man and his society within the varied ecosystems (Mukhopadhyay, 1979 a & b, 1980 a & b). The second part concerns an analysis of more specific examples of changed attitudes in acceptance and tackling the vital issues and challenges to human use of the natural environment, both on meso and micro-scales in the Tista Himalayan terrain, Sikkim and Darjeeling, West Bengal (Fig 16.1). Practically, this is an applied geographical study with a view to correlate between the components of physical and cultural of the Tista Himalaya which is given to a number of environmental hazards like large-scale inundation and floods, frequent landslides and accelerated soil erosion, severe local storms, moderate seismic shocks, dense fog, occasional droughts in some parts along with the effects resulting from human interferences (Mukhopadyay, 1981, 1982, Embleton 1985, Vaidyanadhan 1984). The present study is based on pre-field, field and post-field methods. The physical features and landuse patterns are not included here because of the publication of such information, data etc. already in the previous works by this author (Mukhopa-

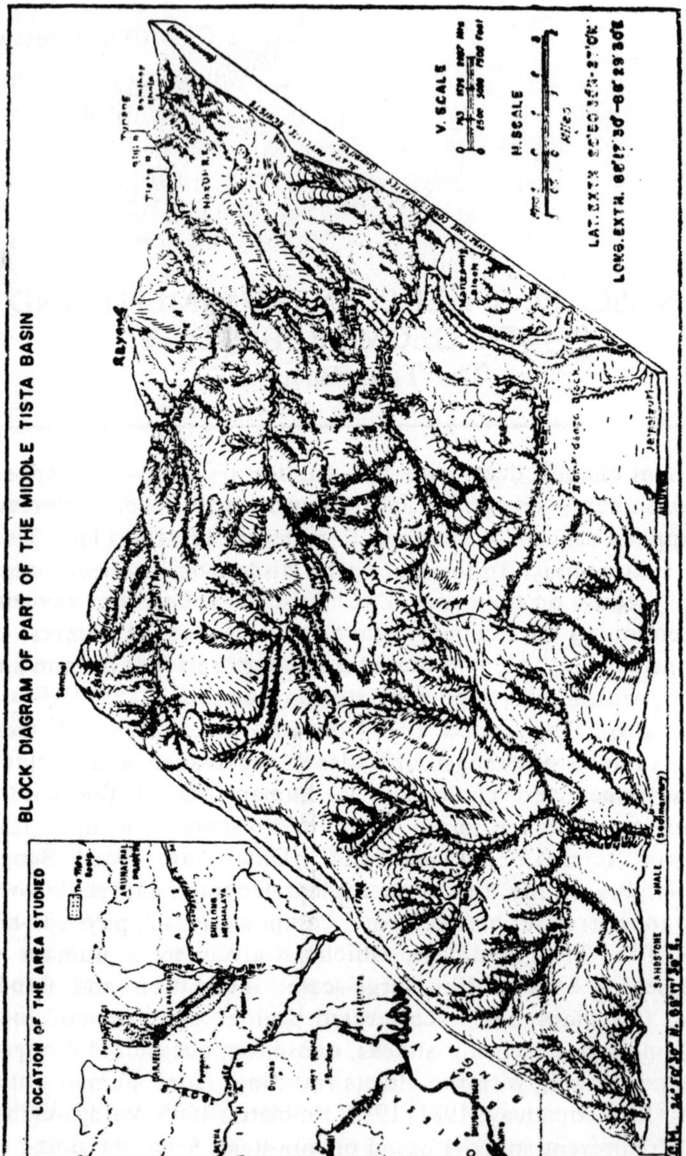


FIG. 16.1 : *Block diagram of part of the middle Tista Basin*

dhyay, 1982, 1983).

**Importance of the environment of case-study area**

The Tista Himalayan terrain is important from the physical and cultural point of view as it also offers a special treatment to study the interaction of different components of landscape sphere within the Himalayan environment. The present mountainous terrain is delineated by the Tista basin (Fig. 16.1), lying roughly, between 26°30′ N to 28°N and 83°E to 88°50′ E with an altitudinal variations between 150m to 7000m a.s.l. in Sikkim and Darjeeling (West Bengal), Eastern India. Actually, this term *The Tista Himalayan terrain* is synonimous with the Tista drainage basin comprising the entire *Sikkim and major parts of the Darjeelieg Himalaya.* The Sikkim Himalaya (about 7300 sq. km.) with the highest and beautiful peaks (6500 m and more) largest glaciers as well as lesser population densities, in the north. and the Darjeeling Himalaya (about 3,000 km²) with the largest plantation sectors (tea and hill agriculture) and commercial or business centres as well as higher population densities in the south. The Tista Himalayan terrain sustains the notable towns or administrative centres like Gangtok, the capital of Sikkim, Singtam, Rangpo, Tadong, Namchi, Mangan, and Gyalshing in Sikkim; and Darjeeling, Kalimpong and Mongpu in West Bengal which are of particular strategic significance and socio-economic interest in Eastern India. Besides the environmental hazards. this part of the Eastern Himalaya as is especially marked by the higher landuse potential for hill-agriculture, forestry and hydel power production. as mentioned earlier (Mukhopadhyay. 1983, 1984, Bagchi, 1944, Karan, 1984, Bagchi and Mukherjee, 1983). Contemporary geographers have also stressed much on the importance of the study of environments for the proper development of a terrain (Vidyanath, 1985. Plesc Alsal, 1985, Ketzoieal, 1985 and others).

**Concept of landscape sphere**

The concept of landscape sphere is being recently appreciated by the geographers in an attempt of correlation between the landscape elements particularly, the landforms and landuse in the course of the environmental management (Vink, 1983). A landscape-sphere which is also termed as geosphere, relates to the dynamism of natural phenomena of the terrain especially the characteristics of surface of the earth and of the space near to this surface. Generally, these characteristics are not static by any means within the sphere or environment. In an interpreta ive sense, this is the environment which accommodates readily most of those processes moulding the mode of living as well as determining the conditions of life for human being accompanied by other organisms in nature. Practically, a range of processes are found to be active in this sphere where other spheres interact to evolve the variegated elements of landscape—both physical and culturals landscape in response to the changing environments in different parts of the earth.

In a comparative sense, it has now been possible to identify a landscape-sphere like the other spheres with its dynamic entity e.g. hydrosphere, biosphere, lithospheres, pedosphere, atmosphere, anthroposphere etc. The landscape may be visualized as the continuously changing result of the interactions between processes in the several types of spheres at different levels. Under a certain environment at a particular region, naturally the landscape thus evolved, relates to the complex relationships among such spheres, as mentioned. Then, the study of a landscape over a certain period of time depicts the processes in a given landscape may be a simple type or complex type with effect of human interferences thus exhibiting heterogeneous features set in a polycyclic landscape developed mainly under multiple cycles of erosion. The human beings are found to have established peculiar cultural landscapes assoaicted with advanced civilization applying the knowledge of science and technology mostly in gently undulating riverine plains or valleys under fluvial (humid) environment. Among the various landscape patterns i.e. Glacial landscape, Periglacial landscape, Fluvioglacial landscape, Aeolian landscape, Marine landscape etc., the Fluvial landscape with high reliability affords more facilities of cultural development associated with higher density of population in respect of the other landscapes evolved under arid, marine, glacial or periglacial environments. It is important to refer also the concept of geotope which means a part of the geosphere which is comparatively homogeneous with respect to its landforms and geomorphic processes i.e. form-process relationnship. (Mukhopadhyay, 1979, 1980, 1982). It may be mentioned that the present study concerns the analysis of the relationship between phenomena and processes in the landscape sphere or geosphere containing the communities of plants, animals and men under varied environments in the surface of the earth.

It seems wise to discuss some relevant points on landscape in connection with the concept of landscape sphere. Although the concept of landscape sphere is a recent one, the term landscape has been used by the earth scientists, social scientists, architectures etc. from an ancient time. Actually, landscape has different meanings to varying groups of people according to their interests and disciplines. Examples may be considered from the approaches and observations as presented by the geographer, the landscape architect, the ecological landscape planner or even a landscape painter (Van Gogh). Broadly speaking, the landscape is thus interpreted as the basis of the human environment, and human existence is dependent largely on the plants and animals living in communities which may be influenced, more or less by man. The concept of landscape sphere is closely associated with the discipline of landscape environment or landscape ecology which concerns the landscape (or geosphere) as the key foundation of the environment mainly for man but essentially for plant communities and animals, as discussed in the following paragraphs.

**Landscape-environment complex (landscape ecology)**

With this introductory discussion on the conceptual framework of landscape sphere, it seems reasonable to present some comments about the landscape environment. Actually the landscape-environment is almost synonimous with the terms landscape ecology which was introduced in 1938 by Troll who later used the term *Geo-ecology*. According to his opinion, the birth of landscape ecology as the outcare of the marriage of geography (landscape) and biology (ecology), where the geographer's point of view stresses that concern is not only with natural landscapes, but also with landscapes including man. It emphasises on the applied geographical problems connected with the cultural landscapes and socio-geographical aspects. Such idea of landscape ecology can be fruitfully applied to human purposes like area or land development, integrated resource planning including regional planning as well as urban planning on the basis of modern methodology i.e. airphoto interpretation, land-use surveying etc. Obviously, it has initiated and promoted a more holistic approach to the geosphere or landscape sphere.

Till now the important points on the concept of landscape sphere are discussed, it seems wise to present some views about the environment in a geographical contex mentioning also the need of an inter-disciplinary approach as made by Vink (1983).

**Environment in a geographical context**

According to Vink environments, as very complex systems, can only be properly studied using a multidisciplinary, and better still with an interdisciplinary, approach. This involves the systematic cooperation of several sciences on an integrated basis. In practical daily life we try wherever possible from the backgrounds of our own discipline to leern of the complex problems involved. In this way separate branches of science such as environmental biology, environmental chemistry and environmental geography have been developed. Incidentally, landscape ecology may be considered to belong to environmental biology or to environmental geography, depending on the particular viewpoints of the scientists involved in the project. In this paper the author considers mainly the geographical viewpoint. It will be made clear in the following pages relating to the Tista Himalayan landscape : a case study that the morphological and the biological aspects are also essential. Now, it seems reasonable to consider briefly the meaning of environment as proposed by Vink (1983).

Vink argues that the environment means different things to different people. Environments as complex systems can only be partially understood and in order to arrive even at a limited understanding, a particular approach has to be chosen. This implies, however, that the environmental scientist has to have a clear perception of his research topic, but the corollary is that he is forced to adopt a limited viewpoint and, therefore, can never obtain more than a limited comprehension of any environment. This limited interpretation has to be supplemented by the restricted interpretations of other scientists to arrive at an acceptable understanding of the whole. Complete expertise on any environment in one person or even in one discipline does not exist. But it is also possible to begin the sequence with the environmental factors and then lead to complexes of these factors which form the environments of the organisms. The following two examples indicate this approach :

1. Drop of water—small pond—small stream—lake—river—sea—ocean. 2. Soil aggregate—soil horizon—soil (pedon)—soil series—soil region or soil landscape.

Vink further suggests that ; within each of these environmental ranges, organisms may be considered to range from micro-organisms to complex associations and communities of micro, meso and macro organisms. The various components and types of landscapes have particular significance not only for man but also for most other organisms. The grouping of the complexes of environmental factors in land units and landscapes or physiographic regions then leads again to landscape environment complex or ecology. The geographical or landscape approach to environments is indeed an excellent way to give full value and extensive treatment to the various attributes of environments. This is done by producing different kinds of maps (for example, maps of vegetation, soils, water, physiography) at a number of scales. Such scales can range from 1 : 100 (very large scale) to 1 : 500 000 (very small scale) and to 1 : 20 000 000 (scale of maps in atlases). With this basic concept on the environment in a geographical context as made by Vink, let some discussion be made in some details regarding the influence of geomorphology particularly in the development of cultural landscape in the Tista Himalayan environment in the following pages.

**Influence of geomorphology in the Tista Himalayan environment**

In the Tista Himalayan terrain, especially in the basin area, Sikkim and northern West Bengal (a case study), geomorphology is so dominant a factor also in the cultural environment that its influence on the pattern and density of agriculture, trade, communication (landuse patterns) etc. is immense. Presenting a geomorphic contrast to the middle (from Tista-Rangit confluence zone to Sevoke area) and lower parts (from Sevoke downstream) the upper part (from glacierized zone to Tista-Rangit confluence) in Greater Himalayan terrain, majestically stand in their variegated landscapes with accompanying glacial-periglacial-fluvioglacial processes set in a framework of severs simplicity, primeval glory, uncommercialized nature, very high and strong relief, landslides and flood hazards and neotectonic movements. Geomorphology has set the foundation and background of various economic and related activities like agriculture, trade and associated communications being discussed in the present paper which can be adopted or can be feasible in a particular area both on macro- and micro-regional scales. The author believes that enhancement of communication system would be helpful in the attempt of regional development relating to eradicating zonal economic imbalances and conflicts in National level with a special reference to Eastern India, a zone of critical significance.

**Geomorphology and terrain communication and transport**

The influence of geomorphology on trade is mostly seen through the lines of communications like roads both metalled and unmetalled, light railways (New Jalpaiguri to Darjeeling), waterways, airways, bi-cable ropeway (Darjeeling) which are the carriers of trade. In regard to the geomorphic permissibility of the Tista terrain, the roads first lead the way as the roads can negotiate steeper slopes and attain greater altitude even in the upper mountainous and hilly tracts (The Eastern Himalayas), Sikkim and Darjeeling. Geomorphology exerts influence on the patterns, characteristics and problems of communications and transport of the varied environmental domains or *ecological sites* at various altitudes within the Tista terrain made of varied types of rocks and soil structures. In the upper part where roads, settlements, villages etc. cover only about 15 per cent of the total area in particular, geomorphology has a deterministic control on communication. The modern technological implication on the progress of economic development relating to conquest of nature is found to be confined mainly in the lower and middle Tista basin area covering parts of Southern Sikkim and northern hill districts (Darjeeling and Jalpaiguri) of West Bengal with lower degree of tectonic movements (neotectonics). As a whole, in the lower part (Southern Darjeeling, moist Tarai belt and Jalpaiguri districts) where the ground slope is very gentle to the south of Sevok area the roads are along the gradient. In contrast, roads of the upper and middle parts, are transverse to the local slope or almost parallel to the mountain of hill side contour as far as possible by engineering skill. Such geomorphic adjustment (geomorpho-engineering) of roads extending both in a valley specially along its flanks (or terraces) and divide crest alignment of the higher Sikkim and Darjeeling Himalayan terrain has been studied.

Even in this modern age, there are unfailing influences of geomorphology on terrain communication, not only for the developing sites relating to tourism (Mirik Singtam etc.) settlement and other projects but also for the developed sites in several selected sections of the Tista basin area. Study of the geomorphological map in conjunction with that showing the distribution of land communication lines (composite geomorphological map) would make it clear. Actually, alignments and various aspects of road structure are also subject to adjustment with a more detailed study of macro, micro and meso regional geomorphology within the Tista basin. Only the major examples are cited here, without going into much details of such problems on geomorphology and communications.

The *geomorphic control* of roads is more strikingly visible in the northern mountainous terrain, for example. One may look at a road map of the Tista basin, Sikkim and parts of hill districts of West Bengal, and compares how the road pattern of the terrain has been influenced largely by geomorphology. Like the Tibet and other parts of Greater Himalayas, the elevated mountainous terrain of the Sikkim and Darjeeling Himalayas are mostly devoid of roads. Normally, the present roads (highways) find their ways through suitable geomorphic corridors formed either by geomorphic agents like the Tista-Rangit Rangpo chu Lachung chu rivers and Pauhunri Zemu-Telung-Lhonak glaciers or litho-techno-structural influence. Like the important highways of the Himalayas, the Kalimpong-chumbi valley routes also display notable geomorphic control in their alignments. The highway in the Himalayas follow the valleys along the terraces, passes, relatively level uplands, ridge tops (divides) or crest alignment depending upon the nature of geo-ecological adjustment at several sectors (Bagchi, 1944; Mukhopadhyay, 1979a, 1982) obviously, there is a close connection between nature of relief (easy or difficult) and alignment of roads in the Tista basin area. Rongphu, Melli Bazar, Naya Bazar, Rongli, Singtam etc. towns (major regional centres) and larger villages (minor and smaller regional centres) like Dentam (Kalet chu), Lachen, Lachung, Dik chu, Tsunthang, Singchuthang, Likship, etc., located in valleys and relatively level uplands being typified by areas of comparatively easy relief. The roads like Kalimpong to Gangtok via Singtam,

Dik chu to Lachen via Tsunthang, Dik chu to Beh via Mangan, Tsunthang to Yumthang via Lachung, Sevoke to Tista Bazar and Gangtok partly the Hill cart road upto the Tibet via Sikkim etc. passing through such zone of comparatively *easy relief* have the advantage of traversing populated and cultivated zones having towns and larger villages which need good roads and thus guarantee the economic viability of the communication lines. Again, the mode of transport is so eminently a function of geomorphology that it cannot be over-emphasised. Besides, it is not always the valleys and low lands that carry roads. Some examples may be cited from the upper and parts of the middle Tista basin where roads or parts of roads for considerable length (10 km and above distance), Gangtok to Darjeeling Majhitar to Tarku via Damthang and to Yuksam via Gyalzing, Naya Bazar to Helay via Fambong and to Uttare via Dentam Renak to Gangtok via Pakyong and Bhusuk, Kalimpong to Natu Lajelepla-Batang La via Rongli and Kyangnosa La, Gangtok to Dik chu and Phen sang etc. have adopted summits (ridge-tops) or divides in many sections. These ridge-tops built up on the distinct divides of various (sinuous) shapes and over the older moraines in parts at different elevations of the Sikkim-Darjeeling Himalayas and surroundings. These are marked by their peculiar situations with an advantage of providing commanding panoramic view even of the distant mountainous landscapes of Sikkim, Darjeeling (West Bengal) and the adjacent countries (Nepal, Bhutan, Tibet and China). Such unique features are observed in its mountainous terrain with scenic splendour, glaciers, shimmering lakes of various sizes and shapes, alpine medows, birch-rhododendron-juniper (sub-alpine and Alpine) forest, terraces along the Tista-Rangit-Lachung Talung chu-Rangpo chu etc. rivers.

Morphologically and in weathering status these sub-elements like ridges (ridge and vales) in particular, of the variegated landscapes are found to be distinct from other elements or sub-elements in the constituent geomorpoic units of the upper and middle Tista basin. These divides, except some parts of the eastern rugged Dongkya Range area, are marked by lower range of dissection index, (below 0.5) roughness index, (below 40), ratio index. relative relief (below 30) and erosional potential, but with higher values of constant of channel maintenance, for examples (Mukhopadhyay, 1979). A ridge-top road (as above) following the divide in the upper environmental domains with lower degree of weathering. Status and erosional potential is associated also with economic advantages. Normally, degree of dissection index is directly proportional to the budget expenses of road-building and maintenance throughout the year. Mainly because of its situations beyond the reach of direct valley development particularly in terms of valley widening and valley deepening (with negligible operation of valley lengthening by head ward erosion), these elevated ridge-top roads have attained the lowest values of dissection index, roughness index etc. thus conforming to the least dissected parts of the mountainous terrain. It should be mentioned that other things being equal the flow of trade adopts the lines or directions of least resistance and not across the geomorphic barriers like the intervening uplands elevated ridges or horsts.

Incidentally, of a close examination of the trade routes at all levels from the village (Rangma, Rahong, Panong, Rabong, Kau etc.) to the larger territorial units (Mangan, Gangtok, Gyalzing, Singtoon, Rangphu, Namchi etc.) in the upper part, unmistakably reveals the geomorphic control. Of course, in the plain level area in the lower part (southern Darjeeling and Jalpaiguri districts) there is little of such geomorphic control but in the high lands and rugged terrain of the upper and middle parts such control is unmistakable. For example, while communicating with the surrounding villages as above, the easier terrains are adopted. Above Lachen or Yumthang in the north and Chumigchen and Namlung in the west etc. there are no metalled roads or railways but only kachcha roads and further uphills in the absence of even there are trails. These northern and marginal hills and mountains are negative or shadow zones as regards intercommunication and micro-regional trade such as the supply of agricultural products of the villages or the products of the household industries which aggregate in the village hats or markets. Geomorphology influences lines of communication and mode of transport, and trade is determined by communication and transport. It is apparent that geomorphology first determines the feasibility and direction of movement. The commodities and carriers of trade flow on land along the path ways determined by geomorphology. This is true of the mountainous terrain of the upper and lower parts and to a considerable extent even in level plains of the lower Tista basin.

However, roads following the divides are conspicuously devoid of spectacular erosional features like nitches, scars or depressions which reduce the possibility of frequent engineering involvements or human interferences relating to the repairing and other works. As a whole, it presents comparatively a low-cost project through the gentlest and lowest part of the surface without, of course, making the detour too lengthy and costly. It is preferable for its relatively simpler structure and permissible extension conforming to the ridge-alignment, as per distribution of the major landscape elements in the upper basin area. Consequently, it be free from welts and ridges which need incision, tunnelling or cutting, and free from depressions which require regular filling or bridging, specially during rainy season.

**Concluding remarks**

The present layout is the outcome of cultural landscape as evolved out of morphological adjustment in the Tista Himalayan terrain as mentioned above. Over the years the environmental character will continue to be more or less same or the whole thing might change as a result of environmental hazards, especially the seismic shocks and its related phenomena (land-slide, floods etc.) apart from the human interferences in the area. The long term planning should keep this in mind while considering environmental issues and challenges of the Tista Himalayan terrain.

## REFERENCES


Bagchi, K. (1944) : Kalimpong—Its Land and People. *Geographical Review*, Calcutta, 3.1, 37-43.

Bagchi, K. and Mukherjee, K.N. (1983) : Diagnostic Survey of West Bengal (North), Department of Geography, University of Calcutta, Calcutta, 8-52 & 74-97.

Cooke, R.U. and Doorn Kamp, J.C. (1977) : *Geomorphology in Environmental Management*, Clarendon Press, Oxford, 8-127 & 326-60.

Embleton, C. (1985) : Notes on the Geomorphology of Kumaon, with particular reference to the Nainital area, in Mukhopadhyay, S. C. (ed.), *Geographical Mosaic*, Professor K. Bagchi Felicitation Volume, Modern Book Agency Pvt. Ltd., Calcutta, 235-43.

Karan, P.P. (1984) : Response to environmental hazards on Sikkim Himalayas,

*Himalayan Research and Development*, Himalayan Research Group, Nainital 2. II., 45-46.

Kreteoi, M., Marton, P., Pecsi, M., Schweitzer, F., Voros, I. (1982) : Pliocene-Pleistocene piedmont correlative sediments in Hungary based on lithological-geomorphological, paleontological and paleomagnetic analysis of the exposures in the open-cast mine at Gyongyosvisonta in Pecsi, Marton (ed), *Quaternary Studies in Hungary*, Inqua Hungarian National Committee, Geographical Research Institute, Hungarian Academy of Sciences, Budapest, 43-73.

Mukhopadhyay, S.C. (1979 a): Some techniques and recent trend of geomorphological studies, *Recent Trends and Comcepts Geagraphy*, vol. I Sponsored by Geographical Research Centre, Patna, Concept Publishing Company, Delhi, 274-272.

Mukhopadhyay, S.C. (1979 b) : The morphological analysis of landforms of the Tista basin in the Eastern Himalaya, Eastern India, *Geographical Review of India*, 41-4, 247-356.

Mukhopadhyay, S.C. (1980 a) : *Geomorphology of the Subarnarekha Basin*, Burdwan University, Burdwan, 63-98, 108-71.

Mukhopadhyay, S.C. (1980 b) : On the morphology, materials, mechanism and movements of the glacierised and adjacent area of the upper Tista basin. *Indian Journal of Landscape Systems*, Calcutta, 3, (1 and 2), 92-101.

Mukhopadhyay, S.C. (1981) : An appraisal of the Tista basin from the perspective of applied geomorphology. *Landscape Systems*, Calcutta, 4, 57-69.

Mukhopadhyay. S.C. (1982) : *The Tista Basin : A Study in Fluvial Geomorphology*, K.P. Bagchi and Company, Calcutta, 97-192, 217-32.

Mukhopadhyay, S.C. (1983) : From-process and tectonics in the middle Tista basin. *Indian Joural of Landscape Systems and Ecological Studies*, Institute of Landscape. Ecology and Ekistics (ILEE), Calcutta, 6, 55-98.

Mukhopadhyay, S.C. (1984) : Evaluation of the geoecological parameter of the upper Tista Basin. *Indian Journal of Landscape Systems and Ecological Studies*, Institute of Landscape, Ecology and Ekistics (ILEE), 7, 87-95.

Mukhopadhyay, S.C. (1985) : On the diversity of geomorphic process and suites of landform development in the polygenetic and polycyclic upper Tista basin. The Sikkim Himalaya, in Mukhopdhyay (ed.). *op cit.*, 316-26.

Pecsi, M. and Retvari, L. (1985) : Concept of the geographlical environment, in Mukhopadhyay (ed.), *op cit.*, 1-8.

Vaidyanadhan, R. (1984) : Emerging concepts and trends in geomorphololology, *Current Science*, Bangalore, 20, 53, 175-178.

Vaidyanath, V. (1985) : A new method to determine arable landuse efficiency in Andhra Pradesh in Mukhopadhyay (ed.), *op. cit.*, 77-87.

Vink, A.P.A. (1983) : *Landscape Ecology and Land Use*, Longman, London and New York, 1-18, 35-76.

MEHDI RAZA
AND
HANIFA BANO

SEVENTEEN

# ECOSYSTEM CONCEPT AND REGIONAL RESOURCE MANAGEMENT

## A Case Study of Dal Lake (J. & K., India)

THE deteriorating quality of the environment and the consequent imbalance and disruption in the natural ecological relationships and the concomitant destruction of valuable natural resources has become a matter of an ever-increasing concern throughout the world. It can hardly be over emphasized that the continued existence of the biosphere as a viable source of essential resources for mankind depends upon a scientific and rational management of the ecosystem. If this growing awareness is not soon translated into positive remedial action, posterity may point an accusing finger towards the present generation for depriving them the means of enjoying the gifts of nature. Scientific management of the environment is, therefore, much more than merely a matter of academic interest.

Like elsewhere in the world, the environmental problems in India, specially in the context of the headlong economic development of the post-independence period, is becoming increasingly serious with the passage of time. The accelerated pace of economic development involved an unplanned and excessive exploitation of renewable and non-renewable resources in many parts of the country. There has been a corresponding increase in man's interference in the ecological balance, beyond the tolerance limit of the natural regenerative processes, seriously imparing the life support mechanism of the ecosystem. It is nobody's thesis that the process of rapid economic development should be deaccelerated. In a developing country like India where war against need and poverty must be accorded top national priority slakening of developmental efforts will be tantamount to condemning a vast majority of our people to remain below the poverty line. What is suggested here is that the process of rapid economic development must be compatible with capacity of regional ecosystem to absorb developmental stresses so that ecological imbalances and environmental degradation are kept within ecologically permissible limits. This calls for a scientific, rational and well-coordinated strategy of economic development that should seek to reconcile the developmental issues with the needs of environmental preservation and management.

**The Dal Lake : A case study**

The present study is rooted in the conceptual context outlined above. The Dal lake is an ideal ecosystem for an environmental impact analysis of rapid economic development on a micro scale. The lake is situated in the valley of Kashmir, in the state of Jammu and Kashmir. Though a small one, this lake plays an extremely important role in the socio-economic life of the valley and has traditionally been the heart of Kashmiri culture since times immemorial. This enchanting lake, the life blood of Kashmir's economy and way of life, has been showing signs of serious entrophication. Already the lake has shrunk by half during the last 50 years and the present open water area, about 12.4 km$^2$, is but a remnant of a once extensive lake. It has been estimated that the lake receives about 64 acre feet of silt annually, thus further aggravating the siltation problem. This sorry state of affairs has been brought about by purely anthropogenic factors, specially excessive and irrational exploitation of its resources. Deforestation of catchment hillsides and irrigational land use practices thereon, increasing agricultural activities on the floating gardens and the resultant heavy nutrient loading, extensive soil erosion in the surrounding hillsides leading to siltation of the lake basin, increasing pressure of population requiring reclamation of more and more area of the lake, hectic urbanization of the foreshore area and direct discharge of raw sewage from the houseboats and lake hillside hotels, luxuriant and rapid weed growth and increasing bacterial activity are some of the contributory causes responsible for the present ill-health of the lake. The quality of water has become considerably polluted on account of unsanitary conditions. The lake has been turned into a veritable "sewage tank" because of the high intake of solid and liquid waste from the surrounding areas. Besides chemical properties of water such as DO, COD, BOD, phosphorous and nitrogen loading etc. are also undergoing changes leading to imbalances in the ecosystem. The lake water also suffers from a high degree of micro-biological pollution. The problem of the lake's entrophication is so serious that if corrective therapy is not undertaken without further delay, the lake may be totally entrophicated within less than a century, leaving in its place stretches of swampy land and pools of stagnant water.

In the above context the present study aims to analyse the state of the entrophication of Dal lake in terms of some aspects of its physical, chemical, biological and human parameters that have a direct bearing on the economic and ecological viability of the lake. The indicators that would be taken into account are: (a) water quality as reflected in temperature, dissolved oxygen, transparency, pH, nitrogen and phosphorus, (b) sedimentation and (c) biological status as reflected in bacteria, phytoplanktons and zooplanktons and macrophytes. An attempt will be made to determine the present state of pollution and degradation in terms of each of the selected indicators.

The methodology of the study is rooted in the ecosystem approach. Every lake has its own entity, deriving its specific characteristics from the geology, morphometry, climate and the biota in and around it. The components of a lake's ecosystem, for that matter any ecosystem, are so finely balanced in nature that any change anywhere leads to disproportionate cumulative changes, which if not checked and reversed in time may lead to a collapse of the ecosystem.

**Geographical setting of Dal Lake**

The Dal lake is situated in the heart of Srinagar city, lying between two hills. On the eastern side of the lake lies the famous Shankaracharaya hill, formerly known as Takhte-Sulaiman while on the west is the hill known as Hariparbat. Though the Dal is a small lake, it nevertheless falls into four sections : Hazratbal, Bod Dal, Lokut Dal and Nagin Lake (formerly it was connected with the Dal Lake).

Several streams feed the Dal lake, the largest being Telbal mullah, which drains Dachigam into the northern end of the lake.

The outflow of the lake is through the Dal gate via a lock system and via the Nagin lake and Amir Khan Nalla into Anchar lake. At the southern end, the lake is fringed by lagoons and a great portion of it is covered by the famous floating gardens (Fig. 17.1). The average depth of the lake, as recorded in 1982, varies from 2.2 metres to about 4 metres, though at many places it reaches more than 6 metres (Table 17.1). With the growth of Srinagar city the lake has been extensively reclaimed to form manmade islands separated by narrow channels and areas of water covered

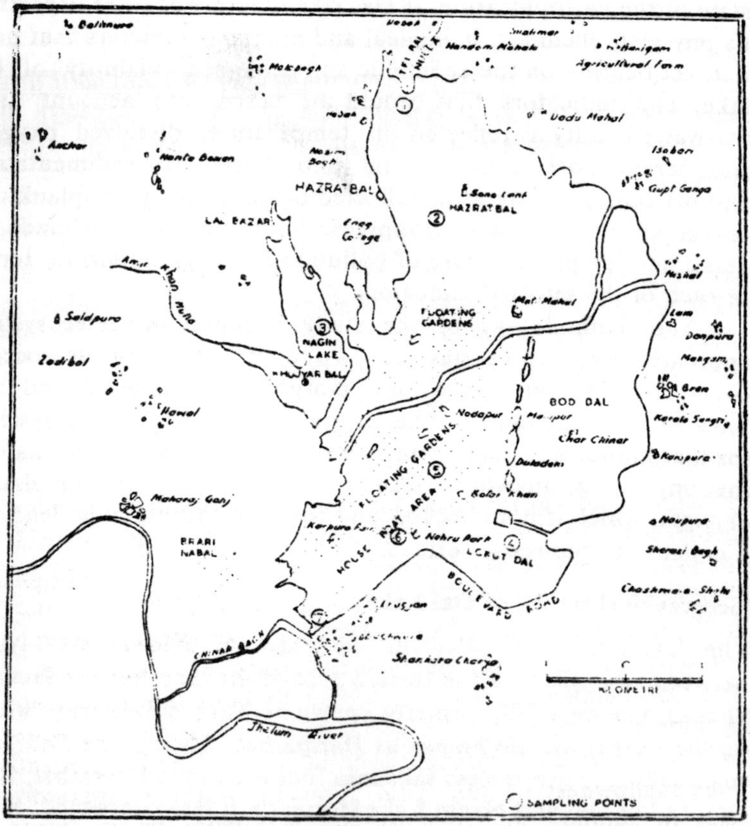


FIG. 17.1 :*Dal Lake and its Basins*

TABLE 17.1 : **Geographical subdivisions of Dal Lake**

*(Area in km²)*

| Subdivisions | Average depth/(m) | Open water | Floating gardens | Marsh | Land |
|---|---|---|---|---|---|
| Hazratbal | 1.5 | 5.6 | 1.5 | 1.3 | 0.4 |
| Bod Dal | 3.0 | 4.5 | 2.4 | 1.3 | 0.01 |
| Lokut Dal | 2.5 | 1.6 | — | — | 1.00 |
| Nagin Lake | 3.4 | 0.7 | 0.3 | 0.3 | 0.1 |
| Total | — | 12.4 | 4.2 | 2.9 | 1.51 |

by floating gardens. A large number of houses and cottages have come up on these manmade islands. The gently sloping terraces around the lake are extensively used for paddy cultivation, orchards and market gardens.

The total catchment area of the Dal lake is about 317 km² and can be divided as follows :

| | *Catchments* | *Area in km²* |
|---|---|---|
| 1. | Dachigam and Telbal | 228 |
| 2. | Lake Hillsides | 47 |
| 3. | Srinagar (North) | 12 |
| 4. | Srinagar (City) | 9 |
| 5. | Dal lake | 21 |

**Hydrological characteristics**

Water temperature plays a crucial role in all aquatic ecosystems as together with the type and quantity of dissolved substances it determines the temperature and density gradient between epilimnion and hypolimnion. The thermal conditions play an important role in the biological organization of lakes by controlling the rate of photosynthesis and other elements necessary for organic life.

The Dal, being a shallow lake, is usually homothermal except during summer when a slight temperature stratification may develop for a brief period. Permanent thermal stratification is prevented by the shallowness of the lake. Typically water temperature increases in summer and reaches the maximum figure during August. Not much spatial variation has been recorded either in the higher temperature of the summer months or the lower temperature of the winter months (Table 17.2).

TABLE 17.2 : Dal Lake temperature

| Stations | | April | June | August | October | December | Mean value | Standard deviation |
|---|---|---|---|---|---|---|---|---|
| I | A | 17 | 16 | 21 | 15 | 11 | 16 | 3.2 |
| | S | 15.5 | 14 | 19 | 10 | 10 | 13.7 | 3.43 |
| | B | 15 | 12 | 18 | 8 | 8 | 12.2 | 3.91 |
| II | A | 19 | 20 | 25.5 | 21.5 | 9 | 19.0 | 5.46 |
| | S | 17 | 19 | 23 | 18.5 | 8 | 17.1 | 4.96 |
| | B | 14 | 17 | 21 | 17 | 7 | 15.2 | 4.66 |
| III | A | 22 | 24.5 | 26 | 21 | 11 | 20.9 | 5.25 |
| | S | 21 | 23.3 | 25.5 | 18 | 9 | 19.36 | 5.74 |
| | B | 19 | 21.0 | 24 | 17 | 8 | 17.8 | 5.4 |
| IV | A | 25 | 26 | 30 | 25 | 10 | 23.2 | 6.8 |
| | S | 24.1 | 22.5 | 28 | 23 | 9 | 21.3 | 6.4 |
| | B | 21.1 | 21.6 | 25 | 20 | 7 | 18.94 | 6.19 |
| V | A | 23 | 24 | 29 | 24 | 10.5 | 22.1 | 6.16 |
| | S | 22 | 20 | 27 | 20.5 | 9 | 19.7 | 5.89 |
| | B | 21.5 | 15.5 | 22 | 19.3 | 8.5 | 17.36 | 4.98 |
| VI | A | 23 | 24 | 25 | 23 | 10 | 21 | 5.55 |
| | S | 22 | 20 | 25 | 20 | 8 | 19 | 5.80 |
| | B | 21.5 | 16.5 | 23 | 19 | 7 | 17.4 | 5.65 |
| VII | A | 23 | 23 | 24 | 22 | 12 | 20.8 | 4.44 |
| | S | 21.2 | 21.5 | 22 | 20 | 10 | 18.94 | 4.51 |
| | B | 20.5 | 18.5 | 19 | 19 | 8 | 17.0 | 4.55 |

A=Atmospheric temperature
S=Surface water temperature
B=Bottom water temperature.

*Note* : Temperature of the lake water was recorded at seven stations in every alternative month in 1982.

## Siltation

The principal reason for the increasing shallowness and areal contraction of the Dal lake is the heavy siltation to which the lake has become prone. Much of the siltation takes place through the Telbal nullah. The irrigation and flood control department has collected average monthly silt concentration data and according to them the total volume of settled silt deposited in the lake by Telbal nullah alone is about 36,200 m³. Table 17.3 shows the silt load in the lake. If this rate of siltation, which no doubt has been greatly accelerated due to anthropogenic factors, is not checked and reversed in time by corrective measures the lake will soon be converted into a swampy, marshy stretch of land.

TABLE 17.3 : **Silt load of the Dal Lake in CC/lit**

| Month | Coarse silt | Medium silt | Fine silt | Total silt | Total flow of Telbal Nalla in month × $10^6$ m³ | Settled volume silt m³ |
|---|---|---|---|---|---|---|
| January | .001 | .001 | .068 | .070 | 6.2 | 434 |
| February | .002 | .002 | .102 | .106 | 8.1 | 859 |
| March | .001 | .002 | .119 | .122 | 16.7 | 2037 |
| April | .001 | .002 | .153 | .156 | 39.7 | 6193 |
| May | .001 | .001 | .113 | .115 | 45.9 | 5278 |
| June | .003 | .003 | .136 | .142 | 36.6 | 5197 |
| July | .002 | .002 | .136 | .140 | 32.2 | 4508 |
| August | .004 | .004 | .170 | .178 | 35.3 | 6283 |
| September | .001 | .002 | .153 | .156 | 21.2 | 3307 |
| October | .002 | .003 | .102 | .107 | 9.9 | 1059 |
| November | .002 | .002 | .102 | .106 | 6.8 | 721 |
| December | .002 | .002 | .068 | .072 | 4.4 | 317 |
| | | | | | Total | 36193 |

## Chemical composition of water

Besides a large number of trace elements, quite a few major elements such as nitrogen (N), phosphorus (P), potassium (K) calcium (C) and magnesium (Mg) etc. cycle through the lake's water. These chemical substances are either derived from the weathering of the rock formation or human inputs in the form of fertilizers and raw-sewage. Since the various components of an ecosystem are in a perfect state of equilibrium, changes in the chemical composition of a water body inevitably leads to ecological disequilibrium as a result of changes brought about in the ecosystem's productivity at different trophic levels. This may eventually cause complete entrophication of a waterbody. In the life history of an aquatic ecosystem, therefore, nutrients, such as nitrogen and phosphorus, dissolved oxygen, pH, specific conductivity and transparency are crucial elements in the preservation and maintenance of the ecological balance.

Nutrients are chemical substances necessary for plant growth. These are generally phosphorus (phosphate, polyphosphate and organically bound phosphorus) and nitrogen (ammonia nitrogen, nitrate, nitrite and organically bound nitrogen). Though nutrients are essential for the biotic components of an aquatic ecosystem, their high concentration is extremely harmful as it leads to formation of extensive beds of macrophytes and algal bloom.

The two principal nutrients in the Dal water are nitrogen and phosphorus. Their total concentration is important as they have an important bearing on the tropic level of the lake. In the lake water nitrogen in the form of ammonia nitrogen is always present in almost all sections of the lake. However, the level of concentration varies from season to season and from place to place (Table 17.4). Both in the surface as well as in the bottom layers, the highest seasonal input of ammonia nitrogen is during the month of October in Dal gate section. The recorded value for the surface and bottom waters being 1.65 mg/lit and 1.93 mg/lit respectively. The lowest level both in the surface and bottom water was recorded in June in the Gagribal section. A perusal of Table 17.4 reveals that ammonia nitrogen input is high during the period August—November while during the summer and rainy seasons the values are much less. There is, therefore, considerable seasonal variation in the level of this nutrient.

In natural water ammonia nitrogen occurs in very small concentration, usually less than 0.1 mg/lit. Concentration above this level is indicative of pollution, specially from untreated sewage. For public water supply ammonia concentration should not exceed 0.5 mg/lit (Berry, 1977). In the Dal lake ammonia concentration is exceptionally high with an average value of 1.01 mg/lit., the standard level of concentration being exceeded at all the recording stations. The level is very high during the tourist season,

TABLE 17.4 : **Seasonal variation of ammonia nitrogen Dal Lake (mg/lit)**

| Stations | | April | June | August | October | December | Mean | Standard deviation |
|---|---|---|---|---|---|---|---|---|
| I | S | 0.64 | 0.70 | 1.47 | 1.41 | 0.91 | 1.02 | 0.35 |
| | B | 0.73 | 0.61 | 1.36 | 1.31 | 1.00 | 1.00 | 0.30 |
| II | S | 1.10 | 0.52 | 1.08 | 1.05 | 1.25 | 1.00 | 0.25 |
| | B | 1.08 | 1.05 | 1.41 | 1.45 | 1.23 | 1.24 | 0.16 |
| III | S | 0.71 | 0.81 | 1.14 | 1.05 | 0.90 | 0.92 | 0.15 |
| | B | 1.00 | 0.58 | 1.11 | 1.80 | 0.94 | 1.08 | 0.40 |
| IV | S | 1.00 | 0.48 | 1.08 | 1.45 | 0.72 | 0.94 | 0.33 |
| | B | 0.75 | 0.44 | 1.42 | 1.34 | 0.94 | 0.97 | 0.36 |
| V | S | 0.92 | 0.63 | 1.14 | 1.52 | 0.91 | 1.02 | 0.30 |
| | B | 0.76 | 0.52 | 1.41 | 1.45 | 0.61 | 0.95 | 0.40 |
| VI | S | 0.76 | 0.50 | 1.00 | 1.32 | 0.73 | 0.86 | 0.27 |
| | B | 1.30 | 0.51 | 1.26 | 1.30 | 0.77 | 1.02 | 0.32 |
| VII | S | 0.73 | 0.58 | 1.11 | 1.65 | 1.23 | 1.06 | 0.38 |
| | B | 0.80 | 0.82 | 1.30 | 1.93 | 0.77 | 1.12 | 0.44 |

S=Surface water.
B=Bottom water.

August to October (Table 17.4). Areas of particularly high concentration are Dal gate and Houseboat sections with August—October reading of 1.11 mg/lit and 1.65 mg/lit for surface and 1.30 mg/lit and 1.93 mg/lit for bottom water and 1.00 mg/lit and 1.32 mg/lit for surface and 1.26 mg/lit and 1.30 mg/lit for bottom water respectively, the average values being 1.50 mg/lit and 1.22 mg/lit respectively. The Dal water on the whole is, therefore, under considerable pollution stress in terms of ammonia concentration.

Nitrite and nitrate are also important components of the nitrogen cycle since their high level may lead to entrophication. The average concentration of nitrate-nitrogen in Dal lake is 0.19 mg/lit, the range of variation being from 0.00 mg/lit to 0.85 mg/lit (Table 17.5). This shows that so far as nitrate concentration in the water of the lake for drinking purposes is concerned, the picture is entirely satisfactory.

**Phosphorus**

Phosphorus is an other important nutrient necessary for aquatic vegetation. In water bodies the principal sources of this nutrient are : leaching of soil and rocks, normal decomposition of plants and animals, agricultural inputs and run-off from the surrounding areas, untreated sewage and industrial affluents. Undesirable level of phosphorus is always indicative of pollution and may lead to entrophication of the waterbody by causing excessive algal bloom and loss of transparency. It is generally believed that phosphorus loading must be regulated in order to eliminate many undesirable properties from water. So far as the desirable level of phosphorus in water is concerned, a reference level of 0.5 mg/lit has been suggested (E.P.A., 1973). The maximum level suggested is from 0.5 mg/lit to 0.007 mg/lit for different water bodies. Concentration of phosphorus above this level is considered high enough to cause entrophication (Table 17.6).

The spatial and temporal variations in the distribution of phosphorus in Dal lake has been presented in Table 17.6. The average level of phosphorus concentration is 1.31 mg/lit. The level is generally high during summer months. Excessive concentration has been recorded during the months of June to August. This may be because the fertilizers are usually applied to agricultural fields on the floating gardens and surrounding areas during this

TABLE 17.5 : **Seasonal variation of Nitrate – Nitrogen in Dal Lake (mg/lit)**

| Stations | | April | June | August | October | December | Mean | Standard deviation |
|---|---|---|---|---|---|---|---|---|
| I | S | 0.22 | 0.53 | 1.04 | 0.04 | 0.09 | 0.38 | 0.37 |
| | B | 0.19 | 0.30 | 0.09 | 0.02 | 0.16 | 0.15 | 0.09 |
| II | S | 0.03 | 0.19 | 0.01 | 0.09 | 0.06 | 0.08 | 0.06 |
| | B | 0.64 | 0.63 | 0.09 | 0.21 | 0.22 | 0.03 | 0.22 |
| III | S | 0.05 | 0.44 | 0.09 | 0.15 | 0.04 | 0.15 | 0.15 |
| | B | 0.05 | 0.60 | 0.23 | 0.24 | 0.05 | 0.23 | 0.19 |
| IV | S | 0.00 | 0.76 | 0.09 | 0.09 | 0.02 | 0.19 | 0.28 |
| | B | 0.03 | 0.18 | 0.13 | 0.34 | 0.10 | 0.15 | 0.10 |
| V | S | 0.12 | 0.40 | 0.05 | 0.06 | 0.07 | 0.14 | 0.13 |
| | B | 0.05 | 0.46 | 0.15 | 0.19 | 0.13 | 0.19 | 0.14 |
| VI | S | 0.00 | 0.56 | 0.13 | 0.15 | 0.07 | 0.18 | 0.19 |
| | B | 0.34 | 0.74 | 0.24 | 0.25 | 0.14 | 0.34 | 0.20 |
| VII | S | 0.02 | 0.20 | 0.15 | 0.41 | 0.09 | 0.17 | 0.13 |
| | B | 0 03 | 0.85 | 0.21 | 0.22 | 0.11 | 0.28 | 0.29 |

S=Surface water
B=Bottom water

TABLE 17.6 : Seasonal variation of Phosphorus in Dal Lake (mg/lit)

| Stations | | April | June | August | October | December | Mean | Standard deviation |
|---|---|---|---|---|---|---|---|---|
| I | S | 1.00 | 1.72 | 2.00 | 0.66 | 1.07 | 1.30 | 0.49 |
| | B | 0.73 | 1.92 | 2.66 | 0.63 | 1.10 | 1.40 | 0.77 |
| II | S | 0.66 | 1.55 | 2.03 | 0.47 | 0.77 | 1.09 | 0.59 |
| | B | 0.85 | 1.77 | 2.29 | 0.29 | 0.60 | 1.16 | 0.75 |
| III | S | 0.66 | 1.84 | 2.16 | 0.40 | 0.74 | 1.13 | 0.74 |
| | B | 0.74 | 1.40 | 3.10 | 0.48 | 0.67 | 1.27 | 0.96 |
| IV | S | 0.65 | 1.88 | 2.21 | 0.32 | 0.62 | 1.14 | 0.75 |
| | B | 0.77 | 1.59 | 2.60 | 0.48 | 0.93 | 1.27 | 0.75 |
| V | S | 0.62 | 1.51 | 2.40 | 0.53 | 0.82 | 1.18 | 0.70 |
| | B | 1.92 | 1.92 | 1.84 | 0.57 | 0.97 | 1.44 | 0.56 |
| VI | S | 0.74 | 1.84 | 2.15 | 0.42 | 0.62 | 1.15 | 0.69 |
| | B | 0.77 | 1.47 | 2.60 | 0.25 | 0.94 | 1.21 | 0.79 |
| VII | S | 1.10 | 1.56 | 3.25 | 0.55 | 1.15 | 1.52 | 0.92 |
| | B | 1.59 | 2.07 | 3.34 | 0.81 | 1.83 | 1.93 | 0.82 |

S=Surface water
B=Bottom water

month. This is also the season of tourist influx which adds considerably to the quantum of sewage into the lake. Spatial analysis indicates that high concentration of phosphorus occurs in the Houseboat, Dal gate and Telbal nullah sections. Examples of entrophication as a result of excessive phosphorus loading is already evident in many parts of the lake, specially in the southern sections and major parts of Hazratbal and Telbal basins which have become so much chocked with floating and rooted macrophytes and dense algal population that it is causing interference in the movement of water crafts. The lake water has a repugnant odour and is generally undesirable for swimming, except small parts of the Gagribal basins.

**Dissolved oxygen**

The level of dissolved oxygen is perhaps the best indicator of the quality of a waterbody. In sluggish waterbodies, such as a lake, pollution leads to rapid depletion of dissolved oxygen (DO) by causing rapid oxygen loss. A low level of DO not only hampers the breakdown of waste material but also attracts to the waterbody, certain undesirable bacteria which flourish in the absence of oxygen and which release in the water foul smelling gases such as methane, ammonia and hydrogen sulphate. Level of dissolved oxygen, therefore, acts as an indicator of status of the water body. The maximum possible oxygen concentration in water is 14.56 mg/lit at 0°C. The average temperature of the Dal lake is 17.4°C where as the average DO level is 8.06 mg/lit (Table 17.7 and 17.8). As per generally accepted standards DO level in the Dal lake should be 9.7 mg/lit (yoznaya, 1981). It means that in the lake as a whole DO level is not low enough to seriously endanger aquatic life but still is much below the desired level. The position is comparatively better in Hazratbal, Nagin and Gagribal sections with average DO levels of 8.46, 8.84, and 8.37 mg/lit respectively. The lowest level of DO (6.56) has been recorded at the Dal gate station which can be regarded as heavily polluted in terms of DO level.

Oxygen depletion is generally caused by heavy input of raw sewage over and above the seasonal capacity of the waterbody to assimilate such wastes. In order to prevent serious degradation of the quality of the water, advanced methods of waste treatment should be introduced as a pollution control measure.

TABLE 17.7 : **Seasonal variation of dissolved Oxygen in Dal Lake (mg/lit)**

| Stations | | April | June | August | October | December | Mean | Standard Deviation |
|---|---|---|---|---|---|---|---|---|
| I | S | 8.90 | 8.30 | 7.00 | 7.00 | 9.00 | 8.04 | 0.88 |
| | B | 8.60 | 7.30 | 7.00 | 6.20 | 8.50 | 7.52 | 0.91 |
| II | S | 13.50 | 10.00 | 8.00 | 6.60 | 8.50 | 9.32 | 2.35 |
| | B | 9.30 | 7.80 | 5.90 | 7.00 | 8.00 | 7.60 | 1.12 |
| III | S | 11.90 | 10.00 | 9.00 | 7.40 | 8.40 | 9.34 | 1.53 |
| | B | 9.00 | 9.50 | 7.30 | 8.00 | 7.90 | 8.34 | 0.80 |
| IV | S | 9.70 | 9.60 | 9.20 | 8.90 | 8.90 | 9.26 | 0.33 |
| | B | 7.00 | 6.90 | 7.30 | 8.40 | 7.80 | 7.48 | 0.55 |
| V | S | 9.50 | 8.40 | 7.40 | 11.00 | 6.50 | 8.56 | 1.57 |
| | B | 7.50 | 7.30 | 6.20 | 10.80 | 4.90 | 7.34 | 1.96 |
| VI | S | 8.30 | 9.50 | 8.50 | 9.40 | 8.20 | 8.78 | 0.55 |
| | B | 7.50 | 8.80 | 8.30 | 8.90 | 7.30 | 8.20 | 0.60 |
| VII | S | 7.70 | 6.90 | 6.50 | 8.00 | 5.10 | 6.84 | 1.02 |
| | B | 7.50 | 6.90 | 6.00 | 7.50 | 4.40 | 6.28 | 1.15 |

TABLE 17.8 : **Mean temperature and mean DO in the Dal Lake**

| Station | Mean temp.°C | Mean DO mg/lit |
|---|---|---|
| Telbal | 12.95 | 7.78 |
| Hazratbal | 16.15 | 8.46 |
| Nagin lake | 18.58 | 8.84 |
| Gagribal | 20.12 | 8.37 |
| Floating Gardens | 18.53 | 7.95 |
| Hauseboat Area | 18.20 | 8.49 |
| Dal gate | 17.97 | 6.56 |

## pH

pH value indicates the level of concentration of hydrogen ions expressed logrithmically. Its value ranges from 1 to 14 with 7.0 representing neutrality (neither acidic nor basic water). High values of pH are indicative of pollution by industrial waste and may produce adverse effect on aquatic life by altering the availability of plant nutrients. Standards have been suggested for acceptable pH levels for different uses. For drinking purpose the pH should be in the range of 7 to 8.5 (Prasad & Ayer, 1983). pH level less than 6.5 or more than 9.2 makes the water excessively acidic or alkaline for human consumption. A range of 5.0 to 9.0 has been suggested for primary contact recreation and public water supply intake. For fresh water aquatic life a range of 6.0 to 9.0 and for areas extensively used for bathing a range of 6.5—8.3 have been recommended. That the Dal suffers from high value pH is borne out by the fact that its mean value varies from 7.77 in Telbal to 8.88 in Gagribal. The mean value is, however, no reliable indication of pH pollution as there is a marked seasonal and spatial variation in its distribution. An examination of Table 17.9 reveals that pH is within permissible limits for drinking water during all seasons in the Telbal and Hazratbal sections. But in Gagribal, floating gardens and Houseboat sections the water has pH in excess of the permissible limit of 8.5 throughout the year. This water, specially in the southern part of the lake, is regularly consumed by the Hanji population who live on the lake and for whom it is the only source of drinking water. Though the Dal water is still within permissible limits for primary contact recreation, it should not be used for extensive bathing and swimming purposes except in Telbal and Hazratbal sections.

TABLE 17.9 : **Seasonal variation of pH in Dal Lake**

| Stations | | April | June | August | October | December | Mean | Standard Division |
|---|---|---|---|---|---|---|---|---|
| I | S | 7.74 | 7.80 | 7.80 | 7.72 | 7.80 | 7.77 | 0.03 |
| | B | 7.70 | 7.75 | 7.60 | 7.68 | 7.75 | 7.72 | 0.03 |
| II | S | 8.20 | 8.00 | 7.90 | 8.60 | 8 00 | 8.14 | 0.25 |
| | B | 7.54 | 7.75 | 7.70 | 8.06 | 8.00 | 7.81 | 0.19 |
| III | S | 8.12 | 9.15 | 8.45 | 8.52 | 8.00 | 8.44 | 0.40 |
| | B | 8.05 | 8.90 | 8.10 | 8.20 | 7.60 | 8.17 | 0.42 |
| IV | S | 9.08 | 8.70 | 8.60 | 8.92 | 9.10 | 8.88 | 0.20 |
| | B | 9.00 | 8.55 | 8.50 | 8.68 | 9.00 | 8.74 | 0.21 |
| V | S | 8.90 | 8.60 | 9.20 | 8.82 | 8.05 | 8.71 | 0.38 |
| | B | 8.56 | 8.55 | 8.25 | 8.12 | 8.20 | 8.33 | 0.18 |
| VI | S | 9.30 | 8.65 | 8.70 | 9.02 | 8.40 | 8.81 | 0.31 |
| | B | 9.20 | 8.60 | 8.55 | 8.10 | 8.10 | 8.51 | 0.40 |
| VII | S | 7.60 | 8.75 | 8.50 | 7.60 | 7.45 | 8.00 | 0.55 |
| | B | 7.55 | 8.80 | 7.40 | 7.20 | 7 30 | 7.64 | 0.58 |

High pH value also adversely affects many species of fresh-water aquatic life, specially if the pH value exceeds 9. From this point of view the problem in the Dal lake has not yet assumed critical dimensions. But generally in all southern stations pH values throughout the year are on the higher side (more than 8.5). It seems that if the problem is not tackled now, high pH values may seriously endanger aquatic life in the near future, specially in the southern parts of the lake. Only species like the mirror carp and leather carp, which are pH tolerant can, and do flourish in the lake.

**Transparency**

Secchi disc values (in metres) of the Dal water, as recorded by junior author in 1982, is given in Table 17.10.

TABLE 17.10 : **Secchi disc readings in Dal Lake (m)**

| Station | Maximum | Minimum |
|---|---|---|
| Hazratbal | 1.70—1.95 | 0.64—0.42 |
| Nagin | 4.00—4.95 | 2.00—2.50 |
| Bod Dal | 1.00—1.50 | 0.87—1.00 |
| Lokut Dal | 2.50—3.00 | 1.95—2.00 |

A perusal of the table reveals that Nagin water is the clearest and most transparent. The waters of Hazratbal, Gagribal and Bod Dal basins have very low transparency as these are the most polluted parts of the lake. There are also significant differences in transparency values between winter and summer months. Seasonal changes in Secchi disc transparency would indicate that maximum light penetration occurs in December. A dramatic decline in transparency occurs in mid to late summer. Transparency level during the latter is explained by the appearance of spring zooplanktons, increased algal production and production of blue-green phytoplanktons.

The Floating garden section of the lake is so weed infested that transparency is almost nill. The same is more or less true of the Dal gate and Houseboat sections, where transparency is negligible because of highly polluted water.

**Biological state of the Lake**

The Dal lake is facing the menace of an ever increasing population by phytoplanktons, zooplanktons and aquatic macrophytes. This sorry state of affairs has been brought about by extensive land use in the lake's catchment resulting in an accelerated rate of nutrient loading and the consequent weed growth to such an extent that the lake has become useless for aquatic sports, games and fisheries to a great extent. An increase in the population of phytoplanktons and zooplanktons, if not controlled, inevitably will lead to the lake's entrophication by causing increased hypolimnetic oxygen deficit and changes in species composition.

The Dal lake is a productive one, having undesirable quantities of algae during certain seasons, specially during the summer. Predominance of species of blue-greens and their abundance have been noted during the summer season. A study, confined to the density of blue and green algae as well as the density of total phytoplankton in the period June to July gives the following spatial picture (ENEX, 1978).

Total phytoplankton : Bod Dal 3696.6/lit, Lokut Dal 1060/lit ; Gagribal 1638.3/lit and Dal gate 9003.3/lit. For blue and green algae species the respective figures are : Bod Dal 463.3/lit ; Lokut Dal 267.4/lit ; Gagribal 518/lit and Dal gate 1071/lit. The present concentration of phytoplankton in the lake cannot be regarded as too high. This may be attributed to the low concentration of soluble nutrients in the lake water.

High concentration of zooplankton has also not been reported except seasonally in the Houseboat area. This is understandable because zooplankton population depends, to a large extent, on the phytoplankton which mostly provide the available food.

The dominant species of zooplankton in the Dal lake are diatom and protozon rotifers whose population fluctuates seasonally in response to a number of environmental parameters which define the ecological niche of these species. Some rough idea of the density and spatial distribution of diatoms in the month of June-July is available (Berry, 1977). Bod Dal 6470/lit. This spatial picture may be related to the availability of algal food.

**Macrophytes**

Macrophytes play an important role in the ecology of water

bodies by acting as modifier and diversifier of organic matter. Thick beds of macrophytes hinder water movement and cause accumulation of silt and organic matter within the lake, thus causing serious hazards to aquatic life.

In the Dal lake an area of $11,673 \times 10^3$ m² is occupied by different types of macrophytes in which the rooted and free floating categories predominate. The maximum area under macrophytes is found in Lokut Dal ($5040 \times 10^3$ m²), followed by Hazratbal basin ($2700 \times 10^3$ m²) and Bod Dal ($850 \times 10^3$ m²).

**Bacteria**

Caliform and streptococci are important indicators of microbiological pollution of water. The presence of the latter in water is the result of pollution caused by recent addition of faeces and urines of warm blooded animals. The presence of faecal streptococci in excess of the acceptable limit is a matter of serious concern because they are generally the carriers of water borne pathogens. Total coliform is indicative of long term pollution.

Standard levels for both total coliform and faecal streptococci have been suggested : 20,000 total coliform per 100 ml at any time for livestock water supply and 10,000/100 ml total coliform for public water supply. For faecal streptococci limit of 200/100 ml for primary contact recreation, 1000/100 ml for agricultural irrigation and 2000/100 ml for public water supply have been suggested.

Microbiological analysis of Dal water reveals that the concentration of total coliform ranges from 640 MPN/100 ml to 24 MPN/100 ml and that for faecal streptococci from 0.00 MPN/100 ml to 2400 MPN/100 ml (NEERI, 1972). The maximum number of coliform group is found in the Hazratbal and Gagribal areas. Basinwise the level of concentration of coliform is : Hazratbal 1500 MPN/100 ml and Houseboat area 2400 MPN/100 ml. For the same basins the level of concentration of faecal streptococci is 640/100 ml in Hazratbal ; only traces in Bod Dal, 240/100 ml in Gagribal and 2400/100 ml in the Houseboat area.

Looking at the faecal streptococci distribution we find that extensive areas throughout the lake suffers from highly polluted water, so far as primary contact recreation is concerned. Extensive area of the lake contain faecal coliform in excess of the commonly used reference level. Extremely polluted faecal streptococci levels for public water supply have been reported from the southern part of the lake, comprising the Houseboat and the Dal gate areas.

No part of the lake contains highly polluted water in terms of total coliform but the level of faecal streptococci in usually high. This pattern can be attributed to pollution from human sources and can be related to the high degree of urbanization that is taking place in areas surrounding the lake.

**Impact of Man**

Man is an integral part of the web of life and functionally interrelated to the organic as well as inorganic components of the ecosystem. But man is not merely an other element in the ecosystem. Being ecologically dominant, he acts both as an instigator of new systems and a component of different trophic levels within existing systems. Man's role, therefore, in destabilization of the delicate ecological equibrium, specially when human number increases beyond ecologically desirable limits, can hardly be over emphasized.

The Dal lake and its environs, according to the 1971 census, had a population of 1,66,040. This was expected to reach a figure of 2,16.083 in 1981, an increase of over 30 per cent over a period of ten years. The largest single concentration of population is on the Dal lake itself. According to the author's estimate over 17,000 persons permanently reside in houseboats or on the floating gardens, with the number temporarily increasing by 20,000 persons during the summer months. This is a matter of great concern because of its adverse impact on ecosystem of the lake. The possibility of population growth and its absorption within the lake's ecosystem is limited. The question is how and by what will the limit be imposed to keep the increasing population away from the lake so that the ecosystem can be kept balanced. Already much of area of the lake is held under legitimate titles and these have been encroached upon by converting water areas into floating gardens which pose one of the greatest dangers to its ecology. They have not only drastically reduced the area of open water but are also directly responsible for deterioration of the quality of water, being the principal source of nutrient intake of the lake. From the aesthetic point of view they are also undesirable as they mark the beauty of the lake. Their removal will not only restore the lake to its original size but will also improve the quality of water. Therefore, from the aesthetic as well as the ecological points of view the removal of the floating gardens and other man made obstructions must be accorded top priority.

An other human activity which is posing serious problems to the lake's ecology is tourism. With its vast expanse of placid water ringed by picturesque hills, the Dal lake has always been a source of attraction to a large number of tourists. The importance of tourism in the local economy is very great. But it is a mixed blessing. No doubt it gives a boost to the regional economy, creates jobs and accelerates infrastructural development but at the same time it also puts the natural ecological balance under severe stress. Since 1940 tourist arrival at Srinagar has shot up by over 2100 per cent. As a natural concomitant of this increase the Dal lake and its environs over the years have acquired a wide variety of infrastructural facilities to cater to the expanding recreation industry, the principal among them being houseboats and other types of tourist accommodation. Over the years the number of houseboats and hotels on the lake as well as along it's periphery has greatly increased. Not many people realize that the pristine charm of the lake is being rapidly eroded by such infrastructural development. The greatest damage is being caused by the unplanned expansion of the accommodation industry. A large number of multistoried hotels have come up along the Boulevard, besides innumerable houseboats on the surface of the lake. This is leading to the lake's pollution as an enormous quantity of construction and other types of garbage, estimated to be about 50,000 tons, is being dumped in the lake every year. The garbage leads to extensive weed growth. being rich in organic matter. Hundreds of houseboats, without adequate and safe sewage disposal arrangement, also constitute a danger to the lake's ecology. The quality of water has seriously deteriorated as a result of such unsanitary conditions. If the lake's ecosystem continues declining at the present rate as a result of tourist related infrastructural development, and no corrective measures are undertaken, it is feared that a point of near entrophication would be reached in a couple of decades.

Enough has been said in the preceding pages to highlight the fact that the Dal lake's ecology is under severe stress. Indeed, if the present level of pollution goes on unchecked, the lake will be lost within the next 100 years. The factors, briefly mentioned above, responsible for the present ill health of the lake's ecology can be categorized as under :

(i) Public encroachment over the lake's surface,
(ii) Prolific weed growth due to large intake of nutrient,
(iii) Silting of basins as result of irrational and unplanned landuse in the lake's catchment,
(iv) Direct disposal of sewage and siltage into the lake, and
(v) Unplanned development of tourism.

The question which naturally comes to mind is, has not the commercialization of the lake been allowed for too long ? The use and exploitation of the lake's resources has been unchecked at the cost of its protection and conservation. If the lake is to be saved from total entrophication corrective measures must be taken on war footing. Suggesting remedial measures is beyond the scope of this article but it is obvious that solution of the numerous problems plaguing this once beautiful lake calls for concerted interdisciplinary effort in which planners, engineers, ecologists, geographers, economists and administrators must pool their resources in order to ensure the survival of the lake as a viable ecosystem, a system which harmonizes economic viability and ecological constraints.

## REFERENCES

Berry, B.J.L. (ed.) (1977) : *The Social Burden of Pollution,* 159.

E.P.A., (1973) : *Guidelines for Developing or Revising Water Quality.*

Yoznaya, N.F. (1981) : *Chemistry of Water and Microbiology*, Moscow, 143.

Prasad, V.S., and Ayer, S.K. (1983) : Drinking water quality aspects, *Science Reporter*, March, 149.

ENEX of New Zealand (1978) : *Study of Pollution in Dal Lake.*

National Environmental and Engineering Institute (NEERI) (1972) : *A Survey of Quality of Water to Srinagar.*

ALOK KUMAR SINGH
AND
S. N. SINGH

# EIGHTEEN

# ENVIRONMENTAL MANAGEMENT AND ECONOMIC REGIONALIZATION IN MIRZAPUR DISTRICT

IN the modern age of science and technology the quality of our environment is deteriorating very rapidly. Pollution problems are becoming more and more serious due to rapid but unplanned urbanization and industrialization. Air, water and soil are the most important components of the environment on which the very survival of human life is dependent. Purity of these environmental components is of great importance and if we continue to pollute them, the human life will be affected adversely.

Pollution has been defined in several ways but the one given by the Environmental Pollution Panel of the U.S. President's Advisory Committee seems to be touching the ground : "Environmental Pollution is the unfavourable alteration of our surroundings, wholly or partly as a by-product of man's actions, through direct or indirect effects of the change in energy pattern levels, physical and chemical constitution and abundances of organisms. These changes may affect the man directly or through his supplies of water and of agricultural and other biological products his physical objects or possession of his opportunities for recreation and appreciation of nature" (Owens, 1965).

## Sources of pollution

The modern world is facing with a wide array of toxic substances including particulates, gases and radioactive materials, which affect unfavourably the general well being of organisms. Pollution is thus an ecological problem affecting both the environment and organisms. No doubt, industries are the most common source of pollutions. The soil is adversely affected by industries in two possible ways. Primarily, due to burning of coal, for power generation, several harmful gases such as sulphur dioxide, carbon mono-oxide, carbon dioxide, fluorine, chlorine etc. are released into the atmosphere. These substances react with the atmospheric moisture forming acids which descend and subsequently pollute the soil. Used water is let out as effluents containing acids, alkalies, toxic, inorganic and organic compounds, colour producing substances, dissolved and suspended solids. When the soil receives these industrial effluents by overflowing from channels or other means, it gets affected adversely. Although edaphic environment has the ability to assimilate a certain amount of waste materials, the constant addition of industrial effluents in large quantities or over a long period above its absorptive capacity pollute the soil and make it unproductive and unhygenic for human health. Unmanaged use of certain chemicals like D.D.T. and other poly-chlorinated compounds produce a wide variety of effects when they are discharged over lands. These chemical pollutants being not easily degraded, accumulate in food chains. After receiving the effluent there is a drastic change in the physico-chemical and biological properties of the soils.

Soils of similar qualities are often occupied by similar biological communities when this natural environment is upset by the addition of pollutants, only the organisms tolerant to those pollutant survive and reproduce whereas the sensitive or less tolerant forms disappear or become restricted. The type of organisms destroyed and the extent of destruction reflect the characteristics and quality of the wastes entering the soil, water, and air. Therefore, many biologists believe that organisms living in such elements of environment provide a fairly sensitive and reliable measure of the change of conditions. Therefore, an indepth ecological investigation of plants growing on such polluted habitants is of urgent need to avoid the health hazards.

**Resource potentials in Mirzapur**

The scientific principles for the planning of environment and rational resource management have not been sufficiently worked out. One factor that has been holding back the planning of resource management, is the inadequate knowledge about spatial differences in the character and magnitude of the impact of human activity on the natural environment. The identification of spatial geo-economic and environmental differences i.e., an environment based geo-economic regionalization, may be useful in order to provide a basis for effective management of regional development. Complete environmental geo-economic regionalization of Mirzapur district has not so far been done. However, attempts that have been made for the purpose of physical, or spatial regionalization of the district as a whole or part of it containing information therein, were to a some extent useful in our environment geo-economic regionalization. Mirzapur district falls in a area virtually dominated by power, cement factory, building stone, coal mining, aluminium and chemicals industry. In analyzing agriculture from the viewpoint of regionalization, used such attributes as the area of cultivated land and overall agricultural landuse as well as the ratio of cultivated land to population and total area.

It may be pointed out that the pressure of population on land is higher than the carrying capacity of the land. While the carrying capacity ranges from 1 to 2.5 persons per ha of food crops in 1980-81, the population density varies from 150,500 persons/km$^2$ of agricultural land. Thus, the population, in general, cannot get the daily per capita required calorie intake. The low nutritional intake is vivid from their and poor health. The higher density of population obviously burdens the land adversely, and of the health of people as well. The poverty of the socio-economic milieu can be further observed from poor literacy, and low level of perception of technology of the local people. However, the region having such physical and socio-cultural environment has been exposed to modern technological and industrial developments during the past three decades.

## Economic development and environmental deterioration

Various developments were initiated especially in the southern part of the district with the undertaking of the Rihand Dam project at Pipri in 1952. The huge reservoir covering an area of 344 km$^2$, known as Pant Sagar, has engulfed a vast forest area with a good piece of cultivated land and settlements. The displaced population moved to adjoining uplands resulting in further deforestation and increase in soil erosion and siltation. It is estimated that about 50 thousand persons were displaced and were consigned to Muirpur and Babhani Uplands. About 12 thousand persons were displaced due to coal mining and establishment of thermal power station at Kota (Shaktinagar). The effected population had to move away, as well, settling wherever they could. The Kanhar irrigation project is likely to cause further ecological damage, and also displace population. The development process does not take into account of the environmental and socio-economic aspects, and suffers from a serious planning lacunae. There has been well marked pollution in the area around Churk and Dala Cement Factory. In deposition of cement dust on soil and vegetation has entrusted the soil with an impervious layer, and damaged foliage which has reduced transpiration and photo-syntheses processes. Soil in the neighbourhood has become alkaline and due to degradation it is becoming unfavourable for cultivation and natural regeneration of plant and forest species. The pollution in the surrounding area of Obra Thermal Plant is clearly noticeable. The daily consumption of large quantities of coal leads to emission of sulphur dioxide, carbon mono-oxide, nitrogen, and oxide besides a huge quantity of solid materials. This causes not only pollution of air but also affects the soil, of nearby areas, water bodies and settlements in a radious of about twenty km. Similarly, Shaktinagar and Anpara Power Projects are likely to cause further environmental deterioration. They would also cause water pollution of Pant Sagar Reservoir and damage organic life. HINDALCO aluminium factory, causes noticeable fluoride pollution which has much effect on water, soil and vegetation. The fluoride concentration is injurious to animal and human health (Pandey, 1978). The maximum and minimum fluoride level, 423.2 and 20.5 ppm were observed at respective northeast distances of 0.5 and 15 km from the factory (Singh, A.K. 1984). These together with effluents from the Kanoria Chemicals, make the industrial zone and adjoining areas polluted and unhealthy.

Environmental deterioration through soil erosion is becoming widespread in the region. Increasing deforestation, road and rail construction without proper consideration for arboriculture, cultivation in marginal and steep lands without contour terracing, over-graving and other numerous undesirable practices have led to phenomenal development of soil erosion over the region.

'Large areas of severe erosion are observed in Lower Rihand-Kanhar Interfluve and west of Rihand-Son confluence (Fig. 18.1). In order to effectively counter soil erosion and encourage conser-

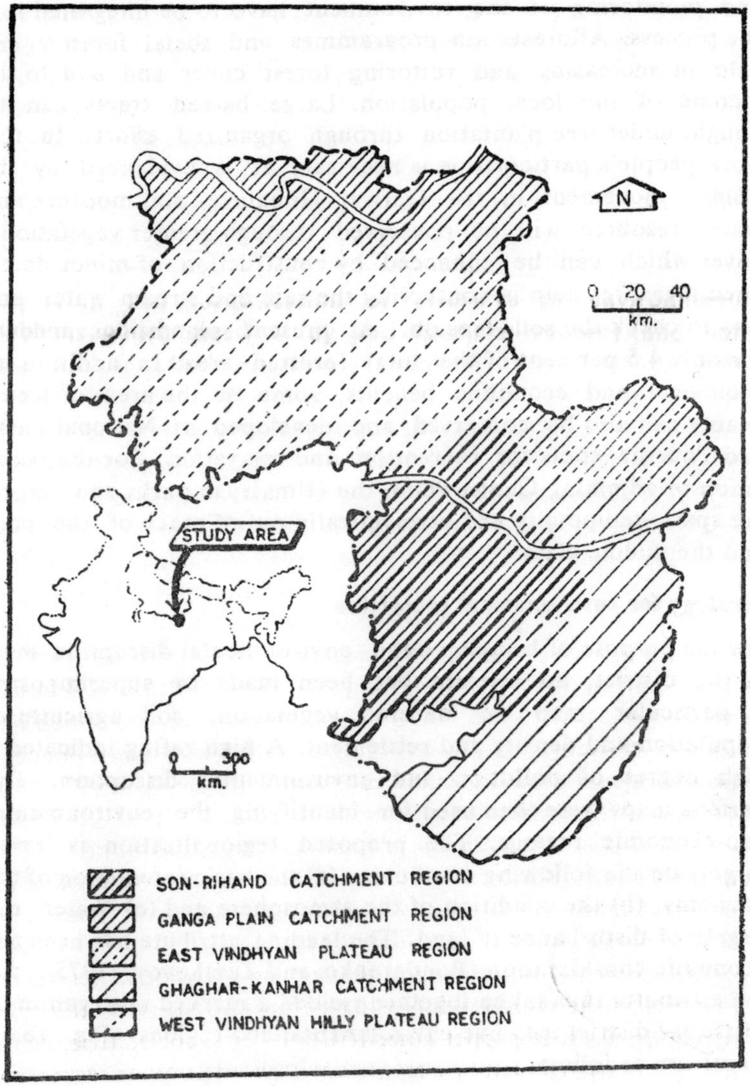


FIG. 18.1 : *Economic regionalization Mirzapur district*

vation, it would be essential to work out an integrated approach for soil, water and forest management. The integrated river basin approach, with watershed units would help in rationalizing development and conservation. In this task, the welfare of the people and monitoring of the environment have to be integrated into the process. Afforestation programmes and social forestry can help in increasing and restoring forest cover and add to the income of the local population. Large barren tracts can be bought under tree plantation through organized efforts. In this work people's participation is necessary as is evidenced by the *Chipko* movement in the Garwal Himalaya, soil moisture and water resource will be conserved through greater vegetational cover which can be enhanced by construction of minor dams. Such measures help in conserving the rain and stream water and also to check the soil erosion. At present plantations account for only 4.5 per cent of the total forested area. In addition to ecological and economic benefits, some of the areas of scenic beauty can also be conserved and developed as National Parks and Tourists spots for recreation and relaxation. For the population of adjoining Ganga Plain, the Himalayan parks and resorts are spatio-temporally and economically out of reach of the poor and the middle class.

**Strategy for environmental protection**

For the purpose of bringing out a environmental disruption map of the district, an attempt has been made by superimposing a particular map on natural vegetation, soil agriculture, population and density and settlement. A high rating indicated a high degree of pollution and environmental disruption. The various maps were then used for identifying the environmental geo-economic regions. The proposed regionalization is based largely on the following attributes : (a) the basic orientation of the economy, (b) the condition of the atmosphere and (c) water the degree of disturbance of land. The leading attribute has been the economic consideration (Bondarenko and Zamkovoy, 1975). All the attributes thus taken together yielded a marked sagregation of Mirzapur district into five environmental-eco-regions (Fig. 18.1). These are as follows :

*Son-Rihand catchment region*

The environment of this region is polluted mainly by the cement power and coal mining industries. The cement dust contains mixture of Ca, K, Si and nitrogen oxide. The fallout of cement dust on soil and vegetation has influenced the environment which constitute an integral source of livelihood. The industry runs largely on power obtained from coal, oil and natural gas the fossil fuels—the burning of which releases carbon dioxide ($CO_2$), sulphur dioxide ($SO_2$), toxic wastes and heat. The daily average coal consumption of Obra Thermal Plant alone is about 9,720 metric tonnes. The enormous quantity of various gases released in the atmosphere and added every day can well be imagined. The effect of pollution has been well observed up to a distance of twenty-five km. The industrial development in the region has led to a substantial increase in water of Pant Sagar, Rihand and Son rivers. The problem of the wastes of the thermal power plant which is discharged in river water is similar in several respects to that of the wastes of the atmosphere. Both the practices are hazardous to human beings. But, more importantly, the effect of both diminish with time.

The process of industrial and agricultural production and other human activity in the region have also laid bad impact on quality of land. The ratio of arable land to population is 0.13 ha per capita, while the ratio of overall agricultural land to population is 0.18 ha per capita. The heavy load on the environment is expected to continue even in the near future. Measures must therefore be initiated to reduce the negative human impact on the biosphere through optimal selection and location of industries in the district.

*Ganga plain catchment region*

In this region, consisting of Mirzapur and Chunar tahsils, environment is polluted mainly by the cement, stone crushers, handloom and other small industries of iron and aluminium and copper rolling mills. The region has mean population density of 540 per km$^2$. Many industries tends to change the use of land, thereby making it unsuitable for agriculture. More than 0.9 per cent of all the land has been effected in this way. The region has 0.07 ha of arable land and 0.10 ha of over all agricultural land per capita. The loss of fertile agricultural land is fairly noticeable. The heavy pollution load in the region affects the atmosphere, which has been receiving the waste products of cement dust and claypot chimenies.

The existing situation is creating substantial difficulties for management and for the future economic development of the region. Further development will be possible only if existing enterprises in stall effective waste treatment devices. New industries are equipped with recycling technologies and further industrial location should be limited mainly to light industry with negligible impact on the environment.

*East Vindhyan plateau region*

Presently the environment of the region is not being polluted by any minor industrial complex. The total population density is 178 per km$^2$. The region is least industrialized with predominating suburban dairy vegetable and agriculture farming but the lack of irrigational facilities limits, their production. The principal constraint of further economic development in the region is lack of transportational facilities. Most part of the region is inaccessible by metalled road. In the rainy season the region becomes virtually isolated from the rest part of district.

*Ghaghar-Kanhar catchment region*

In this region pollution stems mainly from agriculture and transportation. This region occupies the south-east part of Mirzapur district including the block of Nagwa, Dudhi, Babhani and Kon Nyayapanchayat except for the right bank of the east Kanhar river. The population density is over 182 person per km$^2$.

The main economic activity is agriculture and for which accounts for 49 per cent of the total population. Since the region contains less number of industries, pollution tends to be small. The only significant air pollution occurs through industrial plants and near highway with intensive motor traffic. The polluted winds may also affect from nearby industrial region. The soil of the region is very fertile but due to the lack of irrigational facilities, agriculture is still in a premitive stage.

### *West Vindhyan hilly plateau regions*

In this region pollution is associated with some stone crushing industries. This region is situated in the western part of the district with population density of more than 85 person per km². The surface is being disturbed mainly due to mineral extraction (mostly building material with construction of industries residential houses and agricultural facilities etc). The region is characterized by poor soils, lack of water resources and undulating rough terrain. The irrigational potential is quite low and even drinking water becomes scarce especially in the dry season.

## Conclusion

After having traversed the whole gamut of role of water, soil and air in the environment, the very question of management becomes quite relevant. The country has a great variety of environmental conditions. We have the desert semi arid areas, vast fertile alluvial plain, densely forested regions, coastal tracts, mountaneous area etc. Also we have a great variety of development opportunities by way of land, water, mineral, energy and human resources. We have to contend against tremendous natural hazards in the form of floods, droughts, landslides, cyclones, coastal erosion, desertification, siltation etc. and the induced problems of soil erosion, increasing soil salinity water-logging etc. On one hand, rich agricultural terrains are being lost due to various factors, and on the other, deforestation are depriving the community a vital resource of forest and pasture.

Firstly, the question of drought management may be taken up. As already pointed out, there are no typical meteorological drought in Mirzapur except a few arid pockets. Economic shocks are mainly resulting from agricultural failures following subnormal precipitation, when crop practices artificially rise the demand of water requirements. Here long term plans have to be undertaken for scaling down the irrigation practices. Cropping practices have to be restricted to least water intensive varieties. It must be remembered that ground water is available everywhere and at varying depth in varying quantities.

While on this subject, the question of safe drinking water supply for the public health system can be elaborated a little more. As indicated above, in the given dry conditions of the country there can not be any dearth of drinking water (except in a very few selected and limited area of southern arid zone) to the tune of 2.5 litres per capita per day.

Finally, on the question of irrigation, drinking water and cropping practices, it may be emphasised that the ecological balance can be rightly maintained by correct and conjunctive use of surface water and ground water on the basis of an objective assessment of every watershed. With better management and extension of vegetative covers, soil erosion air pollution and other environmental degradations can effectively be arrested.

## REFERENCES

Owens, M. (1965) : Environmental pollution, *United States Presidents Advisory Committee*, 16.

Pandey, S.N. (1978) : Effects of Coal Smoke and Sulphur Dioxide Pollution on Plants. Unpublished thesis, Department of Botany B.H.U. Varanasi, 62.

Singh, A.K. (1984) : Geographical Analysis of Pollution and Settlement in Mirzapur District, U.P., India, Unpublished Ph.D. thesis, Department of Geography B.H.U. Varanasi, 215.

Bondarenko, Ya I and V.P. Zamkovoy (1975) : *An Economic Regionalization of the Donets Basin*, Geograficheskiye vrsledovaniya V Donbasse (Geographical Investigations in the Donets Basin) Donetsh branch, Geographical Society U.S.S.R., 35.

SECTION IV

# ASSESSMENT, MAPPING AND MANAGEMENT

S. R. GANGULY

NINETEEN

# ENVIRONMENTAL EDUCATION

LAST two decades witnessed a surge of interests in environmental research. The two facts which seemed to have emerged into salience from this trend are : Firslty, a developed awareness of the need for using the sum of knowledge that is now available to interpret an organism's relationship with the environment; and, secondly, more systematic processing of data by objective methods requiring sophisticated retrieval system and advanced methodology. The former is indicated by the growth of literature in environmental studies as evident from numerous indigenous, organised and state sponsored activities in the field ; the latter envisage development of skills for designed studies in the field to be able to handle the observable facts efficiently without loss of information.

The last observation underlines the importance of a type of training that environmental education could impart from a new perspective. The demand that the nuclear era related to environment-education, are wider and of newer dimensions than before. It is now commonly agreed that the present curriculum in environment education should be recast in order to introduce modus operandi for more dynamism in the process of studying environment. The principle should be introduced even at an early stage of education as preparation for more mature exploratory studies at an adult stage.

**Objectives of environmental education**

It is admitted that objectives of teaching environment ought to focus attention onto formulating a synthesis from a balanced knowledge of an area of habitation, or discovering spatial relationships between objects or natural phenomena interacting with the conditions of living, and in predicting the nature of future problems taking into account of the present events explained. These objectives need be spelt out in pragmatic terms in the classrooms so that a youth is capable of disposing propositional thinking in interpreting data.

Present education ought to be able to gear the system in meeting the needs of the time. Structure of courses and methods of instruction should adapt to the requirements which demand the teaching concepts to be stated with precision, objectives to be set out in detail ; and the guidelines should be formulated in advance. Strategies in environment education ought to lay emphasis on exploratory learning, training for investigation and interpretation of facts in the surroundings.

It is acknowledged that learning is the outcome of a chain of interactions between an organism and the environment. Interactions develop through responses made to physical objects in the surroundings, through social relationships and by social influences. These interactions contribute to the development of an organism, and development takes place through a network of systematic processes. The realization of this fact lays emphasis on the importance of, and the need for, the knowledge of the things which compose an environment. It is the primary task of environment, education to develop in an individual a sense of awareness of the importance of environment. This awareness can develop by the elements of dynamism which a curriculum in environment education can incorporate.

A meaningful awareness of environment ought to provide an individual with a clear concept of physical space, an ability in grasping relationships between phenomena, and an ability in interpreting the observed facts by some underlying casual factors. The approach visualises some form of structured problem solving method of instruction. The method should enable a beginer environmentalist to assimilate a body of information by applying appropriate yardsticks of accuracy, verifiability and coherence in the whole. This logic of scientific approach to knowledge should be accepted as the governing principle in training programme in environment education.

**Cognitive processes**

The concept learning and development model used by Jeter, Jordan and Ingison (1976) joostulates and earliest beginning of discrimination learning of objects at the pre school years extending right through to higher level until concept is attained. The model visualises a course of cognitive development which makes inference about the developmental phases of an individual through the stages of environment education. This could be an operational definition in concept formation through systematic training in a natural setting.

Concept is a label attached to any action, object or a setting which exists in an individual's object world A minimum set of discrimination seemed necessary if only to avoid confusing the objects. The act is performed by language which has a potential economy in defining and categorizing events and things. Being involved in a task like this an individual is given the facilities in which he begins to discover a relation of pattern between the objects, or notice or note the things similar or dissimilar in which group of precepts are also named. A learner would then move a step forward to analyse the data by grouping them into constituent elements, to synthesize by forming parts to a whole, and to apply methods and principles all that is known to him. The learner in this task is engaged in mental activity in regard to the physical objects present and affecting the sensory processes resulting in perceptual thinking. This embodies the dynamic processes in learning leading to concept formation.

It needs be borne in mind, this approach is distinct from a system when the entire learning content is presented to the learner who internalises or incorporates materials to reproduce on recall. Formal setting can provide instantaneous materials which could call for classification and formulating propositions. But a meaningful discovery learning would seem more complex because it involves problem solving stages preceding any meaningful information that emerges before the learner and that can be internalised. This procedure calls for a direct contact with the phenomena under observation, and most often, the field work in environment education.

**Direct contact method**

Direct contact method and discovery approach emphasize upon developing exploratory skills from an early stage. Exploratory skills in environment education could be developed by learning how to use skills of observation, skills in drawing up and using questionnaires, applying survey methods even though in an elementary form. It is characteristic of the young ones that whenever they notice anything unusual, peculiar, or something new to their experience, they start talking about it, collect information about it in bits and pieces. This impulse can be used by sending them individually, or preferably in groups, out to see their surroundings. They then, with some guidelines could be asked to sort out the data into groups or labels selecting a common trait or feature to identify a group. This procedure provides them with an introduction to enumeration and tabulation. It would be interesting to observe and make a schedule of the questions they ask each other on the spot, and how they thrive for solutions to the questions. Later, some guidelines would be of great help to them to phrase the questions, put the questions in sequence, and record their collections methodically.

**Observation techniques**

The techniques of observation constitute the basis of scientific method of enquiry. It is one of the most basic and direct methods of collecting data. The accumulated wealth of information in biology, astronomy, to name a few, is built on systematic observation. Observation may not be applicable in every field, yet the method where applicable draws attention to a problem. The learners must be trained how to observe a phenomenon in an environment in regard to both its physical features and human activities. The young surveyors will of their own ask questions in any new situations, but adequate guidance would help them how to phrase questions, to determine the content of a question, to recognise a factual question. They must know which group of people to visit or objects to observe which make up a sample.

The training should include such devices as note-taking and record keeping, and simple enumeration leading on to complex methods of classification. Aim of a piece of research is description which notes the elements making up a situation that draws attention. Enumeration, classification complete the description by bringing an order into the elements observed. As a systematic enquiry, a pre-requisite to observation is the identification of a problem or a phenomenon. It is just possible that a problem can be identified from an observation of a systematic occurrence of some behaviour. It can often start with questions like : 'What is the true colour of water in the Hooghly'? or 'What is the effect of throwing tonnes of effluent into a tributary near a densely populated area' ? or 'How was the life or living some 4.5 billion years ago in the Antartic now that gas rich meteorites were found by the expedition team whose records of collections tell us of the use of solar wind and solar flares in those days ? Answer to any of these questions may not readily be found by the young environmentalists. What is important is that the training ought to arouse a questioning mind. A question usually starts from curiosity which is invoked from a situation. This impulse then seeks to obtain more complete knowledge of the situation, object or fact that aroused it, often compelling approach and examine its source. At an early stage of environment education this approach is most useful to be utilised by a teacher. The crux in organising a project lies in choosing, objects or situations and designing instructional procedures so that a lesson unit utilised this instinct of the learners of environment.

A prima facie distinction is necessary between an observable fact and stimule both of which occur in the same context or environment. A learner who is observing is usually keen on collecting objective facts and quite often ignores the signs and symbols associated with the facts observed. Signs and symbols are meaningful in that these constitute components of cells which add meaning to the observed phenomenon. An observer, therefore, has to be perceptive, because meaning is obtained through perception of the environmental stimuli. An individual knows his environment through perception, by using his senses. To develop this mental activity learning resources should include those exercises which are based on discrimination and estimation. The method suggested visualizes development of cognition through phases which are interrelated by graded tasks involved in the direct contact.

**Assessment of the environment**

The direct contact method in environmental education facilitates a learner in cultivating perceptual skills to analyse data for himself. The method aims to stimulate an interest to recognise an environmental fact, and to realise that an environmental fact is an aggregate of a multitude of factors. This fact has not taken place in isolation of the rest of the inter-related observable and covert facts. There is evidently a casual relationship between the events which underlie the fact that has happened and been observed. One who proceeds to describe the environmental fact would choose to do so either by making logical deductions or by some empirical procedures. This in fact is the key-note to graded training toward research which should be embodied in the curriculum of environment education. Environment education should prepare one for using some methods of enquiry if only to draw attention to a problem needing solution, or gather information for further interpretation soliciting follow up studies, or even to reinterpret the existing data to gain further knowledge.

Knowledge of fact can be gained from experience which alone can establish the causes underlying an effect observed as fact. A relief can be caused by soil erosion or by many other influences on geological structure ; this fact is likely to affect the habitation, also occupational pattern and welfare of the settlers. For instance, installation of a thermal power station would have some impact on the local weather conditions over the years which might have affected the health and living conditions in the adjoining areas.

Selection of areas of studies can often open up wider issues calling for attention from related fields. For instance, Vaigai river in Madurai which has been polluted due to sewage flowed into the river the banks of which are known to be agriculturally rich.; Cropping pattern in Tungabhadra with consequential changes in the living conditions of the inhabitants or in the Andamons ; or impact on health and hygiene of vegetation grown in a valley where effluent water is used. Instances can only highlight the need for an inter related approach in environment education to meet the research needs of the day.

The single thread which binds environment education with

other disciplines is the knowledge of demographic characteristics which compose the social environment. Training in collecting information on social environment can begin with a small sample survey, may be as a part of the project led by a small group.

**Environmental education for children**

Familiarity of a child first with the immediate environment, then with the wider environment, at an early stage in environment education seems to be essential as an objective from pragmatic point of view. Concept of space in the classroom between the children, between objects, communication, breathing and hygiene, internal environment of the school and its effect on performance and interaction, all these factors provide the child with the knowledge about his surroundings. Realization of spatial relationships between an organism and the objects in a specific surroundings would lead to his concept formation. He would then be able to use his knowledge to interpret the specific environs as now defined in applied climatology. The recent symposium on the subject at Andhra University stressed on the need for this awareness about the effects of physical environment on human beings.

Project work on this type provides the groundwork of training for research in environment education. A nucleus form of this type of activity could even be, for example, know your neighbourhood organised as a small project by the young learners. The locality can be around the school or an area where some children of the same school live nearby each other. They would first specify the area they would like to go about. The group would be expected to draw maps of the streets they pass through, name them, indicate some contour features, locate churches, mosques or temples, post offices, banks, busy shops and markets, typify their residents and their food habits, clothes, culture, and life etc. all or some of these could make up an acceptable project of interest. The leading idea herein is to broaden their outlook, widen their interests, and to acquaint them with procedures regarding how to find facts for themselves. In an activity like this in which observation of facts has been the guiding principle to look for a solution to the stated problem, young learners learn the limitations of personal judgement although individual experience is of great value.

**Aspects of curriculum planning**

The underlying principle in this approach is to engage a learner with activities based on hypothetico—deductive reasoning. An individual is active through his interactions with the object world, when sensory organs and mental activity—subsumed under perceptual skills operate by selection, processing and assimilation. An activity is an aggregate of responses to numerous stimuli in a given task or situation at a given time. Responses of an individual are elicited in hierarchies in an organised activity, like a laboratory experiment, and the concept which is acquired is gained through chain learning. Embedded in a planned project work is the structure of a designed activity in which a group of individuals find solutions to a problem through principled actions reflected in probabilistic thinking. The setting allows an observer to make a supposition by observing a phenomenon and conjectures a number of consequences, and pursue until subsequent observations or some factual evidence can prove or disprove the supposition.

A planned project activity provides the framework of integrated studies. A project based on keeping a school garden is likely to stimulate interests in botany, language, numbers, some history weather conditions, also recreation from local people. In an activity like this individuals learn some cause and effect relations, develop thinking, develop linguistic skills through naming, indexing and writing which could be broken down into grammatical parts.

This type of activity is centred on a problem, and the entire activity is performed in a natural setting. There is an element of theoretic in this, also much of it is practicum in that it involves chain of activities in which principles are applied. Curriculum planners have to take account of the factors which influence decisions on farming objectives and content areas. Curriculum ought to suit local needs, and this premise would allow enough variations within a geographical boundary. The needs for environment education in an urban setting are not always the same as in the rural setting.

Both teachers and curriculum planners in environment education have great responsibilities in this regard. Lessons should be given in such a way shat a learner would have freedom in experimenting with information given him before an assimilation is likely to take place towards building up a proposition. The exercise should end up with preparing a report of activity, in which they would, with guidance, be able to set out in detail how they started the project, describe what was studied, spell out how the study was organised, specify what their conjectures about consequences were, and state clearly what was found out. Teaching in this way ought to indicate in a learner self-confidence in discovering the facts for himself which is uncharacteristic of formal classroom lessons.

## Conclusion

Activities in environment education make it evidently clear that training in research can begin early, so that maturity is attained at a later stage to take up a formal study. This must be viewed as development through the stages of complexity of knowledge. The method discussed above described how research topics can arise and research interest can develop at a nascent stage, even from things which are not usually cared for. Activities in the natural setting which is stipulated in the direct contact method may be seen as a systematic attempt to gain preparatory knowledge of what could be expected of a researcher in a serious study. The problem which is to study at this level may even remain more vague but thinking is translated through operational activities. It is on this premise that research at an advanced level can build on skills already attained. At the elementary level learners are able to propose some formulations or even make some conjectures which are transformed through experience at an adult stage to be more refined as hypotheses. Application of deductive reasoning is implicit in a designed activity, as it is characterized by argument for a solution to a problem which is 'a priori' in research. A clear cut viewpoint on the methods of teaching environment is imperative. A method that postulates organizing knowledge from simplified wholes through the complex wholes by discovery approach should be given priority in the light of the current explosion of knowledge.

## REFERENCES

**Ausubel, D.P. (1968),** *Educational Psychology: A Cognitive View,* Holt,

Rinchart and Winston, Inc. New York.

Jeter, J., A. Jordan, (1976), "Environment Education : Needs and Specifications", Research and Development Centre for Cognitive Learning. Working Paper No. 150. University of Wisconsin.

Helmstadter, G.C. (1970), *Research Concepts in Human Behaviour*, Appleton-Century-Crafts, New York.

K. S. GOPALAKRISHNAN
AND
M. SAMBASIVA RAO

TWENTY

# ENVIRONMENTAL ASSESSMENT AND MONITORING

MANY of the natural environmental problems that have plagued man are flooding, erosion and siltation. Because of the rapidly expanding population and increase in modern technology and industries, the quality of evironment is suffering great deterioration. The population explosion also has caused rapid and widespread urbanization and has forced revolutionary changes on the human scene. Problems of population and waste disposal have emerged. Energy requirements have grown even faster than population and its generation often changes the natural landscape. The production of hydroelectric energy requires construction of dams and the impounding reservoirs which are responsible for man-made changes in topography. The development of extensive road network in the modern time is another major change. Natural drainage is blocked or diverted into low lying areas.

In coastal regions heavy withdrawal of groundwater has given way for encroachment of seawater in coastal freshwater aquifers. The depressions formed in the Bay of Bengal during monsoon periods have been the main sources of flooding in river valleys, delta plains and coastal plains in southern parts of the India. The misuse of land and deforestation have accelerated erosion in uplands and hilly terrain. The artificial structures constructed by man across major rivers to generate hydroelectricity and to provide assured irrigational facilities have blocked the natural flow of water in rivers and the suspended sediments carried over with water have created siltation problems in reserviors. The fertile valleys have been impounded to accommodate blocked water. The stagnation of water in reservoirs have made it as a place of breeding of mosquitoes which in turn have led to typhoid and malarial diseases. Likewise many problems have appeared because of mans interference with natural ecosystem. Therefore, in the present study an attempt is made to study in brief various environmental problems with special reference to Cauvery basin.

The Cauvery basin* covering an area of 89,000 km$^2$ lies in the states of Karnataka. Kerala and Tamilnadu. The river Cauvery is met by 16 major tributaries and flows almost ENE-WSW in direction from its inception point to Stanley reservoir and then takes a turn towards N-S, WNW-ESE, E-W and NE-SW direction before emptying into Bay of Bengal. The river debouches into the sea with two major distributories namely river Coleroon and river Cauvery. River Vettar, Vennar and Kudamkatar are other minor distributaries. Physiographically the basin is composed of ridge and valley topography on western and central parts with plateaus in between and undulating terrain, rolling, plains, fluvial plains, delta plains and coastal plains on the east. The maximum elevation of 2,637 m is noticed in Nilgris. The Karnataka plateau has an average elevation of about 900 m above Mean Sea Level. The eastern Tamilnadu plains lie below 300 m.

The main objectives of the present study are : to map major geomorphic units of the basin, to delineate zones of erosion, and flood affected areas, to study the nature of distribution of urban centres, and to assess the impact of man's role in changing the face of natural ecosystem of the basin.

**Methodology**

Modern remote sensing techniques have been adopted to delineate various geomorphic units based on graytonal variations, alignment parallel to river courses, hilly terrain or coastline, textural characteristics, landuse and various geomorphic processes involved in their formation. The erosion appraisal of the basis also has been studied from LANDSATS based on drainage and soil formations, gullying, degree of weathering, geomorphic processes and agents involved, vegetal cover, slope, human inter-

* The basin lies within 75° 45′ to 80° 0′E latitude and 10° to 13° 45′ N longitude.

ferance and intensity of precipitation. The flood prone zones are identified based on relief in river valleys, delta plains and coastal plains, location of old or ancient channels, discharge of water during peak flood periods, severity of depressions etc. The major urban centres located in river basin are identified from LANDSATS and their geographical locations, are plotted. The data on population has been taken from provisional Census Hand Book of Tamilnadu, Karnataka and Kerala for the year 1981. Based on the above data an attempt is made to highlighten the man made changes in the face of River Cauvery and major non-urban environmental problems in the Cauvery river basin.

**Environmental geomorphology**

The recent developments in environmental science is to understand the process and agents responsible for deterioration in quality of environment and to suggest a few remedial measures to conserve and protect the same from further deterioration. An attempt has been made by Verstappen and Zuidam (1975), Rao (1975) and Prabhakar Rao (1981) on morphoconservation using serial photographs, and LANDSATS based on degree of weathering, geomorphic units, processes and agents, vegetal cover, lighology relief, human interference, intensity of precipitation, tectonic activity, thermal efficiency etc. in different parts of the world. Their studies proved to be more useful to identify nature of erosion on different gemorphic units. Coates (1973) has also given a detailed account on the role of geomorphic units, processes and agents in changing the face of earth along with physical and climatic elements and human interference. Very often the misuse of land and human interference may lead to adverse environmental effects and destruction of soil and terrain.

Based on the above said characteristics, a similar attempt has been made to describe the relationship between environment and various geomorphic units in the Cauvery River Basin. The major geomorphic units, based on the processes of their formation have been classified into denudational, fluvial, aeolian and coastal landforms. They include both destructive and constructive landforms. The broad geomorphic units are hilly terrain with ridge and valley topography, plateaus, piedmont plains, pediments or pediplains, creep built plains, wash plains, sandstone lateritic plains, river built plains, leavees, flood plains, delta plains, ancient channels, beach ridges, mangroves, marshes, coastal dunes and aeolian deposits. Each geomorphic unit based on the processes involved in their formation, degree of weathering, altitudinal location, slope, climatic conditions, vegetal cover, tectonic activity and agents responsible for removing the materials is being subjected to erosion. There are two types of erosion. One is normal erosion which induces erosion under normal environmental conditions and the later is accelerated erosion which is caused due to man made changes.

The hilly terrain with ridge and valley topography is being subjected to both normal and accelerated erosion. The normal erosion is noticed in the parts of hilly terrain where forests are well preserved. They are reflected in LANDSATS (bands 5 & 7) in dark graytones. Accelerated erosion is noticed in parts of hilly terrain where deforestation has been carried out for commercial purposes, plantation of crops like cardamum, coffee, tea, rubber, fruits and vegetables. In these pockets in the air photos and LANDSATS deep gulleying pattern has been noticed on debris slopes of hilly terrain. The drainage density is very high. Similarly parts of plateaus (Mysore, Mangalore and Hosur plateaus) with undulating topography is being subjected to accelerated erosion. Very often the misuse of land without adopting soil conservation methods has enhanced the intensity of erosion. The pediments or pediplains, and parts of creep built plains in Coimbatore and Salem districts are also being subjected to severe erosion, possibly due to misuse of land and high thermal efficiency. However, the intensity of erosion on creep built plains and wash plains is moderate due to low degree of slope and even terrain. In the fluvial plains due to low degree of slope and even terrain and which are encountered on either side of the major tributaries and distributaries the erosion is very low. Similarly on the delta plains, located in Thanjavur district the erosion is very low. Aeolion action in coastal plains is responsible for migration of dunes into inland. The sandstone lateritic uplands located in Tiruchi and Thanjavur districts in the lower pockets of the basin are also being subjected to moderate to severe erosion due to misuse of land, low water availability and high thermal efficiency.

In addition to accelerated erosion in hilly terrain, of Nilgiris landslides, mudflows, subsidence are very common because of heavy rainfall during monsoon periods. According to Gopalakrishnan *et al.* (1980), sixty major and 350 minor landslides were occured during the year 1978 alone. Due to frequent occurrence of landslides in this pocket of hills in Nilgiris, there were a few deaths reported. Hindering of traffic, damage of metallic roads and crops are common. Environmental hazards in different land units are possibly due to unprecidented rainfall, high thermal efficiency, deforestation, quarrying, random escavations of silt and clay in colluvial fans, quarrying of kopjes, piedmont plains, debris slopes etc. and misuse of land without adopting proper conservation methods. Therefore a detailed study on airphotos on scale 1 : 60,000 is urged for environmental monitoring so that the available environmental potential are used in a planned manner without causing further environmental imbalance. In order to avoid further deterioration a few remedial measures are suggested.

(a) Quarrying may be carried out in denudational hills for building materials instead of kopjes in piedmont zones.

(b) The debris slopes and dissected piedmont plains and pediplains may be utilised for afforestation, pasture, development and orchards after properly levelling the land.

(c) In pockets where deforestation has been carried out on parts of hilly terrain, afforestation of trees of commercial value (teak, sandlewood, blackwood, Ucalyptus, Kucsamum etc.) may increase the economic potential of the forests. Besides afforestation in phased manner careful soil conversation methods have to be adopted to minimise soil erosion till the plants reach the maturity stage by planting selvi pastures.

(d) Similarly soil conservation practices have to be adopted in piedmont plains, creep built plains, colluvial fills and sandstone lateritic plains, where quarrying is intensive for Kankar and regur.

(e) To minimise siltation in the reservoirs, detailed studies have to be carried out in different watersheds of the Cauvery basin to estimate the amount of sediment deposited into the reservoirs annually. Installation of silt monitoring stations in major watersheds will help to measure the amount of soil erosion accurately.

(f) The valley fills, burried piedmonts, colluvial fills, terrace plains, piedmont plains, ancient channels in fluvial and delta plains form as good sources for escavation of groundwater resources.

Flood is a common feature in the river valleys, delta plains and coastal plains during monsoon periods. No doubt, construction of dams across major rivers has minimised the frequency of occurrence of floods in the river valleys. However, the cyclonic activity during north-east monsoon period is causing heavy damage to the paddy, banana and sugarcane crops in the delta and coastal plains during the months of October, November and December. The loss will be maximum when the depressions crossed the coast in and around Cauvery delta. Therefore, a detailed analysis of rainfall, distribution, intensity, variability etc. have to be carried out to estimate the amount of floods in the lower valley, delta and coastal plains. Flood appraisal studies have been carried out by Sambasiva Rao (1982) on Cauvery delta using air photos and LANDSATS. He has maped the zones subjected to floods and concluded that flood intensity is very sever in an around Coleroon river, Cauvery and Vannar systems. The intensity is low to moderate in western and southern parts of the delta plains. Detailed studies using large scale airphotos and climatic data will help to delineate the zones subjected to floods during monsoon periods. An attempt has also been made by Gopalakrishnan *et al* (1980) to study the flood control and the long-term trends in land and water utilisation in Nilgiris.

**Urban landscape**

An analysis of distribution of urban centres in the river basin depicts that in the lower parts of the basin covering Thanjavur, Tiruchi, Coimbatore, and Salem districts in Tamilnadu the concentration of urban centres is very high. This is possibly due to geographical location of the districts in the plain regions below 300 m with even topography. The concentration is also very high along the river valleys in Karnataka. Due to rugged terrain the concentration of urban centres in hilly terrain is very low. However, the concentration of urban centres in Karnataka plateau is moderate due to undulating nature of terrain. In this plateau the important towns are Merecara, Mysore, Mandya, Hassar, Chikmagalur, Chinna Patnam, Arsikere and Tumkur. The population varies from 20 to 100 thousands. Bangalore is the only metropolitan city in the Karnataka plateau with a population of 2.9 million lying just on the north eastern pheriphery of the basin. The Class I towns in Tamilnadu plains are Coimbatore, Salem, Erode, Dindigul, Tiruchi, Thanjavur and Kumbakonam. In these towns the population ranges from 200 to 950 thousand. The understanding of distribution of urban centres and major industrial units located in the Cauvery basin is important to study the impact of their growth and development along the major tributaries in the river valleys. It has been found that many of the major industrial units have been discharging their effluents into the rivers. Similarly, there is a high discharge of sewage water from various urban centres into the rivers. The discharge of both industrial effluents and sewage water is polluting the water in different zones in the rivers. The waters of Cauvery is highly polluted because of a variety of industries located in its catchment area. Among them, paper, chemical and sugar industries are worth mentioning. The effluent of paper industries near Mysore, Nanjemkad and Erode cause a great deterioration in the quality of water. This was further intensified by a number of sugar and chemical units located in the Mysore, Mandyas, Periyar, Salem and Tiruchi districts. Therefore, the study of geographical distribution of urban centres and major industrial units in the Cauvery basin and the amount of industrial effluents and sewage water discharged into the river in detail will help to monitor the environmental hazards created by man. The further study in this direction is under progress.

**Environmental change : Natural and human**

The rapid growth of population within the twentieth century has forced widespread changes in the natural system. The food and energy requirements have increased enormously. To overcome the problems many artificial structures have been constructed across the rivers, to extend the areas under canal irrigation, to produce hydro-electricity and to control floods. The construction of dams across rivers has led to impounding of rich and fertile river valleys and checked the natural flow of water. Added to this the tribal people were forced to vaccate the river valleys and settle elsewhere. The checking of natural flow of water has deposited huge quantities of silty soils into the reservoirs reducing gradually reservoir's capacity. This siltation has become one of the important problem. In the downstreams there is shrinkage in the size of rivers and very often man has occupied the river bluffs and river islands for cultivation purposes. This process is well depicted in the Cauvery distributary. The large scale deforestation in the hilly terrain in order to meet the fuel demand has also led to change in micro-climatic conditions and rapid depletion of organic rich soils. Erosion has been accelerated. The mismanagement of land for the cultivation of potatoes, fruits and other vegetables on the debris slopes, piedmont plains without adopting soil conservation methods has also accelerated erosion. The widespread deforestation carried out in hilly terrain also disturbed the natural ecosystem.

The rapid urbanization and industrialization processes on either side of the river valley has resulted in deterioration of water quality in the rivers and disturbed the aquatic ecosystems. The sugar industries in and around Mandya in Karnataka, paper manufacturing industries at Pallipalayam near Erode and distillaries in Tiruchi in Tamilnadu are a few other examples located on the banks of the river, discharging the effluents into the river. The rapid growth of population in the Ootacamund town and its tourists significance has also led to shrinkage of Doty lake recently due to encroachment and disposal of waste products into the lake. The eastern portion of the lake was converted into a race course in the past (Gopalakrishnan *et al*, 1980). The rapid development of industries in and around Ootacamund has also resulted to widespread deforestation. The concentration of macro industries in and around Bangalore and rapid growth of population has resulted to various environmental problems and health hazards. The management of sewage water, distribution of sufficient quantity of chlorinated water, solid waste disposal, traffic, congestion, urban poor, housing problem air pollution, water pollution, noise pollution, transportation problem, etc. have emerged and focused the environmentalists to think for remedial measures. Likewise many changes have been made by man in the natural Cauvery system, which further needs a detailed analysis.

## Conclusion

The Cauvery river basin covering an area of about 89,000 km$^2$ has been studied with view to map various geomorphic units and to bring out non-urban environmental problems because of man's interference with natural ecosystem. The broad land systems are hilly terrain with ridge and valley topography, undulating terrain, rolling plains, fluvial plains, delta plains and coastal plains. Based on the studies from LANDSATS (1 : 250,000) of bands 5 and 7, drainage and textural characteristics, lithological and soil formations, degree of weathering, geomorphic processes and agents involved, vegetal cover, slope, human interference and intensity of precipitation the zones subjected to erosion, floods and siltation are mapped. Finally based on distribution of urban centres and industries an attempt was made to describe how to monitor the various environmental problems and man made changes in the river basin.

## REFERENCES


Donald, R. Coates (1973) : *Environmental Geomorphology and Landscape Conservation*, Dowden, Hutchinson and Rose Inc. Stroudsburg, Pennsylvania, III. Non-urban, 483.

Gopalakrishnan, K.S. (1973) : Evolution of settlements in delta, *The Indian Geographical Journal*, XLVIII, (2). 65-75.

Gopalakrishnan, K.S., R. Manner Mannan and P. Sampath (1980) : Flood control in Nilgiris-longterm trends in land and water utilisation, symposium on water resources of the southern states, Department of Geography, Madurai Kamaraj University, Madurai.

Prabhakara Rao, P. (1981) : "Applied Geographic Studies in the Semiarid Tracts of Anantapur district" ; unpublished Ph.D. thesis of Andhra University, Waltair, India.

Roa, D.P. (1975) : Applied geomorphological mapping for erosion surveys ; the example of the Oliver basin, Calabria, *I.T.C. Journal*, 3.

Sambasiva Rao, M. (1982) : Flood intensity and land evaluation in the Cauvery delta, Tamil nadu, applying modern remote sansing techniques : *Photonirvachack*, 10 (3), 23-30.

Verstappen, H. Th and R.A. Zuidam, (1975) : ITC system of geomorphological surveys, in *Text book photointerpretation*, ITC, The Netherlands, III.

Prithvish Nag
and
Anima Dutta

TWENTY ONE

# ISSUES IN ENVIRONMENTAL MAPPING

Most of the geographers consider maps as an important aspect for the completion of a study or research. For the study of environment, the maps play a greater role, more particularly for considering the interrelated aspects in space. Here the model building techniques and statistical methods can only partially explain the issues involved. The International Geographical Union (IGU) has several commissions and working groups on environmental related aspects. The working group on Cartographic de l'Environment et de sa Dynamique and another working groups jointly with the International Cartographic Association (ICA) on Environmental Maps and Atlases have been particularly working in this thrust area. Other IGU and ICA bodies are directly or indirectly involved in mapping the environment. Recently, the International Council of Scientific Unions (ICSU) has identified two major thrust areas. They are (a) thematic mapping, and (b) monitoring environmental change. The issues involved are directly related to environmental mapping. There has been an attempt to work out a cartographic base for environmental mapping at a global level. Geographers and cartographers engaged in the study of the environment can not possibly remain indifferent to the global and national concerns. In the International Seminar on Environmental Maps and Atlases held in Calcutta in 1983, different aspects of environmental mapping were discussed. These aspects can be grouped the following headings :

(a) Rural and urban environment: Assessment and mapping,
(b) Environmental planning in the context of development,
(c) Perception and mapping of the environment,
(d) Environmental hazards—assessment and mapping, and
(e) Remote sensing and environmental mapping.

It was recommended that different agencies collecting data by field surveys and other methods should coordinate their activities which would help in the formation of a data bank. Though, National Natural Resources Management System (NNRMS) exit in the country, the necessity of having a cartographic institution like the National Atlas and Thematic Mapping Organisation (NATMO) as 'nodal agency' was felt.

**What is an environmental map ?**

Several attempts have been made to define an environmental map, or to highlight the issues involved in environmental mapping, or to discuss the contents of an environmental atlas. The aspects related to the study of environment by geographers have been discussed elsewhere. What is an environmental map ? How it is different from other maps ? The definition of "environmental map" is a delicate point, for without some limitation and in the absence of convention, we can run the risk of reviewing virtually the whole scope of thematic cartography (Vazquez Maure, 1980). In a meeting of the I.G.U./I.C.A. Working Group on Environmental Maps and Atlases, this aspect was discussed at a greater length. The environmental map was defined as follows :

> We have understood initially by the 'environment' those geographical circumstances which influence (pressure, ameliorate or degrade) the human ecological system. Atlases of the environment are related to the work of other scientific disciplines and display cartographically the elements in and complexity of the environment for a region, or country or a group of regions or countries (Vazquez Maure, 1980).

Though this definition is imprecise, the emphasis is on the cartographic display of the work of other scientific disciplines related to environment. Environmental mapping should not be considered as a sub-discipline trailing behind the developments in other scientific disciplines, but there is a feedback to such scientific researches from the spatial analysis of the related aspects as well. Such feedback has become important in the critical study of the environment-related disciplines. However, there are general users of environmental maps who are concerned with more basic issues like wildlife, forests, climate and the like.

**Environmental cartography**

The information about the environment is increasing every day and this phenomenon is found all over. Attempts have been made to establish 'environmental information system' for a region or a nation. Remote sensing techniques have become a major and most regular source of information for environmental mapping. The basic issue in environmental mapping is no longer the scarcity of data, but selectivity of data is posing problems for the environmental cartographers. Information is now available on almost all conceivable issues. This brings into the question of depicting the environment in its totality. Whether a single map should be accepted comprehensive for mapping the environment, or a series of maps are required on interrelated aspects in order to make a complete enquiry into the subject ? If one conducts a survey concerning this question, he will find that environmental mapping in a form of atlas is considered more appropriate. However, single type of maps have also been produced but mostly for experimental purposes.

On the other hand, the issues in environmental cartography is not altogether different from other specialized fields of map-making. For example, the choice of scale, generalization, interapolation, symbolization, selection of colour etc. are to be worked out before discussing the design and contents of a map or an atlas. With the help of same environmental data cartographers can produce different map images by selecting different combinations of cartographic parameters. Hence, for the purpose of environmental assessment, the user has to be trained in interpreting these maps. Obviously, the issue of matching the scale of the map to the users' requirements is of fundamental importance. "Environmental assessment is iterative by nature, that is, alternatives are studied in a multi-staged or progressive screening process which compares a large number of alternatives initially (e.g. at a scale of 1 : 250 000), and at each successive stage the number of alternatives being compared decrease while the analysis becomes more detailed (e.g. at a scale of 1 : 2 000). It is well known cartographic rule that a change of scale or purpose of a map usually necessitates going back to the initial compilation stage and starting again" (Ives, 1980). Morrison (1980) has suggested for the improvement and expanded use of feedback loop from the map percipient to the cartographer. Such feedback should be taken into account in environmental cartography.

**Technological aspects in environmental mapping**

Environmental cartography oriented towards production of atlases, obviously need a data bank with computer facilities. The conventional and non-conventional data are now available in large volume and with high precision. An environmental cartographer cannot ignore the availability of such information. For environmental mapping, there is an additional source of data which is based on laboratory work. Regarding the mapping of the air pollution, for example, $SO_2$. CO, oxides of Nitrogen ($NO_2$), ozone (Oz), lead (Pb), dust etc,. components have to be worked out. Similarly, for water pollution, organic and inorganic contents, bio-chemical oxygen demand etc. have to be worked out in scientific laboratories.

Considering the volume of data to be handled, it is becoming inreasingly difficult to prepare environmental maps by conventional methods. Quick processing and plotting of data of all sorts have become essential. Further, the issues for mapping are not necessarily concerned with the static components of the environment, such as relief or climate. Area-specific, problem-specific and time-specific issues have become more significant for mapping. Technological advancements have been made to meet these challenges. However, such advancements are not always free from limitations, e.g. the cloud cover creates problems for collecting information from air photographs or imageries for the mapping of the flooded areas.

Most of the information can now be available in digital form or can be converted into digital form in CCT tapes. The boundaries can be processed by vector or raster method into this form. This technological advancement has increased the potentiality of producing maps at a faster speed. The production of maps within a reasonable time has been considered essential owing to the spatio-temporal issues now important in environmental mapping. In fact, such advancements have been made in the whole processing and finally to printing or producing a hard copy.

**Some environmental issues**

In order to understand the environment one has to make a positive approach and carry out objective studies and analysis of the components that act collectively to create different sets of environmental conditions (Dutt, 1982). According to him, this approach includes : (a) natural environment, (b) natural hazards, (c) man-made environment, (d) population, (e) water distribution and population, (f) air pollution, and (g) fertilizers, insecticides and pesticides. Das Gupta (1982) has worked out a list of maps for a proposed structure of an environmental atlas. These are : (a) terristrial environment, (b) climate, (c) ecology, (d) environment, (e) human environmental hazards, and (g) pollution. Probably, this scheme is more suitable for Indian situation.

Possibly the world's first simple environmental atlas was published in 1973 in Hungary. This atlas includes nine plates and 78 text pages (Popp-Vary, 1980). The *Atlas of the Present State of Environment in Hungary* at a scale of 1 : 450 000 consists of the following subjects :

(a) Territorial division of big factories strongly deteriorating the air,
(b) Estimated division of sulpher dioxide pollution in the air,
(c) Division of medium and large pollution sources by towns,
(d) Division of industrial areas and air pollution observation posts in Buuapest,
(e) Graph : The annual mean of sulpher and dust content of air in Budapest by observation posts,
(f) Quality of superficial waters,
(g) Division of agricultural lands with unfavourable conditions,
(h) Forests having primarily at present and future a recreational function, and
(i) The protected natural and landscape patterns.

Most of the issues are representative factors of the whole zone of consideration of the environment as thought important in a situation like Hungary. Conceptual development in last 12 years has raised several new issues which will now find place in environmental atlas. The Environmental Agency of the Central Government of Japan has been involved in environmental study from a long time. It has a programme to prepare vegetation maps and distribution maps of animals and plants at a scale of 1 : 200 000. The same agency also plans to produce the *Atlas of Natural Environment* for the Japanese Archipelago. Obviously, in this atlas, the physical and natural aspects have been basically considered. Some of the important aspects identified for mapping are coastal changes, edible seaweeds, coral reefs, oceanic pollution etc. (Masai, 1980). National Atlas and Thematic Mapping Organisation is also planning to prepare an *Environmental Atlas of India.* An attempt in this direction is shown in Fig. 21.1.

The preparation of environmental atlases has been a rare phenomenon. Models have been generated which are meant to explain how the environment should be mapped. On the other hand, there has been a lot of study on environment which includes the preparation of maps. The feedback from such studies has been substantial, but not systematic. As a result, different aspects have been selected, say for the assessment and mapping of the urban environment. The methodology and orientation are different as well. The technology used in mapping has also been responsible for producing maps on uncommon aspects of the environment. For example, with the help of remote sensing (MSS imageries), following aspects were mapped for urban environment of Japan : (a) green space ratio, (b) duration of sunshine, (c) sky amount, (d) tree vigor, (e) urban climate, and (f) land conditions. Evaluated values were put into National Standard Grid System ($\frac{1}{4}$ divided mesh : 250×250 m) (Kitani & Fuchimoto, 1980).

Though a few atlases on environment have been prepared, there are maps or plates on environment as a part of a larger mapping programme. Almost all the atlases have been made by the I.G.U./I.C.A. Working Group on Environmental Maps and Atlases regarding environmental maps in nineteen atlases (mostly national atlases). In this study special attempt was made to examine maps on : (a) vegetation, (b) environmental pollution aspects, and (c) fauna. A detailed discussion has been provided

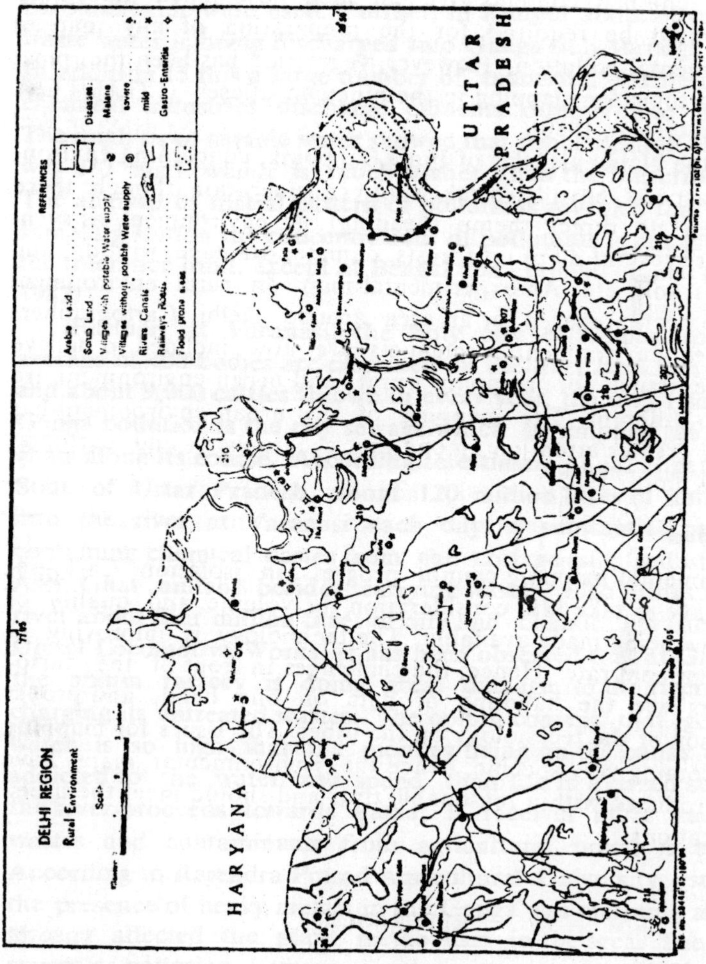


FIG. 21.1 : *Delhi Region : Rural Environment*

about the map contents and description according to different atlases. The above major themes were further classified into several other themes. For example, environmental pollution maps were examined for : (a) water, (b) air, and (c) soil pollution aspects. The map contents give an idea how much laboratory work would be required for the preparation of any map on environmental pollution. However, vegetation has been the most common topic for mapping in the nineteen atlases (I.G.U./I.C.A Working Group, 1980).

In the National Atlas of India, mapping, concerning environmental aspects, have been taken into consideration, though there has been no direct theme (heading) on environment as such. There are several maps on forests, climate, drainage, wildlife and wetlands, soils etc. An experimental map on rural environment has also been prepared for an area around Delhi. Cartographers and other environmental scientists are now inclined to believe that the issues for mapping rural and urban environment are entirely different. The mapping of the urban environment has drawn a greater attention among the geographers and cartographers (Kumra, 1982).

## Conclusion

Environmental mapping cannot progress in isolation. Cartographers have to take into consideration the volume and quality of data now being made available. The technology is improving at an exceptional rate. Hence, the challenges in front of the cartographers are the handling of data on one hand, and proper utilization of the technology on the other. The issues for mapping are becoming very specific, hence the environmental maps have to be produced within reasonable time and should meet the users' specifications.

## REFERENCES

Das Gupta, S.P. (1982) : Structure of environmental atlas, in Das Gupta (ed.), *Essays in Thematic Cartography*, National Atlas and Thematic Mapping Organisation, Monograph 2, Calcutta.

Dutt, G.K. (1982): Mapping the environment, *International Journal on Environmental Studies*, 19, 25-33.

I.G.U./I.C.A. Working Group on Environmental Maps and Atlases (1980) :

Samples of a Documentational Scheme Concerning Environmental Atlases and Maps, Utrecht.

Ives, Laura (1980) : The utilization of maps for environmental assessment, paper presented in the 10th I.C.A. Conference, Tokyo.

Kitani, Yukio and Fuchimoto, Masataka (1980) : Urban environmental survey by remote sensing and representation of the results, paper presented to the 10th I.C.A Conference, Tokyo.

Kumra, V.K. (1982) : *Environmental Problems in Kanpur City*, Varanasi.

Masai, Yasuo (1980) : Environmental atlases and maps of Japan, paper presented to the 10th I.C.A. Conference, Tokyo.

Morrison, J.L. (1976) : The relevance of some psychophysical cartographic research and basic map reading tasks, paper presented to the I.C.A. Conference, Moscow.

Nag, P. and *et al.* (1985): Mapping of the environmental risk zones, in *Assessment and Mapping of the Environment*, National Atlas and Thematic Mapping Organisation, Monograph 4, Calcutta (in Press).

National Atlas and Thematic Mapping Organisation of India (1982) : *National Atlas of India*, Department of Science and Technology, Calcutta.

Papp-Vary, Arpad (1980) : Environmental Cartography, paper presented to the 10th I.C.A. Conference, Tokyo.

Vazqel Maure, F. (1980) : Examples of environmental maps, in I.G.U /I.C.A. Working Group on Environmental Maps and Atlases, *Examples of Environmental Maps*, Institute Geografico Nacional, Madrid.

G. N. Pandey

## TWENTY TWO

# ENVIRONMENTAL POLLUTION
## Monitoring and Control

INDUSTRIALIZATION is considered a barometer of civilization. On one side industrialization brings in wealth and prosperity, and on the other it deteriorates the environment, we live in. For quite some time it was considered a problem of developed nations, but since the Stockhom Conference (1972), it has taken a new dimension with a real concern for effective control on environmental degradation throughout the world including the developing countries.

In 1974, the Government of India also decided to go in for environmental management based on certain standards. It was decided to set up a 'Central Board' and 'State Boards' for the prevention and control of water pollution. Both, the Central Board and several State Boards for prevention and control of water pollution adopted the standards provided by the Indian Standards Institute (I.S.I.). The standards adopted by Uttar Pradesh Pollution Control Board for prevention and control of water pollution are also on a similar pattern. A closer evaluation of the standards clearly indicates that the several pollutants including organic pollutants have not been given proper place in the Act. In 1981, it was decided to enlarge the scope of pollution control in order to include air pollution control as well. In February 1983, ambient air quality standards were finalized and notified by Central Board for Prevention and Control of Air Pollution. The emission standards for different industrial environments are yet to be finalized and notified.

Recently, there is a growing concern to develop multidisciplinary approach to study the environmental degradation. For obvious reasons, the growing population is the main cause for pollution, specially in developing countries. The increasing population necessitates higher production of agricultural products by suitable imputs of fertilizers and pesticides, higher industrial production, to provide the necessities to the growing populats e.g., cloth, medicines, sugar, iron and steel, and energy production, etc. In one of the recent evaluation of pollution loads, specially organic load on river system, it was noticed that 90 per cent is contributed by sewages, and only 10 per cent from industries. Out of 10 per cent, 3 per cent is contributed by sewage and only 7 per cent mainly by large industries. It is, however, essential that even industrial pollution should be controlled on a priority basis.

It is generally observed that there are only a few types of industries which contribute significantly to environmental pollution. For example, in one of the largest states of the country, e.g., Uttar Pradesh, out of 252 major polluting industries about 35 per cent are covered only by sugar, distillery, pulp and paper, tanneries, and textiles.

### The origin of waste

One characteristic, peculiar to man, is that he progressively changes his environment to meet his biological and social needs. On a socially organized basis man provides himself with the materials necessary for life which he removes initially as raw materials from his environment. It is in the provision and utilization of these material necessities, that worthless and sometimes harmful byproducts originate. Such products go to make up wastes of all description and they are the inevitable results of man's interaction with his physical and biological surroundings. If not removed, they reduce the wholesomeness and quality of these surroundings. To separate waste from environment, it has been one of the man's oldest activities.

In primitive nomad societies the problem of waste disposal is at its least. The nomad separates waste from environment by changing his surrounding. It is when man exists in permanent communities that the removal of his waste products must be conducted on an organised basis. It is of interest to note that it was the transition from Palaeolithic society based on hunting and gathering to Neolithic society based on farming and agriculture which heralded the need of waste disposal. But it has been the excess of the industrial revolution and its aftermath up to the present day which has resulted in a qualitatively different type of problem. Although waste disposal must, as a minimum requirement, conserve the environment's ability to sustain human life, the degree to which it practised at levels above this minimum depends on the economic, political, cultural and aesthetic values of the waste producing society.

**Pollutants**

Wastes are not necessarily pollutants in themselves but all have the capacity to be so. In exactly the same way that there is no single taxonomic group of plants which can be considered to be weeds, but rather the designation 'weed' depends on the locality occupied by the plant, so a waste becomes a pollutant when it occurs in the wrong place. It must be appreciated that production of waste is an inevitable consequence of the first and second law of thermodynamics ; the production of actual pollutants is far from inevitable.

The following definition of a pollutant has been given by Werner :

> A substance or effect is normally considered to be a pollutant if it adversely alters the environment by changing the growth rate of a species, interferes with the food chain, is toxic, or interferes with health, comfort, amenities, or property values of people.' Generally, a pollutant is a substance or effect introduced into the environment in-significant amounts as sewage, waste, accidental discharge, or as a byproduct of manufacturing process or other human activity. A pollution substance can be a solid, semi-solid, liquid, gas or sub-molecular particle. A polluting effect is normally, some kind of waste energy such as heat, noise of vibration.

**Pollution**

Pollution is the result of the action or presence of pollutant in a part of the environment where it is considered to have deleterious effects. Pollution comes from getting rid of wastes at the least possible cost. The Royal Commission of Environmental Pollution in its Fifth Report has defined the pollution as follows :

> Pollution occurs when as a results of man's activities, enough of a substance is present in the environment to have harmful effects. Many substances which can become pollutants are present naturally in the environment in lesser amounts and may be beneficial or even essential to it.

**Monitoring**

Monitoring of pollutants for environmental analysis may be carried out continuously or intermitantly, although with the advent of monitoring units which are well provided with funds, the use of continuous monitoring systems has increased. Monitoring systems basically fall into five categories :

*Instrumental systems for measuring physical properties*

The following equipments are used :

(i) Sound level meters,
(ii) Geiger counters,
(iii) Thermometers,
(iv) Suggest replace this with oxygen meter,
(v) Hydrometers, and
(vi) Suggest colorimeter.

*Instrumental methods of chemical analysis*

Until recently these were restricted to the research laboratory, but now they are used quite extensively in the environmental field. Examples of instrumental methods currently used in environmental analysis are :

(i) Atomic absorption spectroscopy,
(ii) Infra-red spectrophotometry, and
(iii) Gas-chromatography.

*Wet method of chemical analysis*

This is an extensive field which is not quite so important today as it once was, as many of the traditional techniques have been replaced by instrumental methods. Three important wet techniques are :

(i) Volumetric analysis,
(ii) Gravimetric analysis, and
(iii) Colorimetric analysis.

*Biological systems of monitoring*

The type and number of living organisms are used to assess the quality of an environment. Biological systems are referred to as environmental indicators (integrators). They sum-up the entire effect of all the pollutants present in the locality. For example, the use of lichens to estimate concentrations of sulphur dioxide in the air.

*Hybrid systems*

In practice, many of the complex analytical systems used employ elements from two or more of the above. Thus, the Biochemical Oxygen Demand (BOD) test is a hybrid of either a wet chemical technique and a biological system, or an instrumental chemical system and a biological system.

**Control**

There are basically four aspects for controlling the pollution :

*Setting of standards*

By observation and experiment, relationships are established between the intensity, duration and frequency of exposure to pollutants and the risk or magnitude of an undesirable effect occurring in the environment. When exposure levels are low and long term, it becomes very difficult to accurately deduce such relationships. In the absence of scientific information it becomes necessary to adopt other approaches based on experience and inspired estimates. Environmental Health standard may be set in one of the three ways.

*Limits set for a pollutant taken up by an organism or a population*

These are sometimes known as primary protection standards. Examples are Acceptable Daily Intake (ADI) of toxic substances and Maximum Permissible Intakes (MPI) of radioactive substances.

*Limits set for a pollutant present in a specified environment, medium, or product (food or consumer goods)*

Such limits are sometimes called 'derived working levels'. These include standards for ambient air and standards for water quality. When applied to products such as food they are frequently referred to as 'product standards'.

*Limits set for discharges or emissions from polluting sources*

These attempt to limit the release of a pollutant into a medium. A well known example is the so-called Royal Commission Standard set for sewage effluent in the State of Uttar Pradesh. Here the effluent should have a B.O.D. of less than $20^{-3}$ gm $1^{-1}$ and suspended solids contents of less than $30^{-3}$ gm $1^{-1}$ and the river into which it is discharged should dilute it eight folds.

**Legislative provision**

As a result of the re-awakening of interest in environmental matters which started in the mid-1960's, many countries have now consolidated and defied previous legislation and introduced new laws which endeavour to deal with the environment as a whole, for example, Pollution Control Act (1974) of the U.K. In the case of pollutants arising from sources, however, two distinct approaches have been adopted in legislation :

*Environment quality management*

Once the appropriate parameters of quality have been isolated, a desired standard can be laid down. Knowledge of the background levels of the pollutant together with quantitative data about its sources can be used to develop a mathematical model with predictive powers. Steps can then be taken to improve the actual quality of the environment. This approach is typical of the U.S.A. and European legislation relating to air pollution.

*Best practicable mecns (BPM)*

The B.P.M. approach involves application of the best available technology which is practicable in the circumstances to reduce the level of an emission by the greatest amount possible. The enforcement agency must therefore make a decision in individual cases regarding the constitution of B.P.M.

**Devising technological abatement processes**

A few types of abatement equipment currently adopted are as follows :

(i) Activated sludge plants,
(ii) Percolating filters (biological filters, trickling filters or bacteria beds),
(iii) Anaerobic digesters,
(iv) Cyclones,
(v) Electrostatic precipitators,
(vi) Centrifuges,
(vii) Fabric filters,
(viii) Countercurrent scrubbers, and
(ix) Controlled tipping.

**Environmental standards**

In view of the comments made above about the movement of pollutants through the media, one would think that it would be sensible to have at least one body responsible for an overall view of the environment. This is certainly not the case in the U.K. where pollution control is extremely fragmented.

**Environmental pollution control**

Environmental pollution control could be analyzed on the pattern of process control. A closer scrutiny would reveal the analogy between process control and environmental pollution control. There are four questions invariably asked about any control which are equally applicable to environmental control :

(i) What is to be controlled ?
(ii) How far is to be controlled ?
(iii) How fast is to be controlled ? and
(iv) How should it be controlled ?

The first question clearly falls within the jurisdiction of medical scientists who could very well identify the main or potential pollutants, viz., solids, liquids and gas which are undesirable to human beings and should be controlled (Appendix I,).

In order to determine how far to take controlling measures, it is essential to know the desired limit or tolerance limit whichever is appropriate for each substance. This can be decided with the help of medical scientists as well. The second aspect of this question is to determine the present level of each pollutant in the environment. This can be done by proper and effective monitoring of the pollutants. The difference between the present level of pollutant and desired or permissible level will decide the extent to which the pollutants should be removed from the environment. It is felt essential that an independent agency should execute the monitoring in order to have the position of the present level of pollutants.

How fast to control these pollutants really depends upon the degree of undesirability and hazards associated with the pollutants. The another factor that determines this process is to see that the place of industrialization is not reduced.

The last question really concerns the jurisdiction of technologists and microbiologists due to the decomposing role of micro-organisms. As far as possible the recycling should be used, which will not only help in controlling the pollution but will also assist the economy of the country by proper utilization of effluents. This question is concerned, with : (a) essential use of all suitable technological developments so far, (b) suitable arrangements must be made for technology transfer, and (c) execution of suitable technological developments without interfering the sentiments of the people.

The pollution control method depends upon the type of pollutants in effluents of various chemical complexes. Basically it is by controlling the source itself. It can be achieved by several methods such as change in raw materials, operational changes, modification of process equipments, and using suitable catalyst to convert pollutants to products, which are not harmful to life. These methods are limited in their application as the destructions of only those wastes are possible which are combustible. In general there are two basic principles involved in pollution abatement :

(a) dilute and disperse, and (b) concentrate and contain. A third method which could be considered of great potential specially due to dwendling of natural resources is recycling.

The first method is to dilute and disperse should also include dispersal of industry. Those places which are heavily polluted should not be available for any further concentration of industries. This method is quite satisfactory for industrial waste as well. Proper distribution of pollutional load well below the assimilation capacity of natural stream will definitely keep the natural stream healthy and available for abatement. This method is equally suitable for air pollution control.

With regard to the second method i.e., concentrate and contain, it is appropriate for radioactive materials and noise pollution. Radioactive pollution is caused due to radioactive elements when they are added to atmosphere on water source. Muller (1928) demonstrated that the radioactive elements are very dangerous to life. Atomic explosion in 1945 destroyed the whole of Hiroshima and Nagasaki. Even after several years the environment was polluted with radioactive emissions causing several diseases, including cancer.

The increasing trends of radioactivities in pharmaceutical and atomic energy research laboratories have caused the dangerous radiation materials polluted in the atmosphere and water sources. Radioactive elements from industrial atomic reactors are the potential sources of pollution, especially when accidental discharge of cooling water is made. Concentration of these radioactive isotopes should not exceed certain limits when discharged to receiving systems. The radiological wastes are not allowed to pollute the atmosphere or water source but they may be dumped under the soil based on the regulations framed by the Atomic Energy Commission. Wastes should only be discharged within controlled areas.

With dwindling resources of metals, pulp and paper, rubber, etc., it is considered essential to optimize their use in day-to-day life. When indiscriminately thrown into the atmosphere, they cause serious pollution, for example plastic and rubber are known to cause cancer, but properly collected they could be used again and again.

The pollution control method depends upon the types of pollutants in effluents of various chemical complexes, automobile exhaust and boiler effluents. Basically, it is controlled at the source itself which can be achieved by several methods such as raw material change, operational changes, modification of process equipment, and using suitable catalyst to convert pollutants to products which are not harmful to life. These methods are limited in their application as the destruction of only those wastes is possible which are combustible.

**International concern for pollution**

Air pollution, however, knows no barriers and, although it normally only affects the country in which it is emitted, it can be carried by winds to other countries. With the greater development in international transport, mobile emitters too have brought their own pollution problems. For these reasons a variety of international committees have been established in order to bring about further exchange of information and research with a view of drawing up of common standards for pollution control. The most important organisations in this field are as follows :

*European Economic Community*

An environment programme has been established to improve the setting and quality of life and people's surroundings and living conditions. The aim of the programme is to promote studies into specific problems and the establishment of proposals for the improvement of the environment. A number of proposals have already been adopted :

(i) Acceptance of 'polluter pays' principles,
(ii) Resolution on energy and the environment, and
(iii) Directives to control the sulphur contents of oil and the maintenance of vehicles.

Directives are binding on member states but the mode of implementation is left to the individual governments. The programme is concerned with the establishment of international emission standards for some pollutants. A measure opposed by the U.K. as it felt that emission standards should be closely related to both what the environment can reasonably be expected to bear and to the use to which the environment is likely to be put.

*Organisation for Economic Co-operation and Development*

This organisation brings together the major western industrialized countries providing a forum to discuss common problems and co-ordination of policies. The Environmental Committee, established in 1970, conducts investigations and proposes solutions to all common environmental problems, but at the same time it considers economic factors as well. It does not involve itself directly in research but promotes ideas which can then be taken up by specialised establishments.

*United Nations environmental programme*

U.N.E.P. was set up in 1972, as a result of the Stockholm Conference on the Human Environment. Its objectives include the establishment of a comprehensive and unified environmental monitoring system and the providing of essential data on environmentally significant chemicals.

*Committee on the challenges for modern society*

Established in 1969 by N.A.T.O. for the purpose of undertaking international studies of specific problems of the human environment. The objective is to bring the exchange of experience and technical knowledge and to put the findings into practice.

*United Nations Economic Commission for Europe*

In 1971, this commission set up involving the E.C.E. Governments. Projects have been undertaken on various problems including air pollution control. In addition to these, there are other bodies which have been established, consisting of non-governmental organisations, with a primary aim of promoting the collaboration and exchange of information. One organisation, 'The International Union of Air Pollution Prevention Associations' was established in 1965 to deal specifically with air pollution matters.

**Geographical considerations for location of industries**

The industrialization has not been uniform throughout the country. In order to have well balanced industrial development several concessions are being given to those who are interested to put up their industries in backward areas. These incentives include loan at a reduced rate of interest as well as subsidies. This could be defended on the basis of socio-economic considerations. But there is a scientific basis for having much higher standard of living in the backward areas which is practically more than half of the total area in every state in this country, due to the fact that total pollution load whether organic or inorganic or both is much less.

On the other hand, there are several places, viz., Kanpur, Bombay, Madras, Delhi, Calcutta and Varansi, where pollution load is much more than the assimilation capacity and consequently the receiving streams are badly affected. Such places should be declared as 'restricted area' for any further industrial development. There should be in-built provision for a time bound action programme. It is, therefore, obvious that a distinction has to be made between the region where there is practically no industrialization, e.g., Ghazipur, Jaunpur, Chamoli, Garhwal, etc., from those which have several industries contributing to a large amount of industrial effluents to such an extent that the assimilation capacity has been surpassed.

In the state of Uttar Pradesh, 252 industries have been identified as major polluting industries which clearly indicate that there are only four districts which have more than ten industries, viz., Deoria (15), Ghaziabad (14), Kanpur (74) and Saharanpur (10). The figures in the brackets indicate the number of industries. But, there are 12 districts in Uttar Pradesh which do not have any polluting industries at all. Similarly, a close analysis of types of industries involved indicates that there are only five types of industries, viz., sugar (87), tannery (77), distillery (26), pulp and paper (22), and textile (10), which contribute to about 83 per cent of total effluent from industries. The figures in each bracket indicate the number of industries.

Obviously, while planning new industrial activities, specially for Kanpur, Ghaziabad, Deoria and Saharanpur, environmental impact assessment must be carried out and only those industries which do not make any significant contribution to environmental degradation should be located, at the same time all incentives should be given for industrialization in other districts. Similarly, efforts should be concentrated to control pollution where assimilation capacity has been surpassed. Regarding the different categories of industries, sugar factaries are normally located in remote rural areas where there are not many other sources for pollution. But location of 61 tanneries at a particular place like Kanpur has definitely created a special situation where the treatment of the effluent to the prescribed limit is absolutely essential.

The World Management Congress in its last meeting held at New Delhi projected the concept of eventual sharing of every thing that is available. This would amount to possible exploitation of all the available resources as far as possible. For example, the effluents from distillery and tannery have fairly high B.O.D. contents. Hence, their collective discharge always leads to serious consequences from pollution point of view. The ideal location for distillery or tannery will be a remote place where there is no additional effluent containing high B.O.D. contents. There is no dirth of such suitable locations in India.

### Combination of industrial effluents

A summary of industrial waste, its origin and character is given in Appendix vii of this volume. The characteristics of effluents from different industries are different. The characteristics of effluent from some selected industries are given in the several appendices which clearly indicate that their properties could be made use for planned industrial development, e.g., liquid effluents, tannery are acidic in nature while the effluents from textile and Jute mills are alkaline. Farther, B.O.D. of tannery effluent is between 6,000 and 10,000 ppm depending upon whether the effluents are coming from chrome tanning or vegetable tanning or a combined unit of both, while the B.O.D. of effluent from cotton textile is only about 350. Hence, combination of these two effluents would involve much less investment. Similarly, the effluents from pulp and paper craft process is generally alkaline and the B.O.D. is about 375. On the other hand, the B.O.D. of distillery is about 40,000 to 70,000 and is acidic in nature. Therefore, a combination of these two would require much less investment. Even a combination of effluents from fertilizer and that of distillery would result in considerable saving.

### Low pollution technology

There has been shift from high technology to low technology which could be used from development of village and small scale industries in developing countries but in view of deterioration of the environment, even in developing countries, it is imperative to go in for development of low pollution technology which could be adopted without much contribution to environmental pollution.

**Use of process chemicals**

There are several chemicals including heavy metals which are used in industry. The degree of harmful effects and nature of pollution are different in different cases. Efforts have been made successfully to develop chemicals used as solvents with low pollution level in a tannery.

**Recycle**

The dwindling resources of materials as well as energy suggest for a suitable action for the conservation of both the materials as well as energy. It is in this connection that recycle is considered essential and pressing at present. The importance of recycle is obvious due to two important considerations. Some of the heavy metals, e.g., cadmium, lead, arsenic, chromium, etc., which are dumped into natural streams, are serious pollutants. Secondly, the resources of such non-renewable materials are in limited supply. For example, the world's known reserves of chromium are about 775 million metric tonns, of which 1.86 million metric tonns are mined annually at present. Thus, at the current rate of use, the known reserves would last about 415 years. The actual world consumption of chromium is increasing, however, at the rate of 2.5 per cent annually. If the growth rate continues in such a way it will deplete the resource stock, not in 415 years but in just 95 years. If we suppose that reserves yet to be discovered, perhaps the present known reserves would last from 95 to 154 years. Even if it is possible from now onward to recycle 100 per cent of the chromium so that none of the initial reserves were lost, the demand would exceed the supply in 235 years. Chromium is definitely one of the most serious pollutants amongst all. Most of the natural streams do not have any further assimilation capacity and hence cannot be used for disposal of such materials any longer.

**Disposal of industrial effluents**

It is general observation that most of the industrial effluents either join the river directly or through aqueduct. Hence the introduction of the effluent into any natural stream is only through laminar sublayer where the flow is practically longitudinal which does not permit any radial mixing. This is the only reason why a large number of fishes are found dead with the input of concentrated amount of effluent either organic or inorganic. A suitable arrangement for disposal of effluent especially away from the bank of a river which is accessible for drinking purposes would definitely eliminate the possibility of such a large scale killing of living organisms. Even those animals which provide milk also consume water for drinking purposes from such polluted source. During festive days when tens of thousands of people use river water for drinking purposes from this laminar zone which is heavily polluted. Hence proper disposal of industrial effluents at appropriate points in the receiving stream could go a long way to fully utilize the assimilation capacity.

Recently, there is a growing concern to develop multidiscipli[nary] approach to study the environmental degradation. [illegible] is the main cause for pollution [illegible] countries. The increasing popu[lation] [illegible] higher production of agricultural products [illegible] of fertilizers and pesticides, higher industrial [illegible] to the growing population [illegible] iron and steel, [illegible] evaluation of pollution loads, [illegible] it was noticed that 90 per [cent] [illegible] only 10 per cent from indus[tries] [illegible] is contributed by sewage and [illegible] industries. It is, however, essential [illegible] should be controlled on a priority [basis]. [illegible]

It is generally observed that there are only a few types of industries which contribute significantly to environmental pollu[tion]. [illegible]

V. K. KUMRA

TWENTY THREE

# ENVIRONMENTAL POLLUTION AND MANAGEMENT

ENVIRONMENTAL pollution has become a problem of major concern in recent years especially, in developing countries like India. The continued growth and expansion of existing industries, the development of latest technologies and products and increasing use of motor vehicles coupled with population growth especially in large urban areas are emitting varying nature of pollutants in large quantities into the atmosphere. Consequently, so many questions are arising to which degree the living organism can adjust to the polluted environment and thus it has become a growing concern of the nation to save the natural environment from pollution.

All over the planet. urban centers are expanding by migration and natural increase of population. About 20 per cent of the world's population live in cities of more than 100,000 persons. By 2,000 A.D. more than 60 per cent of men and women would be perhaps be town and city dwellers. Growth of towns and cities in India both in terms of population and areal coverage is phenomenal. Industrial growth is an adjunct to these features in making this problem more complex.

Man lives in the environment of air, water, soil, tolerance, it is called as 'polluted'. The water pollution may be caused by any one or more of the following factors such as : (a) atmospheric dissolved gases, (b) decomposition of animal and vegetable materials, (c) weathering of soil and rock materials, and (d) industrial effluents and sewage discharge into water bodies.

The main water pollutants are suspended solids. The disposal and dilution of sewage and industrial effluents, bio-chemical oxygen demands, phosphorous, heavy metals, nitrates, chloride, carbonates, bacteria and radioactive substances. The ratio and extent of water pollution depend upon several factors, viz. : (a) waste water disposal system including static and dynamic methods for treatment of domestic and city sewage, (b) hydrological condition of diluting bodies, extent of dilution and self-purification capacity of the stream, (c) chemical and biological characteristics of different types of effluents discharging from certain areas, and (d) socio-economic condition of the communities, generating toxic wastes.

The most dangerous type of pollution to our surface water bodies is caused due to untreated discharge of domestic wastes. It has caused severe damage to water quality mainly in river banks near the cities. A fast increasing number of industries and rapid growth of urban population has resulted in the disposal of highly contaminated wastes into the main sources of water. It has been observed that wastes of almost every industrial centre, viz., Bombay—Kalyan, Bombay—Pune, Calcutta—Hooghly—Asansol, Durgapur, Kanpur, Ahmedabad, Delhi and other industrial areas have been badly polluted as the discharge of untreated water into the main stream is rather a rule than exception. Rivers are practically turning into sewers with their discharge depleting due to massive withdrawal for irrigation, industry and water supply and with increasing quantities of filth, sewage and toxic industrial wastes finding their way into them. In India, most of the rivers like Ganga, Yamuna, Mahanadi, Cauvery, Hooghly, Gomati, Godavari, Chambal etc. have become disposal sites for industrial and domestic waste water of cities and towns, losing their water quantity and aesthetic value. Among all the rivers Ganga is the worst affected from pollution.

River Ganga enters in plains at Hardwar. It is the first spot where 'organised pollution' is felt. The IDPL plant at Rishikesh and the Bharat Heavy Electricals Limited unit at Hardwar, besides a few private industries throw in effluents without treatment. During the 2,000 km course of the holy river, from Hardwar to the Bay of Bengal, a want on discharge of wastes plays havoc with health, agriculture and marine life. The BHEL factory at Bhadrabad, about seven kms from here discharges the wastes into the river at Jawalapur. The 'nullah', carrying greasy and oily discharge from the heavy electricals, is dangerous as it has been felt by the villagers of the concerned areas. Almost half a kilometre of the nullah stretch caught fire and flames up to 15 feet damaged the vegetation. The fire was put out only after tonnes of sand and carbondioxide were used. Further, in Kanpur about 40 m.g.d. of waste water is being discharged into Ganga only through sewers. In addition to this a large number of tanneries, textile mills and chemical industries discharge effluents directly into the river. The analysis of sample water showed that concentration of B.O.D. was 270 mg/l which is much higher than the prescribed limit. The analysis of spatial pattern of pollutants revealed that at every sampling station the concentration of pollutants is not more than the tolerance limit, except at Bengal and Jajmau ghats (Kumra, 1982).

Further, at Varanasi, the holy city of Hindus, though an average of 400 bodies are cremated on the river banks every day and about 9,000 cattles thrown in every year, the major source of Ganga pollution is the city sewage which is emptied at various ghats along its course. According to estimate of Jal Nigam of the State of Uttar Pradesh, about 120 million litres of sewage gets into the river at Varanasi each day, a substantial part of it containing chemical wastes from the cottage silk industry. The Assi Ghat nullah, besides carrying wastes, also brings into the river about four million litres of oily and chemical wastes from Diesel Locomotive Works. It has been observed that at Ghazipur, the opium factory is doing great damage to the river by discharging its untreated wastes. The opium content that gets into water is so high that the monkey living on the banks become addicted to the water and spend their life in a dazed state. As the river proceeds towards Patna, it receives more and more wastes and contaminants from agricultural practices as well. According to Rajendra Prasad Agricultural University estimates, the presence of heavy metals in the Ganga was alarming and has already affected the plant metabolism in the area. The biggest source of pollution, however, is the city sewage (30-50 million litres) though a substantial part of it is utilized for irrigating farms. Ironically Patna is the only city along the course of the river which has extensive arrangement for the treatment of city sewage so that river is kept clean but now its capacity is not enough tc treat the huge amount of city waste. Among all the cities on its bank, water of Ganga is heaviest polluted at Calcutta, because of the row of the industries on the river bank. The industries whose number in the basin is 285. discharge their toxic wastes directly into the river. Total volume of liquid wastes discharged through the various outfalls is 252 m.g.d. of which 77 m.g.d. is contributed only by industries. The daily average pollutional load contributed in terms of B.O.D. total solids, suspended solids and dissolved solids are 106, 2308, 1058 and 1251 tonnes, respectively of which 66, 904, 415, and 482 tonnes are discharged only by industries. In addition, discharge of various toxicants by over one thousands ships calling at Calcutta Port every year affect the quality of water. Thus, the study revealed that the quality of Ganga water is deteriorating very rapidly and if appropriate measures already proposed by the Government of India in the Seventh Five-Year Plan are not taken in time, the river water will not be suitable for irrigation to fields.

Yamuna, another important river of the country is also being polluted due to discharge of domestic and industrial wastes at Agra, Mathura, Delhi and other towns situated at its bank. Delhi, discharges 0.96 million kilo litres of toxic waste water into river Yamuna every day. This contaminated water contains 250 thousand kg of dissolved solids, 125 thousand kg of suspended matter and 125 thousand kg of B.O.D. The real assault on the river by all factors starts at downstream near Wazirabad waterworks. Sewer pipes discharge about 120 million gallons of waste water and filth daily without biological treatment. After that another 100 million gallons of water is discharged by the dreadful Nazafgrah drain which carries industrial waste water of west and north Delhi. The matter of utmost concern is that only 46.4 per cent of the effluent water and sullage flowing back into Yamuna through 17 drains is being chemically treated before discharge but it is of primary nature only. Further, it is notified that the Hindon river which joins Yamuna at Okhla is the most polluted river in the country. Therefore, it can be estimated that like Yamuna and Ganga rivers, other rivers and water bodies are facing similiar problems which will not only deteriorate water quality but will also create underivable affects on human health, thus overall management of our water resources is required.

**Noise pollution**

Noise is a twenty four hours bug and it is getting increasingly hazardous. Noise pollution may be both natural and artificial. However, the artificial noise pollution, the subject pertinent to the present discussion, has existed and grown with advancement of civilization with no interregnum and has assumed a proportion which is perhaps unsurmountable with modern human ingenuity. The sources of noise may be non-industrial or industrial in origin, man-made and natural as well. Among the non-industrial sources, street noise due to hawkers, recreation, thunder, public loudspeaker, political and religious propoganda, etc. While the industrial and others include noise from various industrial operations, transportation, vehicular movement, such as car, motor, truck, train, motorcycles, aircraft and, explosion. Though this problem has been a problem since the beginning of human civilization but it has recently been considered as a health hazard.

Enough work had not been done to study the spatial pattern of noise levels and its effects on health in cities. Random surveys revealed that it led to an increased pressure and heart beats and affected the hearing capacity of man. People in noisy areas have higher blood pressure and suffering from fatigue. There reflexes become somewhat 'edgy' and their 'efficiency has been severely affected.

A study conducted by E.N.T. Department of Maulana Azad Medical College in Delhi and by the author at Kanpur revealed that the level of noise varies from 64 dB at Teen Murti Marg to 92 dB at Daryaganj in Delhi, while it varies from 40 dB at H.B.T.I. west campus to 89 dB at Parade Crossing in Kanpur (Table 23.1). In another study carried out by Environmental Monitoring Laboratory of the Industrial Toxilogical Research Centre, Lucknow, the noise levels in various localities in Lucknow varies from 88 dB at Kaisarbagh to 100 dB at Hazratganj Crossing during the peak hours (Table 23.1).

Bombay has been characterised as the noisiest city in India with an average noise levels varying between 75 to 100 dB. At Santa Cruz Airport, it reaches 104 dB. Owing to operation of electric trains inside the city, noise intensity is high along main rail routes of the city, which has badly affected the hearing

TABLE 23.1 : Noise levels at important crossing of Delhi, Kanpur and Lucknow cities (in decibles)

| Delhi* | | Kanpur** | | Lucknow*** | |
|---|---|---|---|---|---|
| Crossings/Localities | Noise levels | Crossings/localities | Noise levels | Crossings/localities | Noise levels |
| Daryaganj crossing | 92 | Parade Crossing | 89 | Hazratganj Crossing | 100 |
| Subjimandi | 79 | LIC Building Crossing | 82 | King George Medical College | 94 |
| Fatehpuri | 84 | Charles Crossing | 81 | Nur Manjil area | 90 |
| Sadar bazar | 80 | Gumti No. 5 | 82 | Kaisarbagh Crossing | 88 |
| Chandni Chowk | 88 | Collectorgunj Crossing | 79 | Royal Hotel Crossing | 95 |
| Lady Harding Hospital | 81 | Bara Chauraha | 84 | Charbagh Rly. Station | 94 |
| Teen Murti Marg | 64 | H.B.T.I. | 40 | Lalkuan | 96 |
| Minto Bridge area | 88 | Canal Crossing | 80 | Aminabad | 95 |
| Cannaught Place | 80 | Eye Hospital Crossing | 68 | Sadar Crossing | 90 |
| Safdarjung Airport | 81 | Rawatpur | 71 | | |

* E.N.T. Department of Maulana Azad Medical College, Delhi.

** Survey by the author.

*** Industrial Toxicological Research Centre.

capacity of people residing near the rail track. In other cities noise levels are not monitored but most of the million cities are suffering serious problem of noise pollution.

The human beings may be affected in many ways when exposed to higher levels of noise than the tolerance limit. Studies have shown that a person could be safely exposed to 90 dB for 8 hours but if a person is exposed to 100 dB of noise for only one and half hours or 115 dB for 45 minutes it is likely to cause irreversible damage to his hearing capacity. People living in cities become prone to deafness at the age of 45 years, whereas those residing in quieter places have nothing to fear until they are 65. Loud noise can also make the pupil of the eye dilate which requires a very quick change of eyes focus seriously imparing the workers ability to fine and delicate works. Therefore, noise pollution control is urgently needed especially in big cities. For this purpose, appropriate management techniques should be adopted and adequate standards for controlling outdoor and indoor noise should be made available. In addition, Government should also pass the Noise Pollution Act similar to Water and Air Pollution Acts which have already been approved by the Government of India.

**Solid waste disposal**

In addition to the above-mentioned pollution problems, solid waste disposal is also a significant source of environmental pollution. It includes a large number of old and discarded articles, ranging from small pins, glassware, plastic containers, polythene bags, waste papers, ash and plant leaves. Generally, waste materials are classified into two major categories : (a) household solid waste, and (b) commercial waste. Household solid wastes are collected locally and the amount depends upon the size and consumption habits. It is disposed off on either public or private contractors' sites. The other type of wastes are generated by commercial establishments but are usually disposed off by the available public facilities. The quantity and quality of refuse depend on various factors, such as geographical location, climatic conditions and nature of community, i.e. commercial, residential or industrial.

Population growth and greater affluence and unrelenting commercial enterprises all are the root cause of problem of solid waste generation and disposal. The average urban garbage disposal rate is 0.35 kg./day/capita with a maximum of 0.57 kg/day/capita in Calcutta, though it is much less than the cities of U.S.A. (2.6 kg/day/capita). But if the same rate of garbage generation continues, it will be a difficult task for the planners to manage it. In an earlier chapter of this volume, a detailed discussion has been made on this problem already.

**Environmental management**

The above analysis of environmental problems reveals that there is an urgent need to check the continuously increasing human needs as it will be very difficult to cope up with the complex problems of pollution in 2001 A.D., when population is likely be doubled. For this purpose it is essential to evolve a comprehensive strategy for environmental mangement of Indian towns and cities. "Management implies a conscious choice from a variety of alternative proposal and furthermore that such a choice involves purposeful commitment to recognised and desired objectives. Whenever possible, management implies to the deliberate adoption of strategy or number of strategies designed to meet realistically short-term objectives, yet specifically providing sufficient flexibility for the preservation of longer term option" (O' Riordan, 1971). Environmental management is extremely a complex affair as the recognised and desired objectives are varied and often conflicting. The objectives of the environmental management are varied as each individuals or groups values, perception, needs and preferences varies. The objective is often generalised as the maintenance of 'environmental quality', a term which is defined differently by variety of people. The objectives of comprehensive environmental planning have been divided into three areas ; (a) protection of physical and mental health, (b) enhancement of economic values, and (c) preservation of sensory and participatory pleasure. However, all this involves a number of programmes, such as education, information and monitoring system, legislation, standards and policies, issues of developments and necessary safeguards for proper planning, and management. Some of these aspects have been discussed in different chapters of this volume.

Environmental education and information helps in developing environmental awareness. The environmental education and training is based on the distinction that education aims at establishing senstivity to environmental problems, raising the levels of awareness and generating commitments; while training aims at developing and mastering skills, solving political problems and refers to specialized actions. The process of issue-creation through public awareness is likely to go through three stages : (a) identification of the problem, i.e. the existence of the problem is identified, (b) realisation of the problem, i.e. the scope, intensity, causes and consequences of the problems are realised by the citizen, and (c) mobilisation of awareness, i.e. public awareness is expressed through visible and formal channels, like interest groups or political parties. At the first stage, educational and research institutions can play a very useful role while at the second mass media can make a significant contribution. The task of adding a course in environmental preservation in the school and colleges curricula is not easy. Some efforts has been made by the National Council of Educational Research and Training, New Delhi collaboration with International Union for Conservation of Nature to make environmental education as part of the 10 plus 2 school system. But this step is not enough. It will not bear fruit unless environmental education is taught in a multidisplinary perspective. Moreover, there is paucity of source materials for such a course as well as non-availability of teachers to teach problem and action-oriented environmental programmes. Thus, courses on environmental education should be planned in such a manner that they involve schools, universities, social voluntary and rural organisations or bodies, including panchayats, scientifie institutions and above all the common man.

In order to know the present level of pollution, there must be constant monitoring of pollutants in air, water and soil which reflect the environmental situation of an area. This is essential for both, human adjustment and environmental management. Prediction of possible changes is an most essential pre-requisite for planning and management processes. The modern methods and techniques of automatic recording and transmission of data, remote sensing through satellite are valuable for rapid monitoring of the environment. The task of observation and monitoring should be much ahead for the purpose of forecasting, planning and management. In addition to this, environmental impact and cost benefit analysis should be made .n order to know the results of implementation of environmental pollution controlling measures. There should be a scientific analysis about the benefits of installing a waste treatment plant in a textile mill and how much cost of production will increase.

To control pollution, there is a need of appropriate standards suitable to the Indian condition. The standard means the comparison of a quantity to a scientific or arbitrary standards or to a 'specified base'. Therefore, environmental indices mean tools used to monitor, and quantitatively report, environmental status and trends based on the specified standards. Whenever the measurements are longer than two standards the index value is high, which indicates problem (Inhaber. 1976). Many standards dealing with pollutants are set on the basis of danger to human health. For example, Indian Standards Institute has established noise level standards for outdoor and indoor locations for residential and official buildings of the city and rural areas (Table 23.2). But these standards do not fit with every locality of the cities as their functions varies from one to another. The validity of the standards should be checked regularly, and the standards should neither be too lenient nor too strict, that is they are not economically defeating. The spirit of and purpose of the standards should be to safeguard the environment. It is also necessary to follow the standards prescribed by the Government by every citizen and industries. To follow standards for air, water and noise pollution,

TABLE 23.2 : **Acceptable noise levels in dB (A)**

| Acceptable outdoor noise levels in residential areas | | Acceptable indoor noise levels for various type of buildings | |
|---|---|---|---|
| Location | Noise levels | Location | Noise levels |
| Rural | 25-35 | Radios and T.V. Studios | 25-30 |
| Suburban | 30-40 | Music Room | 30-35 |
| Residential urban | 35-45 | Hospital, classroom and auditorium | 35-40 |
| Urban Residential cum commercial | 40-50 | Appartments, hostels, Conference rooms, Office | 35-40 |
| City | 45-50 | Court rooms, Private Offices, Library | 45-50 |
| Industrial area | 50-60 | Restaurants | 50-55 |

*Source* : I.S.I. Publication UDC 534 : 83 : 7144, March 1969.

there must be legislations.

It is necessary to provide legislative measures to ensure the protection of environment. The Indian Government passed the first national level Act to control water pollution in 1974. The Act took fourteen years to pass through the hurdles of parliamentary procedures and lobbying. Passing of the prevention and Control of Water Pollution Act was an indication of Government's awareness. The Act also created a Government Body for the enforcement of the legislation. The second legislation passed by parliament was the Air Pollution (Prevention and Control) Act of 1980. The Department of Environment is now responsible for implementation of Water and Air Pollution Act. These measures are however not enough if not backed by executive authority. There is still lack of technical competence to assess the environmental standards and the administrative machinery does not appear to be quite enthusiastic about all this for their own reasons. Legislation has to be made more comprehensive and precise. This is not possible through administrative measures alone but peoples participation and cooperation is also necessary. It is true that no control measure or planning can get success without the support the voluntary cooperation of the people concerned. For example, improper disposal by garbage may not be controlled unless public of a locality do not dispose their daily waste material at proper sites and the local authority collects and dispose it properly. To implement that planning measures, there is a need for the formulating of a rational and national environmental management policy.

For preparation of a national environmental policy Department of Environment (D.O.E.) has to play a watchdog's role and to study and attract the attention of the government about the causes and conseqeuences of environmental deterioration. The D.O.E. did not have an administrative responsibilities, similar to its counterparts in U.S.A., Japan and Sweden. So far the agency has only played advisory role.

The major question to examine is what has been the net result of all these efforts? Has environmental pollution been checked in the waters of the Ganga and the Yamuna? Are residents of Greater Calcutta, Bombay, Delhi, and Madras breathing cleaner air? It may be here argued that five years is a short time to judge an agency's perfermance. But, the creation of a federal agency is only a preliminary measure. For years, Government of India claims to be committed to protect the environment. Except in a few cases, the government has not been able to forcefully implement its commitment. Sewage and industrial waste water is still being dumped into rivers without treatment. People are still dying without knowing that pollution is partly responsible for their death. For this purpose, the government should take quick actions against pollution, implement rules and regulations, and impose heavy taxes on the offenders.

The Indian Factories Act, Rule 18 clearly provides legal powers to State Inspectorate of Factories for ensuring proper treatment of industrial wastes before they are drained out of the factories or other industrial units. As observed, the existing laws have proved to be ineffective because the provisions are vague. Similarly, Section 241 of Delhi Municipal Corporation Act, 1957 has also a similar provision in prosecuting industrial units draining out industrial wastes. But, the fine is too small and hence the factory owners have been indulging in violating the legal provisions with impunity. Therefore, the administration should immidiately move for amending the existing laws so that industrial units whether small, medium or large responsible for water pollution should be effectively prosecuted.

Very recently the government has shown keen interest too in saving Ganga from pollution. Rupees 250 crore project of the Central Government envisages the renovation and recommissioning of the existing sewage treatment plants and installation of new ones as a major step towards pollution control. A report prepared by the D.O.E. in 1984 says that the main by-products of the process will be methane, biogas, algae and fish. It is estimated that on an average it should be possible to get an annual income of Rs. 37 per capita contribution of sewage by the sale of these products and against that the expenditure of pumping and sewage would be Rs. 23 per capita per year. The plan which is to be carried out during the period 1985—1990, will be overseen by an apex body at the Centre. Though the task is very difficult, but it will not only maintain the quality of Ganga water and its paucity but also provide by-products worth millions of rupees.

The issues of environment and development are of much importance. There is a wrong notion that environment and development are incompatible. The national environmental management policy has to be environment-development-oriented rather that being by and large, development-oriented (Kayastha, 1985). It is the need of the day that we must maintain, protect and preseve the environment not only for ourselves but also for future generations and survival of mankind at large. The stage has come where clean environment is not a luxery but an essential aspect for survival. A sense of moral code among citizens is the only hope for making life worth living, both in rural and urban environments.

A. B. MUKERJI

# TWENTY FOUR

# FOREGLEAM

THIS paper aims to discuss on pedagogy of teaching of environment in geography with emphasis on a detailed discussion of its two basic aspects : contents and methods. Related to these dimensions, depending upon ones perception of their significance in the total understanding of environment and inclination and competence the issues might differ. In so far as content is concerned. I will follow the general principle : what shall we teach, at least partly, would determine what we shall not teach or what we should not teach. Similarly, there needs to be a discussion on at least a modicum of methods of studying the facets of environment which are relevant to the sustenance of life.

Some of the issues I would like to highlight are : (i) definitional doubts, (ii) kinds of environment, (iii) environment : an inter-related entity, (iv) environment and development, (v) resources, (vi) debate of doom, and (vii) priority identifications (beer cans on Himalayan snows).

## Definitional doubts

Etymologically, environment simply means that which surrounds. Doubts arise on two counts : what would all environment include and how is it different from ecology. The environment, the surrounding, which is the subject of definitional doubts, means many things to many people, from the administrative conditions prevailing in an office to the economic system obtaining in a country plus, of course, everything that has been created by nature. Few terms are so widely used, misunderstanding or complication arises from an ignorance of the essential, scientific fact and this is also derived from its etymology that environment must connote both context and resource. Water, air, soil and forest, are at the same time both context and resource. Also, although created by nature, through natural processes independent of human beings, environment is the environment of humankind not of stars. planets, rivers, forests and terraces. All of them have surrounding, environment, but that is not the environment of our discussion. Environment has its significance only in relation to humankind. All the components of environment are used by us : 'have dominion over the fish of the sea, and over the fowl of the air, and over every living thing that moveth upon the earth'. (*Bible*, Genesis, Chapter 1, 28). Both the contexts and resources are significant to humankind. Resources occur in a context and their exploitation and conversion also are founded in characteristic contexts. In their own turn resources constitute context as well.

The term environment we have thought is natural environment and of humankind. How is it different then from ecology ? The Greeks used the word *oikos* to describe a home which has the intimations of surrounding or environment and which stands in interactive relationship with the inhabitants. The German biologist, Haeckel, introducing the concept of ecology in 1866 defined it as the study, of inter-relationship between an organism human beings, plants and animals and its environment. The emphasis here is on relationship and not on the components of environment. Ecology and environment are, thus, inter-related but neither synonymous nor inter-changeable concepts.

**Kinds of environment**

From the wide spectrum of its use in diverse contexts, it is obvious that the term environment must have several kinds, each having a different meaning endowed on it by the user. At least five kinds of environment can be identified from the published literature in geography and biological and social sciences . natural (physical), biotic, social, political, economic, and cultural. It is necessary to convey the particular kind of environment involved in discussion. Mainly, we are concerned with natural (physical) environment in both learned discussions and common parlance. This environment is composed of two subsets, the animate and the inanimate, and the two individually and in combinatian constitute the environment of humankind. Depending upon the prevailing fads and exigencies of survival one is emphasised at the cost of the other. At one time it is the smog of the impossible metropolitan centres, at another it is the muddy waters of the polluted Ganga, yet another is the scene of beer cans on the Himalayan snows. Logic of definition would include forests and wildlife forming the biotic complex as the animate component while water, air, sunlight, and ozone constitute the inanimate. Both the animate and the inanimate components constitute the life-support system of humankind.

**Environment : An inter-related entity**

The natural environment is constituted of elements, water, atmosphere, soil. and vegetation and innumerable identified and undiscovered ones. These components are casually and functionally inter-related. Their origin. placing and functioning are not fortuitous but genetically ordered. Vegetation types have their characteristic habitats. Because the components are different and their interrelationships are characteristic of them, the environments are different. Mainly, the properties, the values of the attributes, create major differences in the environments. The Himalayan and the Thar Desert environments are different because of the differences in the attributes of the components which are common to both. The podzol soil, the coniferous forests. and cold and humid climate. all characteristic components of the Himalayan environment are as much inter-related as the interrelated components of xerophytic vegetation. sandy soils, and desert climate are in the Thar Desert environment.

Thus. one kind of inter-relationship refers to the natural elements among themselves. The other kind of inter-relationship, of urgent relevance to humankind, is that which operates between humankind and natural environment. Lest the gentle ladies are annoyed I am not using the easier and, far more popular term man-environment relationship. The crucial point to be emphasised here is the bilateral nature of the relationship, symbiotic and parasitic, between human groups and the natural environment. They form an interacting system, also termed man-environment system. Changes in one component will produce changes in the other. Even though there are environmental changes independent of human actions, there are no such changes which do not affect human life. Similarly, the most urgent environmental problems are those which have been created by human actions, deliberately or unwittingly.

**Environment and development**

It must be exiomatic that environmental problems are to be considered as an integral part of the overall processes of economic and social development. But in the development literature and in the development processes one can easily discover a dialectical relationship between environment and development. While one school of development strategists shoots the bullet on the target of development by all logic and at any cost, regardless of the long term devastating consequences on environment, the other obivious of the compulsions of existence shoots the development processes down s) that environment might be protected. In this dialectical battl the advocates of the two schools tend to take extreme, irreconciliable positions, both totally unrealistic. The pessimistic, helpless notion is that the dialectics of development and environment is beyond resolution, is based on an entirely wrong premise. We need both development and environment.

Conceptually, development is a function, a result, of the interplay of three basic variables : institution, technology and environment. Development must involve economic processes of production, consumption and distribution. Production is contingent upon utilisation of natural resources not exploitation which, prima facie, has a negative connotation. The natural resources, animate and inanimate, directly or indirectly constitute our environment. There can, thus, be no development, in the long run with the destruction or without the conservation of environment. It is necessary for the environmentalists to emphasize clearly that the terms utilization and destruction are not synonymous or interchangeable. Through every period of mankinds long history of terrestrial occupance beginning from the neolithic, depending upon the institutionally created needs and technological equipment, environmental resources have been utilised. With the progress of technological, material civilisation, the volume and the manner of utilisation have undergone sea changes, the former has impossibly increased, the latter has become ruthless exploitation. Undoubtedly, it is the insatiable consumerism which has led to the destruction of environment. Any development that is environmentally neutral is, by the very logic of dialectics, destructive of environment and development must, therefore, be symbiotic and not parasitic and, preservation of environment be considered an integral component of all development processes which are, one way or the other, related to environment.

**Environmental resources**

Earlier it was pointed out that two basic dimensions or connotations of environment are context and resources i.e. environmental contexts and environmental resources. Many of the contexts and also the components are resources. Forest is a context, a component, and a resource. Slope is more a context and a component and not a resource.

We all know that resources which are all natural, are at the first stage utilised or extracted by primary activities and are then converted into manufactured products by secondary economic activities. But all the components of environment are not resources. In order that they constitute resource they must be cultural appraisals. This has been from the beginning of human history. As cultures have become increasingly complex, a larger number of components have been identified as resources and increasingly their exploitation and utilisation have become diverse, their destruction imminent, and their protection difficult.

It is in this context that I suggest a re-examination of the concept of renewable resources, that of non-renewable resources being already understood with convergent connotations. To put a straight question : how renewable is a renewable resource. The property under question is not that of renewability itself but the degree of renewability. It is being submitted that the degree of renewability is a function of the rate at which a resource is being created and the rate at which it is being exploited. Obviously, under non-planned, *laisez faire* economy the two rates would not be equal and there would be a timelag between the two. For most renewable resources the rate of recreation is slower than the rate of exploitation. Hence, during the period extending between the time of complete exploitation and that the recreation in the form which can be exploited, the resource simply does not exist. The renewable resource, during the period, is non-renewable. By this argument neither soil nor forest can be considered renewable to a complete degree of renewability or renewable in an unqualified connotation. A *sheesham* (*Dalbergia sissoo*) tree takes eighty to hundred years to fully grow and be considered ready to be felled for building construction. A patch of forests comprised of this species once felled would need at least eighty years to be fully recreated and during these eighty years there would be no *sheesham* for the community depending only upon this forest tract. This concept of renewabilitiy can help us to understand the message of UNEP, or the Club of Rome. or Limits to Growth. Coal is non-renewable because its recreation involves geological time span while forest whose recreation is limited to human time is, to begin with, renewable but becomes non-renewable. While we cannot do anything with the nature of coal as a resource we can, by adopting a time staggered felling, convert the non-renewable forest into a renewable resource.

**Debate of doom**

In the literature on environment there is an almost unbearable pessimism on its survival heralding the doom of humankind. The most frightening part of this literature tries to convince us that we have destroyed our environment beyond all possible hopes of regeneration, recovery, and re-creation and hence our own destruction is inescapable. This fear is not founded in facts. There is no question but that there has been massive environmental degradation since the Industrial Revolution, more in the western industrialised countries and less in the agricultural countries. Of the third world, more in the industrial parts of the latter countries than in the agricultural regions inhabited predominantly by simpler societies with low levels of consumerism. Also, there is enough monitoring of the destruction of both environmental context and resources to frighten us beyond our limits—the pollution and degeneration of the lite support system, Notwithstanding this declaration of an imminent loss of paradize and the portrayal of the scenario of a vast annihilated earth the massive efforts invested by national and international agencies in the portection and regeneration of environment and the impressive successes they have yielded cannot fail to reassure us of our capacity and self-confidence in creating adequate survival strategies. Smog can be removed, ozone can be protected, soil erosion can be prevented, barren slopes can be afforested and water pollution can be eliminated. For each task there are suitable technologies available, only that for their implementation needed are finances and political will and a responsive institutional structure. Teaching of environment that continually harps on hopelessness and helplessness can ultimately paralyse all motivations for thinking and action.

**Beer cans on Himalayan snows**

While on the one hand there is a worldwide alarm sounded for an environmental doom there is a callous and ridiculous misplacing and identification of priorities of environmental problems which need study and financial support and political will for their solution. There is, according to some learned environmentalists, environmental pollution of the Himalayas by the bear cans dumped by the tourists and the mountaineers, and they insist that this is one of the major urgent problems in environmental protection. In an area of upwards of 30,000 miles$^2$ of snow, annually regenerated mass, some hundreds of beer cans are polluting the environment on a scale that must create panic in UNEP. As it is the number of tourists, Indians and foreigners, who will ever tread on the Himalayan snow is a miniscule and the number of beer cans they will throw on the snow cannot be so large as to hurtle the Himalaya down on the plains. I submit that the priorities must be decided on the basis of the segment of human population that is involved or affected—the first priority in India must go to those environmental problems which endanger the survival of the vast majority of our population, poverty and quality of life, mode of living, health, and sustenance activities. Nor can one forget the marginal populations socially and spatially in formulating the priorities. Forest, flood, soil, and water are our first priorities. There can be no relative ranking among them. All are linked in an interactive system. We cannot remove forests and eliminate flood, protect soil, and conserve water. It is a chain of destruction. Hence, all belong to the top priority. Also these four components of environment, in varying measure, are necessary for different groups. Forests are needed for a large majority of population but they are absolutely necessary for tribal life. There would be no tribe without forests. There are complaints of high decibels and murky air and dusty sky : clean air and blue sky are necessary not for an aesthetic environment but for humankinds, health and survival, work efficiency and quality of living. Our enthusiasm for any environmental problem that does not focus upon poverty of life is, prima facie, unethical. Freons destroying ozone should be our concern but they cannot be our first priority.

Environmental studies in geography, as in Physical Geography, need to be conducted mainly in the field. It must be axiomatic that there is no substitute for field measurement, observation, and enquiry. The internal logic of the environmental processes cannot be formulated by deductions alone. Both for data gathering and monitoring field component becomes a sine qua non. Undoubtedly, no one would question the proven utility of the widely used techniques based on aerial photographs, landsat imageries, and topographical sheets but greater insights to the site relationships can be obtained only through micro scale studies involving field observation. In any case it not only generates fresh data but also supplements that collected by other agencies. Generally, the government agency data are totally inadequate, unreliable, more so in the case of data relating to land use covers and of the relatively little accessible areas. Lastly, the perception and use of environment can be documented only through field generanted data.

## Concluding remarks

Practically every environmental issue has a potential for generating debate. All the issues are multi-faceted and are involved with the entire gamut of human existence. Basically, environment is natural environment and has inter-active relationship with humankind, to which it provides both context and resources for development. While there is no inherent conflict between environment and development, the priorities in the identification, study, and solution of environmental problems must be realistic and related to the environmental perception of the majority of the population. Environmental problems are also intimately related to class structure of the communities, political economy and government and these aspects should be gradually introduced in our discussions.

# APPENDICES AND INDEX

number of alternatives being compared [illegible] while the analysis becomes more detailed [illegible] scale [illegible] and [illegible] a change [illegible] of emphasis [illegible] back to the initial formulation stage [illegible] (1981) [illegible] has suggested [illegible] feedback [illegible] the main [illegible] in the [illegible] feedback should be taken [illegible] in environmental cartography.

**Technological aspects in environmental mapping**

Environmental cartography [illegible] towards production of atlases [illegible] hand in hand with computer facilities. The [illegible] and non-conventional data are now available in large [illegible] with high precision [illegible] environmental [illegible] information [illegible]

[illegible]

*Appendix I*

## Pollutants causing air pollution along with their sources and harmful effects and tolerance limits

| Pollutants | Major Sources | Principal effects |
|---|---|---|
| 1 | 2 | 3 |
| 1. Ammonia | Ammonia and urea plant | Adverse health effect, toxicity to fish and aquatic life. |
| 2. Arsenic | Gas purification plant in ammonia and urea manufacturing | Causes black foot disease and slow poisoning tc plants and crops. |
| 3. Boron | Manufacture of synthetic borons | Causes tonic effects for living organisms, destruction of plants. |
| 4. Carbon-monoxide | i. Transportation<br>ii. Fuel combustion<br>iii. Industrial processes<br>iv. Solid wastes disposal | Reduction in the oxygen carrying capacity of blood, reduces visual and mental activity. |
| 5. Carbon-dioxide | i. Transporation<br>ii. Fuel combustion<br>iii. Industrial processes<br>iv. Solid waste disposal | Causes eye irritation, suffocation and cloud dust. |
| 6. Colour | Various process plants | Excess colour is objectionable aesthatically to all the living beings. |
| 7. Dust | Various process plants | Reduce visibility. |
| 8. Flouride | Effluents in phosphoric acid and super phosphate plants | Causes dental and skeletal fluoros is affects hatching of offs in fish. |
| 9. Lead | Various industrial and mining effluents | Strong poison and higher contamination may be fatal |
| 10. Nitrogen oxide, nitric oxide and nitrous oxide | i. Transportation<br>ii. Fuel combustion<br>iii. Processing<br>iv. Solid waste disposals | Visibility reduction, sensory irritation and plant damage. |
| 11. Odour and taste | Due to presence of some organic compounds and dissolved gases like hy- | Bad odour and taste make the air, water and edibles unfit for use. |

| 1 | 2 | 3 |
|---|---|---|
| | drogen sulfide and inorganic salts of iron, zinc, copper etc. | |
| 12. Oil and Grease | Lubrication of machinery gasoline filling station and fat manufacturing plants | Suppresses dissolution of oxygen in water, floats on water and causes toxic and synergic effects. |
| 13. pH | Due to pressence of acids or alkalis | Higher or lower pH effects aquatic life and makes it unfit for human use. |
| 14. Phosphate | Phosphoric acid and complex fertilizer units | Phosphates alongwith ammonia causes algal blooms which increases cost of water treatment. |
| 15. Sulfur oxides | i. Transportation<br>ii. Fuel combustion<br>iii. Industrial processes<br>iv. Solid waste disposal | Causes sensory and respiratory irritation, plant damage corrosion, adverse effects on health. |
| 16. Temperature | Industrial cooling | Prolonged exposure may produce undesirable physiological effects, causes adverse effects on aquatic life, taste and odour. |
| 17. Turbidity | Due to presence of suspended clay or dispersed organic and microoganics | Affects aquatic life also objectionable from aesthetic point of view. |
| 18. Urea | Urea plant | Toxicity of ammonia for hydrolysis, entrophieation of water bodies |

*Courtsy* : G.N. Pandey

## Appendix II

### Rubber and plastics wastes causing cancer

| Carcinogen | Site of cancer | Type of cancer | Species |
|---|---|---|---|
| Cobalt | Musscless tissue | Parenteral | Rat |
| Cadmium | Connective tissue | Parenteral | Rat |
| Selenium | Liver, Thyroid | Oral | Rat |
| Chloroform | Liver | Oral Parenteral | Mouse |
| Carbonates, Ethyl-carbonate, isopre-phylnchlorophenyl carbonate | Lung | Oral | Rat |
| Methylated | Skin | Cutaneous | Mouse |
| Napthelene | Brain, hypophysis | Oral | Rat |
| Nitroamine | Nasal, Binus, Lung Oesephagus, Stomach Lever, Kidney, Bladder | Parenteral | Pig |
| Carbon and Silicon | Connective tissues | Parenteral | Rat |
| Polymers | | | Rabit |
| Hydrazine | Lung | Oral | Mouse |
| Vinyl chloride | Lung, skin, bone | Respiratory | Rat |

*Courtsy* : G.N. Pandey

*Appendix III*

## Noise pollution and its effects

| Interference | Sound level | Effect |
|---|---|---|
| 1. Sleep | 40-70 | Period of sleep is shortened |
| 2. Hearing | 80 | Hearing ability deteriorates |
| 3. Structure and equipments | above 150 | Breakage of structure and equipments |
| 4. Health | | Increased heart rate, constriction of blood vessel, digestive spasus |

*Courtsy* : G.N. Pandey

*Appendix IV*

## Recommendated noise criteria for steady background noise in various Indoor functional activities

| Type of spaces | PNC curve | LA LB |
|---|---|---|
| 1. Concert halls, Opera house, recital hall (for listening to join musical sound) | Not to exceed 15 | 26 |
| 2. Broad-cast and recording studia (distance microphone pickup used) | 10 to 20 | 20 to 30 |
| 3. Large auditoriums, large drama threatres and churches (for very good listening conditions) | Not to exceed 20 | Not to exceed 30 |
| 4. Board-cast, television and recording studia (close microphone pick up used only) | Not to exceed 25 | Not to exceed 34 |
| 5. Executive offices and conference rooms for 50 people (No amplification) | Not to exceed 35 | Not to exceed 42 |
| 6. Bed rooms, sleeping quarters, hospitals, hotels (for sleeping, resting and relaxing) | 25 to 40 | 34 to 47 |
| 7. Offices, small conference rooms, class-rooms, libraries (for good listening conditions) | 30 to 40 | 38 to 47 |
| 8. Living rooms and drawing rooms (for conversing or listening to radio and TV) | 30 to 40 | 38 to 47 |
| 9. Large offices, reception areas, retail shops and stores, cafeterias, restaurants etc. | 35 to 45 | 42 to 52 |
| 10. Lobbies, laboratory workspace, drafting and engineering rooms (for fair listening) | 40 to 50 | 47 to 56 |
| 11. Shops, garages. power plant control rooms (for acceptable speech and telephone communications) | 50 to 60 | 56 to 66 |
| 12. For work spaces where speech or telephone communication is not required but there must be no risk hearing damage | 60 to 75 | 66 to 80 |

*Courtsy* : G.N. Pandey

**Different types of pollutants in paper and pulp industry (Types of their control)**

| Source | $H_2S$ | Particulate pollutant lb/ton dry pulp | Type of control |
|---|---|---|---|
| Digester blow system | 0.1-0.7 | Negligible | Untreated |
| Smelt tank | N.A. | 20 | Uncontrol |
| Lime kiln | 1 | 18.7 | Scrubber |
| Recovery Furnace | 3.6 | 150 | Primary stock gas scrubber |
| | 3.6 | 7-16 | Electrostatic precipitator |
| | 0.7 | 12.25 | Venturi scrubber |
| Multiple effect | 1.2 | Negligible | Untreated |
| Evaporator | 0-0.5 | Negligible | Black liquor oxidation |
| Oxidation tower | N.A. | Negligible | Black liquor oxidation |

*Courtsy* : **G.N. Pandey**

# Appendix VI

## Important air pollutants concentration in some of the Indian cities

| No. | Cities | Sulfur dioxide in parts per million | | Nitrogen oxide in parts per million | | Hydrogen sulfide parts per million | | Particulate micro gram/metre | |
|---|---|---|---|---|---|---|---|---|---|
| | | Average | Maximum | Average | Maximum | Average | Maximum | Average | Maximum |
| 1. | Ahmedabad | 0.004 | 0.005 | 0.005 | 0.0109 | — | — | 341.98 | 504.7 |
| 2. | Calcutta | 0.0148 | 0.0664 | 0.17 | 0.0550 | — | — | 317.72 | 663.4 |
| 3. | Delhi | 0.022 | 0.133 | 0.0085 | 0.019 | 0.0002 | 0.001 | 630.79 | 996.8 |
| 4. | Hyderabad | 0.0031 | 0.09 | 0.0037 | 0.0096 | — | — | 150.2 | 219.7 |
| 5. | Jaipur | 0.0019 | 0.005 | 0.0033 | 0.0067 | — | — | 432.0 | 1114.0 |
| 6. | Kanpur | 0.03 | 0.06 | 0.0042 | 0.012 | — | — | 590.6 | 1564.0 |
| 7. | Madras | 0.0114 | 0.018 | 0.0139 | 0.06 | 0.004 | 0.016 | 59.8 | 103.3 |
| 8. | Nagpur Centre-1 | 0.0037 | 0.0119 | 0.128 | 0.0283 | — | — | 334.8 | 671.9 |
| | Centre-2 | 0.0046 | 0.0136 | 0.0136 | 0.034 | — | — | 269.6 | 420.7 |

*Courtsy* : G.N. Pandey

# Appendix VII

## Summary of industrial waste : Its origin and character

| Industries producing wastes | Origin of Major Wastes | Major characteristics |
|---|---|---|
| 1 | 2 | 3 |
| 1. Textile | Cooking of fibres, desizing of fabrics. | Highly alkaline, color high BOD |
| 2. Leather goods | Dehairing, soaking and deliming of hide. | High total solids, hardness, salt sulfides, chromium, pH, precipitated lime and BOD |
| 3. Laundry Trade | Washing of fabrics. | High turbidity alkalinity and organic solids |
| 4. Canned goods | Trimming, culling, juicing and blanching of fruits and vegetables | High in dissolved organic matter, mainly protein, fat and lactose |
| 5. Dairy products | Dillution of whole milk, separated milk, butter milk | High in dissolved organic matter, mainly protein, fact and lactose. |
| 6. Brewed and distilled beverages. | Steeping and pressing grain residue from distillation of alcohol | High in dissolved organic solids, containing nitrogen and fermented or their products |
| 7. Meat and poultry products | Stock yards, slaughtering of bones and facts, residues in concentrated grease and wash water, pickling of chickens | High in dissolved suspended organic matter, blood, other proteins, and fats |
| 8. Animal feedlots | Excreta from animal | High in organic suspended folids and BOD |

| | | |
|---|---|---|
| 9. Beet sugar | Transfer, screening and juicing water drainings from lime | High in dissolved and suspended organic matter containing sugar and protein |
| 10. Pharmaceutical products | Mycelium, spent filtrate and wash waters | High in suspended and dissolved organic matter, including vitamins |
| 11. Yeast | Residue from yeast filtration | High in solids (mainly organic) and BOD |
| 12. Pickles | Lime water, brine, alum and turmeric, syrup, seeds etc. | Variable pH, high suspended solids, colour and organic matter |
| 13. Coffee | Pulping and fermenting of coffee bean | High BOD and suspended solids |
| 14. Fish | Rejects from centrifuge, evaporator and other wash water wastes | Very high BOD, total organic solids and odour |
| 15. Rice | Soaking, cooking and washing of rice | High BOD, total and suspended solids (mainly starch) |
| 16. Soft drinks | Bottle washing, floor and equipment cleaning, syrup storage tank drains | High pH, suspended solids and BOD |
| 17. Bakeries | Washing aud greasing of pans, floor washings | High BOD, grease floor washing, sugars, flour, detergents |
| 18. Water production | Filter backwash, lime soda sludge, brine, alum sludge | Minerals aud suspended solids |
| 19. Pulp and paper | Coking, refining washing of fibres screening of paper pulp | Collodial and dissolved solids, inorganic filters |
| 20. Photographic | Spent solutions of developer and fixer | Alkaline, containing various organic and inorganic reducing agents |
| 21. Steel | Cooking of coal, washings of blast furnace flue gases, pickling of steel | Low pH, acids, phenol and ore, cokes, lime stone, alkali oils and fine suspended solids |

| 1 | 2 | 3 |
|---|---|---|
| 22. Metal-plated | Stripping of oxides, cleaning and plating of metals | Acid metals, toxic low volume mainly mineral matter |
| 23. Iron-foundary | Washing of used sand by hydraulic discharge | High suspended solids, mainly sand, some clay and coal |
| 24. Oil fields and Refineries | Drilling muds, salt, oil and some natural gas ; acid suldge and miscellaneous oils from refining | High dissolved salts from high BOD odour, phenol and sulfur compounds from refinery |
| 25. Fuel oil use | Spills from fuel tank filling waste, auto crank case oils | High in emulsified and dissolved Oils |
| 26. Rubber | Washing of latex, coagulated from crude rubber | High BOD and odour, high suspended solids, variable pH, high chlorides |
| 27. Glass | Polishing and cleaning of glass | Red colour, Alkaline nonsettleable suspended solids |
| 28. Naval stores | Washing solvent recovery, and oil recovery water | Acid, high BOD |
| 29. Glue manufacturing | Linewash acid, extraction of non-specific proteins | High COD, BOD, pH, chromium periodic strong mineral acids |
| 30. Wood preservation | Steam condensates | High in COD, BOD, solids, phenols |
| 31. Candle Manufacturing | Wax spillsm stearic acid condensates, dilute acids | Organic (fatty) acid, Low pH, low organic content |
| 32. Plywood Manufacturing | Glue washing | High BOD, pH, phenols, potential toxicity |
| 33. Acids | Dilute wash water, dilute acids | Low pH, low organic contents |
| 34. Detergents | Washing and purifying soap and detergents | High in BOD and sopanified soda |

| | | |
|---|---|---|
| 35. Corn starch | Evaporator condensate or bottom when not reused or recovered, syrup finol washes, wastes from bottling up process | High BOD and dissolved organic matter ; mainly starch and related material |
| 36. Explosives | Washing TNT and gun-cotton for purification, washing and pickling of cartridges | TNT coloured, acid, odourous and contains organic acids and alcohol from powder and cotton, metals, acid, oi's and soaps |
| 37. Pesticides | Washing and purification products such as 2, 4, D and DDT | High organic matter, benzene ring structure, toxic to bacteria and fish acid |
| 38. Phosphate and phosphorus | Washing and screening of floating rock | Clays, slimes and talls oils low pH high suspended solid phosphorus silica and fluorids |
| 39. Formaldehyde | Residues from manufacturing synthetic resins and from dyeing synthetic fibers | Normally high BOD and HCHD, toxic to bacteria in high concentrations |
| 40. Plastics and resins | Unit operations from polymer preparation and use ; spills and equipment wash downs | Acids, caustic dissolved organic matter such as phenols, formaldehyde, etc. |
| 41. Steam power | Cooling water, doilling blow-down coal drainage | Hot, high volume and high inorganic and dissolved solid |
| 42. Coal processing | Cleaning and classification of coal leaching and sulfur started with water | High suspended solids, mainly coal, low pH, high $H_2SO_4$ |
| 43. Nuclear power and radioactive materials | Processing ores, slaundering of contaminated cloths, research lab wastes processing of fuel, power plant cooling waters | Radioactive elements, can be very acid, 'hot' |

*Courtsy* : G.N. Pandey

## Characteristics of waste waters from some organic chemical industries

| Industry | Flow $M^3$/ton | BOD mg/1 | COD mg/1 | BOD/COD | Other important characteristics |
|---|---|---|---|---|---|
| Dyes and pigments | 180—900 | 200— 400 | 500—2000 | 0.2—0.4 | Heavy metals, solids, colour pH |
| Primary petrochemical | | | | | |
| ethylene | 0.18—5.4 | 100—1000 | 500—3000 | 0.2—0.33 | Phenol, Oil and pH |
| Propylene | 0.36—7.3 | 100—1000 | 500—3000 | 0.2—0.33 | Phenol pH |
| Primary Intermediates | | | | | |
| Toluene | 1.1—11.0 | 300—2500 | 1000—5000 | 0.3—0.5 | |
| Xylene | 0.72—11 | 500 -4000 | 1000—8000 | 0.5 | |
| Methanol | 1.1—11 | 300—1000 | 500—2000 | 0.5—063 | Oil and solid |
| Ethanol 8 | 0.72—7.2 | 500—4000 | 1000—8000 | 0.5 | Heavy metals |
| Ethyl benzone | 1.1—11 | 500—3000 | 1000—7000 | 0.43 -0.5 | Heavy metals |
| Chlorinated H/CS | 0.18—3.6 | 50—150 | 100—500 | 0.3—0.5 | Oils, solids, pH |

*Courtsy* : G.N. Pandey

## Characteristics of waste water from elected industries in India

| Industry | Flow | pH | Suspended solids mg/$l$ | BOD mg/$l$ | COD mg/$l$ | BOD mg/$l$ | BOD load per kg product | Other important characteristics |
|---|---|---|---|---|---|---|---|---|
| 1 | 2 | 3 | 4 | 5 | 6 | 7 | 8 | 9 |
| Dairy | 7 $m^3/m^3$ milk | — | 690 | 816 | | 6.6 | 5.7 | Oil and putrescibility |
| Distillery | 14 $m^3/m^3$ rectified spirit | 4.0—4.5 | — | 40,000 | 80,000 | 0.5 | 560 | High TDS, $C_1$ $SO_4$ colour and putreocibility |
| Cotton textile | 25 $m^3$/1000 metre cloth | 7—10 | 375 | 350 | 525 | 0.66 | 8.75 | TDS 300 mg/$l$% Sodium 90 and colour |
| Viscose R Rayon | 570 $m^3$/tonne | 2.08—4.1 | 200 | — | 210 | — | — | Zinc, 6-18 mg/$l$ polysulfides |
| Tannery | 3.2 $m^3$/1000 kg hide | 8.0 | 1900 | 1,000 | 3,500 | 0.48 | 5.4 | Gr 0.35—0.45 kg/100 kg hide, C 14600 mg/$l$, tannin, colour |

| 1 | 2 | 3 | 4 | 5 | 6 | 7 | 8 | 9 |
|---|---|---|---|---|---|---|---|---|
| Pulp and paper craft process | 310 m/tonne | 6.5—8.2 | 375 | 160 | 610 | 0.26 | 50 | Lignin, colour, suspended fibre |
| Primary polymers polythylene | — | 1.44—5.8 | — | 200 | 4,000 | — | — | Solids |
| Polypropylene | 1.44—5.8 | — | 200 | — | 4,000 | — | | |
| Polystyrene | 1.8—3.6 | — | 1000 | — | 3,000 | 0.05 | 0.025 | |
| Polyvinyl chloride | 5.5—11 | 50—500 | 1000 | — | 2,000 | 0.05 | 0.25 | |
| Cellulose acetate | 0.036—.72 | 500—2000 | 1000 | — | 5,000 | 0.46 | 0.5 | |
| Butyl rubber | | | | | | | | |
| Miscellaneous substances | 7.2—21.6 | 800—2000 | 2500 | — | 5,000 | 0.32 | 0.4 | |
| Isocyanate | 18.0—36 | 1000—2500 | 4000 | — | 8,000 | 0.25 | .31 | Nitrogen |
| Phenyle glycerine | 18.0—36 | 1000—2500 | 4000 | — | 8,000 | 0.25 | 0.31 | Phenol content |
| Parathion | 10.8—28.8 | 1500—3500 | 3000 | — | 6,000 | 0.5 | 0.59 | Solids, pH |
| Tributyl phosphate | 3.6—14.4 | 500—2000 | 1000 | — | 3,000 | 0.5 | 0.67 | Phosphorus |
| Coke $3m^3$/tonne Oven. of coke | 3.6—11.2 | 82 | 1200 | | 2,900 | 0.41 | 3.6 | NH phenol CN $CN^3$ tar and oil |
| Refinery $1.7m^3$/tonne oil processed | — | 300 | 200 | — | — | — | 0.34 | Free oil, 2500 mg/l emulsified oil, 100 mg/l mercaptance 115 mg/l |

| | | | | | | | | |
|---|---|---|---|---|---|---|---|---|
| Phenol 30.0 m³/ tonne formal-dehyde resin | 4.0—7.0 | 23 | 600 | — | 32,160 | 0.73 | 71 | Phenol, 7900 mg/l HCHD 4360 mg/l |
| Parathion | 2.0 | — | 700 | — | 3,000 | 0.23 | — | TS 27000 mg/l parathion 20 mg/l |
| Carbaryl | 7.10 | Nil | Nil | — | 10,000 | — | — | TS 40,000 mg/l No. carbaryl low toxicity |

*Courtry* : G.N. Pandey.

## Works of Professor S. L. Kayastha

1. The Himalayan Beas Basin-A Hydrographical Study. *The National Geographical Journal of India*, (*N.G.J.I.*) Vol. 1, Part I, Sept., 1956. Reprinted in *Indian Journal of Power and River Valley Development*, Vol. VI, No. 1, 1956.
2. Geography and its Place in Education. *Journal of Government College*, Rupar, Vol. I, No. 6, 1956.
3. Organisation of Productive Process in U.S.S.R., *Kashi Vidya Peeth, Publication*, 1952.
4. Demographic Features of the Himalayan Bean Basin, *N.G.J.I.*, March, 1956.
5. Tourist Industry of Kangra, Kulu and Mandi, Presented in the Indian Science Congress, Agra, 1956, published in *N.G.J.I.*, Sept. 1956.
6. The Place of Geography in National Reconstruction. Presented in the International Geography Seminar, Aligarh, 1956, published in *Journal of Banaras Hindu University*, March, 1958. Proceedings of the International Geography Seminar, 1959. Abstract in *PRZRCLAD GEOGRAFICZNY PLSKA AKADENIA NAUM INSTYTUT GEOGRAFII* ; Warsaw, 1956, pp. 557-558.
7. The Himalayan Beas Basin-A River Basin Concept in the Delimitation of Himalayan Districts. Written jointly with Dr. R.L. Singh, Presented as Presidential Address in Historical Geography Section of the International Geographical Congress, Rio de Janeiro, 1956, Published in *N.G.J.I.*, June, 1957.
8. Gaddis of Dauladhar-A Study in Human Ecology, Presented in Indian Science Congress, Calcutta, 1957, published in *N.G.J.I.*, June, 1957.
9. True Azimuths, and Distances of World Capitals from Varanasi. Written jointly with Shri N. Prasad, published in *N.G.J.I.*, June, 1957.
10. Precipitation Characteristics in the Himalayan Beas Basin

Presented in Indian Science Congress, Madras, 1958, published in *B.H.U. Scientific Journal*, Vol. VIII (2), 1957-58.

11. Chirths of Kangra Valley-A Study in Agricultural Society. Presented in Indian Science Congress, Delhi, 1959, published in *N.G.J.I.*, March, 1959.
12. Geography and Geographical Society, Inaugural Address to Geographical Society, published in *Journal of Wesley College*, Azamgarh, March, 1958.
13. Tourism in India. Written jointly with Dr. R.L. Singh. Published in *Journal of Banaras Hindu University*, June, 1958.
14. The Role of Geographer in Town Planning. Presented in the Seminar Indian Council of Geographers, Madras, 1958.
15. Kangra-A Himalayan Town, *N.G.J.I.*, June, 1958.
16. Occupational Structure in the Himalayan Beas Basin, Indian Science Congress, 1960, published in *N.G.J.I.*, March, 1960.
17. A Geographical Study of Transport and Communication in the Himalayan Beas Basin. Indian Science Congress, 1960, published in *N.G.J.I.*, June, 1960.
18. Careers in Geography — Presented in the National Geographical Society of India, August, 1956.
19. Ponyculture and its Problems in the Himalayan Beas Basin. Presented in the Indian Science Congress, Roorkee, 1961.
20. Animal Husbandry in the Himalayan Beas Basin, Indian Science Congress, Roorkee, 1961, published in *N.G.J.I.*, June, 1961.
21. Human Habitations in the Himalayan Beas Basin. Presented in the Indian Science Congress, Cuttack, 1962, published in *the Indian Geographer*. Vol. 6. December, 1961.
22. Suratgarh Farm-A Geographical Appraisal. Presented in the Indian Science Congress, Cuttack, 1962, published in *N.G.J.I.*, June, 1962.
23. Kandla-A Study in Port Development, *N.G.J.I.*, March, 1963.
24. Some Aspects of Landuse in the Himalayan Beas Basin.

Presented in the Seminar of the Indian Council of Geographers, Delhi, 1963.

25. Social Structure in the Himalayan Beas Basin. Indian Science Congress, 1963.
26. Horticulture in the Himalayas. Indian Science Congress, 1963.
27. Climatics Features in Horticulture in the Himalayas. To be published in *Geography Journal*, Bhopal.
28. Some Aspects of Soil Erosion and Conservation in India, Indian Science Congress, Calcutta, 1964.
29. World's Population and Resources. University Extension Lecture, B.H.U., 1964 (in the Press).
30. A Geographical Appraisal of Population and Resources with special reference to food resources, Summer School in Geography, Nainital, June, 1965.
31. Conservation of Natural Resources in the Himalaya-A Vital Need. Seminar on Himalaya, Indian School of International Studies and Delhi School of Economics, 1965.
32. Animal Husbandry and its Development in Himachal Pradesh. Indian Science Congress, Chandigarh, 1966.
33. Some Problems of Development in Lahaul and Spiti, Indian Science Congress, Hyderabad, 1967, published in *N.G.J.I.*, June, 1967.
34. An Appraisal of Water Resources and Need for National Water Policy in India, U.G.C. Seminar on Applied Geography, November, 1966, B.H.U.
35. Some Aspects of Resource Conservation, U.G.C. Seminar, Patna, 1967.
36. Tourism in Himachal Pradesh, Indian Science Congress, Varanasi, 1968.
37. Regions for Planning and Development in Sonpar Region (U.P.) (Conjointly with S.N. Misra) *Proceedings Regional Development and Planning*, Mysore, 1968.
38. Trend of Population Growth in Uttar Pradesh conjointly with Om Prakash. International Geographical Congress, New Delhi, 1968.
39. Growth of Tribal Economy and Impact of Modern Economic Developments on Tribal Economy and Society in Sonpar Region (U.P.).

40. Trend of Urbanization in the Himalayan Beas Basin. International Geographical Congress, New Delhi, 1968.
41. Kangra Valley. Chapter contributed for I.G.U., 1968 in the book "Indian Regional Study" edited by R.L. Singh.
42. Land Resources and Population Problem of India with Special Reference to Food Production : Indian Science Congress, Bombay, 1969 (Abstract Proceedings Pt. IV, pp. 17-18).
43. Development and Conservation in the Himalayan Region Indian Science Congress, Kharagpur, 1970.
44. Landuse Ranking in Eastern Uttar Pradesh (conjointly with J.N. Pandey) Indian Science Congress, Kharagpur, 1970.
45. Himachal Region in India : *A Regional Geography*, National Geographical Society of India, Varanasi, 1971.
46. Introductory Chapter in India : *A Regional Geography*, Jt. Author.
47. Kashmir Region in India: *A Regional Geography*, 1971, Jt. Author.
48. Problems of Northern Frontiers : Himalayan Seva Sangh (Abstract), New Delhi, 1972, pp. 22-23.
49. A Note on Management and Conservation of Natural Resources Symposium on Ecology, Feb. 24-27, U.G.C. and Smithsonian Institute, and Indian Science Academy.
50. Geographical Appraisal of Conservation and Resource Development in B.N. Chakravorty University, Kurukshetra, Extension Lecture, 1971.
51. The State of Environment Post-Graduate Research Unit, Geography Department, Gorakhpur University, 1971.
52. Problems of Development in the Himalaya, National Conference on 'Himalaya' November, 1971, New Delhi.
53. Some Aspects of Transportation in Dhanbad (Jt. Author). Indian Science Congress, Calcutta, Published in *N.G.J.I.* 1972.
54. Agricultural Innovations among Farmers of Mahasu, Mandi and Kinnaur (H.P.) Correlates and Levels : Section I, Agricultural Geography and Rural Settlements, I.G.U. Montreal, 1972, *N.G.J.I.* March, 1973.
55. A note on Rural House in Himachal Pradesh I.G.U.

Symposium on 'Rural Settlements in Monsoon Asia', *N.G.J.I.*, March, 1972, pp. 383-385.

56. Hydro-electric Power Resources and Development in Himachal Pradesh, Indian Science Congress, Chandigarh, 1973.
57. Approaches to Environmental Study and Problems U.G.C. Seminar on Teaching and Research Methodology in Geography, Gorakhpur University, December 24-27, 1972.
58. Environmental Geography : 5 Lecture Papers Proceedings U.G.C. Advanced Level Summer Science Institute in Geography, B.H.U. May-June, 1973.
59. Changing Aspects of Cultural Patterns in Rural India : Case Studies, Paper Presented before Study Team of Professors from U.S.A. led by Prof. A.G. Noble, Chairman, Department of Geography, University of Akron, Varanasi, July, 1973.
60. Development Problems of Himalayan Region Indian Science Congress 1974, Nagpur.
61. Regional Approach Planning at District Level : A Case Study of Allahabad, Uttar Pradesh. (Jt. Author J. Prasad) First Asian Symposium on Regional Planning and National Development, Institute of Development Studies, University of Mysore, July 1-5, 1974.
62. Himachal Pradesh : A Study in Population Characteristics Indian Science Congress, 1975, Delhi.
63. The World Population and Resource Problems and the *World Population Souvenir*, 1974 World Population Year, B.H.U. Varanasi, pp. 21-30, Year 1974.
64. The Problem of Unemployed Amongst the Educated in India, All India Seminar on Reforms in Higher Education B.H.U., Varanasi, October, 1975.
65. Resources, Industrial Development Patterns and Planning in Eastern Uttar Pradesh (Jt. Author M.B. Singh), Proceedings All India Symposium on Resource Development and Planning (November 5-7, 1976), University of Madras, Madras.
66. Indo-Soviet Trade, *N.G.J.I.*, March-June, 1976.
67. Problems of Development in Drought Areas : A Case Study of Karchhana and Meja Tahsils, Allahabad District

(Jt. Author with S. Prasad) Proceedings, Indian Science Congress, Waltair January, 1976.

68. Rural Landuse Changes and Planning : A Case Study of Village Pratappur, Mirzapur (conjointly with A.K. Singh), Article submitted to I.G.U. Congress, Moscow, 1976.

69. Areas Planning and Development : A Case Study of Phulpur (conjointly with J. Prasad), Article submitted to XXIII I.G.U. Congress, Moscow, 1976.

70. Intra Village Spacing : Measurement and Comparison (conjointly with R.P.B. Singh) Article submitted to XXIII I.G. U. Cóngress, Moscow, 1976.

71. An Appraisal of Environmental Constraints on Economic Development in a Hill-Tribal Areas of Southern Uttar Pradesh, Article submitted to XXIII I.G.U. Congress, Moscow, 1976.

72. Human Perception and Adjustment to Environmental Hazards: A Case Study of Three Villages of Ganga Flood Plain (conjointly with R.P. Yadav) in *Man, Culture and Settlement*, N.G.S.I., Varanasi.

73. Size and Spacing of Rural Settlements in South-east Haryana (conjointly with Debal Singh), Proceedings I.G.U International Symposium on Rural Settlements, December, 1975.

74. Population of India : An Appraisal of Growth Characteristics, Problems and Planning Policies, *Journal of Medical Sciences*, New Delhi (Ayurvigyan) September, 1975.

75. Problems of Erosion in a Fertile Agricultural Tract : A Case Study of Affected Area in Phulpur Block (Allahabad Distt.) (with J. Prasad), Indian Science Congress, Bhubaneswar, 1977.

76. Problems of Settlement Location in Flood Prone Area : A Case Study of Lower Rapti Basin (with R.P. Yadav), Indian Science Congress, Bhubaneswar, 1977.

77. The Indian Ocean : An Appraisal of the Significance and Resource Potential to India, All India Inter-Disciplinary Colloquim on 'Law as an Instrument of Development', Law School, B.H.U. Varanasi, February, 4-7, 1977.

78. The Environmental Study and International Policy and

Programmes, S.L. Kayastha, Annual Conference, Association of Geography Teachers, Madras, February, 1977, published in *Geography Teachers*.

79. Forests and Environment in the Himalaya (with B.N. Juyal). Issue paper presented before All India National Seminar on Forests, Environment and People in the Himalaya, New Delhi, December, 1977.

80. The Problem of Economic and Non-Economic Evolution of Man on Environment (with S.N. Misra). Paper submitted for Special I.G.U. Volume of the Commission on Environmental Problems, Praha, 1977.

81. Demographic Characteristics of Small Towns in Mirzapur Region (with V.K. Kumra and M.B. Singh), Origin, Evolution and Development of Small Towns along the River Ghaghra : A Comparative Study. S.L. Kayastha and R.P. Yadav ; Papers 9 and 10 presented before Seminar on "Problems of Small Towns in Developing Countries", Second All India Social Scientists Convention, February, 20-22, 1977.

82. A Study of Preference and Behaviour Pattern of Tourists in Varanasi : A Geographical Analysis. *N.G.J.I.* (with S.N. Singh), Sept.-Dec., 1977, pp. 143-150.

83. Review of Resource Geography, Indian Council of Social Science Research, New Delhi.
Book Review of *The Political Geography of Federalism : A Enquiry into its Origins and Stability* by R.D. Dikshit, McMillan, 1975, *N.G.J.I.*, Sept.-Dec., 1977, pp. 207-208.

84. An Analysis of Rainfall and Production of Wheat in the Drought-Prone Gwalior-Chambal Regions of M.P. (with M.B. Singh), All India Symposium on Drought-Prone Areas of India, S.V. University, Tirupati, January, 4-6, 1978.

85. Effects of Land Utilization and Cropping Pattern in Robertsganj Block : A Case Study of Drought-Prone Areas (with V.K. Kumra and Nar Singh), Tirupati Symposium, 1979.

86. Rural Drinking Water Problem in Drought-Prone Area of Varanasi District (with Ram Bilas), Tirupati Symposium, 1979.

87. Flood Hazard in Lower Ghaghra Plain (U.P.) India : A

Study in Perception and Impact on Socio-Economic Development (with R.P. Yadav), I.G.U. Working Group on Environmental Perception, Ibadan (Nigeria), July, 1978.

88. Environmental Implications of Resource Development in India : An Appraisal (with S.N. Misra). I.G.U. Commission on Environmental Problems, Ile-Ife (Nigeria), July, 1978, Published by the Commission, U*NESCO Ecological Studies*, 1978, Regional Conference, Nigeria, 1978, pp. 44-67.

89. Air Pollution Problem in Kanpur City : (with V.K. Kumra), *Indian Journal of Air Pollution Control*, Vol. I, No. 1, January, 1978. pp. 38-42.

90. Spatial Disorganisation and Internal Migration in India : Some Strategies for Reconstructing the Space Economy and Development (with Shekhar Mukherji), Submitted 1978. *Canadian Studies in Population*, Edmonton, Vol. 6.

91. Eastern U.P. : A Causative Analysis of Backwardness, (with M.B. Singh), Seminar on Factors in Economic Backwardness in Eastern U.P., December, 21-23, 1978, Gandhian Institute of Study, Varanasi.

92. Pattern and Prospect of Service Centres in Spatial Context : Case of the Siwan District. (with Rana P.B. Singh, Ram Babu Singh), Published in *Transformation of Rural Habitat in Indian Perspective, A Geographic Dimension*, Varanasi, N.G.S.I., Res. Pub. No. 19, 1978, pp. 108-116.

93. Regional Development in Special Areas : Mountainous Regions, The Indian Experience, Indo-Soviet Symposium on Regional Development and Planning, Moscow-Tiblishi-Baku (U.S S.R.), October-November, 1978.

98. Scientific Study and Control of Ecological Deterioration in Urban Ecosystem : A Case Study of Kanpur (with V.K. Kumra), Symposium of I.G.U. Commission on Man and Environment, July 1-12, 1979, Moscow-Kurek-Irkutsk (U.S.S.R.).

99. The Centrality Oriented Connectivity and Planning in Muftiganj Block (Jaunpur, U.P.) (with S.P. Mishra), Geographers' Association, Gorakhpur, 18-19 May, 1979.

100. A Spatial Analysis of Manufacturing Industry in U.P. (with M.B. Singh), Indian Science Congress, Hyderabad, January, 1979.

101. Perception and Adjustment in Diara Environment : A Case Study of Three Sample Villages of Dhampur Diara (Lower Ghaghra Region), U.P. (with R.P. Yadav), Indian Science Congress, Hyderabad, January, 1979.

102. Air pollution and its Relationship to the Prevailing Climatic Conditions : A Case Study of Kanpur City (with V.K. Kumra), Paper Submitted to 1979 for 5th International Clean Air Congress Buenos Aires, Argentina (October 20-26, 1980).

103. The Socio-Economic Development of Tribal vs. Non-Tribal Population in South Mirzapur Region (with R.S. Yadav), IGU, International Symposium on Transformation of Rural Habitat in Developing Countries, Varanasi : 22-25, January, 1979.

104. Identification and Determination of Hierarchy of Rural Service Centres in Ghazipur Tahsil, U.P. IGU. Symposium, Varanasi 1979 (with B.N. Singh).

105. Flood Plains and Human Habitations : A Study in Settlement Process and Socio-Economic Changes, (with R.P. Yadav), I.G.U. Symposium, Varanasi, 1979.

106. Problems of Water Pollution in Rural Areas : A Case Study of Kanpur (with V.K. Kumra), I.G.U. Symposium, Varanasi, 1979.

107. Methodological Approach to Identify the Functional Hierarchy of Settlements in Rural Areas : A Study of Middle Ganga Plain (with S.P. Mishra), I.G.U. Symposium, Varanasi, 1979.

108. Geographical Aspects of Education and World Hunger, Second World Congress of International Association of Education for World Peace, January, 1979 (Sectional Presidential Address).

109. Geoscientist Number of National Geographical Journal of India, conjointly Edited. Presented to Eminent Senior Geoscientists at Indian National Congress, Hyderabad, 1979.

110. Eastern Uttar Pradesh : Development Patterns and Potentials of Sugar Industry (with M.B. Singh). *The*

*Natianal Geographer*, Vol. XV, No. 1, June, 1980, pp. 45-51.

111. Noise as a Factor in Environment (with V.K. Kumra), *Philippine Geographical Journal*, Vol. XXIV, July-Aug-Sept., 1980, No. 3, pp. 127-142.

112. Effect of Environmental Pollution and Human Health : A Case Study of Kanpur City (with V.K. Kumra), Geology & Geography Section, Indian Science Congress, Culcutta, 1980.

113. Rural Housing Problem in Flood Environment : A Case Study of two Sample Villages of Middle Ganga Valley (with R.P. Yadava), Geology and Geography Section, Indian Science Congress, Calcutta, 1980.

114. An Analysis of Development Process of South Mirzapur Region (with R.S. Yadav), Geology & Geography Section, Indian Science Congress, Calcutta, 1980.

115. Flood Induced Population Migration in India : A Case Study of Ghaghara Zone (with R.P. Yadava), I.G.U. Commission on Population Geography, International Symposium, Karachi, 1980, Accepted for Publication in Selected Papers by Prof. L.A. Kosinski.

116. An Appraisal of Water Resources of India and Need for National Water Policy, I.G.U./U.N.U. Panel on Use and Management of Natural Resources, I.G.C., Japan, 1980, Accepted for Publication in *GEO. JOURNAL* No, 6 (1981).

117. Floods in Eastern Uttar Pradesh : A Study in Occurrence, Awareness Losses, Adjustment and Amelioration Measures, General Symposia 3 : Natural Disasters as Environment (with R.P. Yadava), Presented Paper in the I.G.U. Congress, Tokyo, Japan, 1980.

118. An Appraisal of Urban Environment and Suggestions for Improvement of Kanpur Region : A Case Study (with V.K. Kumra). Presented Paper in the I.G.U. Commission on Environmental Problems, August, 1980, I.G.C., Japan.

119. Aspects of Flood Hazard in the Middle Ganga Valley : A Case Study of Selected Settlements of Eastern U.P. (with R.P. Yadava), I.G.U. Working Group on

Perception of Environment, I.G.U. Congress, Japan, 1980.

120. Regional Development Through Social Planning : A Micro-Level Case Study from India, (with R.B. Singh), I.G.U. Commission on Regional Systems and Policies, I.G.U. Congress, Japan, 1980.

121. Emerging Dynamics of Integrated Rural Development (with R.B. Singh), I.G.U. Commission on Regional Systems and Regional Development, Nagoya (Japan), 1980.

122. South Mirzapur Region : A Geographical Analysis of Development Processs (with R.S. Yadav), I.G.U. Commission on Regional Geography, I.G.U. Congress, Japan, 1980.

123. *Souvenir, Indian Science Congress, 68th Session, Varanasi.* Co-edited B.H.U. Press, Varanasi, 1981.

124. Settlement Planning for Integrated Rural Development in a Micro-Level Unit : Case Study of Daraundha Block (with R.B. Singh), National Symposium on Regional Planning and Rural Development, G.B. Pant Soc. Sc. Instt. Allahabad, 23-24, March, 1981.

125. An Approach towards Balanced Development in South Mirzapur Region (with R.S. Yadav). G.B. Pant Soc. Sc. Instt., Allahabad, 23-24, March, 1981.

126. Tourism in U.P. Himalaya (with S.N. Singh), Published in *The U.P. Himalaya : Nature, Man and Culture*, Edited by Dr. O.P. Singh et al.

127. Resource Management in Rural Environment : Geographers' Viewpoint (with R.B. Singh), Silver Jubilee Symposium, International Society for Tropical Ecology, Bhopal, Oct. 1981.

128. Population and Environmental Problems, World Environment Day, 6th June, 1981, Varanasi.

129. Approaches to Environmental Study and Problems, Special Science Number, *Pragya*, 1981 (Forthcoming).

130. An Approach to Block Level Planning in Developing Countries : A Case Study of Dhobi Block in Middle Ganga Plain (with S.P. Mishra), Indian Science Congress (Geol. & Geog. Section), Varanasi, 1981.

131. Tourism in Varanasi City : A Functional Analysis (with

S.N. Singh), Indian Science Congress, 1981.

132. Agropolitan Growth Strategy of Rural Development: Reflections on Indian Situation (with R.B. Singh), National Symposium on Agrarian Structure, Agricultural Growth and Rural Development, G.B. Pant Soc. Sc. Instt., Allahabad, Sept. 1981.

133. Mapping Population Intensity in an Urban Environment (with V.K. Kumra), Third National Cartographic Conference, Calcutta, Oct. 19-21, 1981.

134. Regional Planning for Socio-Economic Facilities in Spatial Context in Ghazipur Tahsil (with B.N. Singh), International Symposium on Urban and Regional Development Policies in Developing Countries, I.I.T. Kharagpur, December, 1981.

135. Disease Ecology and Ecological Planning for Health (with V.K. Kumra), International Symposium on Incidence of Infections and Chronic Diseases, & I.G.U. Working Group Meeting of Medical Geography, Madras, December 1-4, 1981.

136. Environmental Problems and Development Constraints in the Himalaya, Centenary Lecture, Guest Speaker, Hosie University, Tokyo, 18-9-1980.

137. Environmental Perception and Problems, edited a Special Issue of the *National Geographical Journal of India* on the Occasion of 24th International Geographical Congress, Tokyo, 1980, vol. 26, Pts. 1-2.

138. Regional Development in the Soviet Union, All India Radio, Varanasi, 22-4-80.

139. Rural Settlements System of Uttar Pradesh, India: A Study of Spatial Characteristics (with Rana P.B. Singh), I.G.U. Symposium on National Settlement System, Hyderabad, March (20-23), 1980.

140. Indian Ocean Islands, *Scientific Encyclopaedia*, Sponsored by Rajasthan Academy (edited: M.N. Nigam), April, 1980.

141. Habitat and Development, edited a Special Issue of the *National Geographical Journal of India*, Sept.-Dec. 1980, Issued for the 68th Indian Science Congress, Session (Jan. 1981), Varanasi.

142. Significance of Forests in the Conservation of Water Resources: An Integrated Ecological Approach with

Special Reference to Ground Water Resources, Summer Institute of Recent Development in Ground Water Geophysics for Soft Rock Terrain, Deptt. of Geophysics, B.H.U., Varanasi, June, 1980.

143. The Role of Forests in the Regional Economy Within the Framework of a National Forest Policy in India, Indo-Soviet Symposium on Rational Utilisation and Conservation of Renewable Natural Resources for Regional Development, Vigyan Bhawan, New Delhi, 10-15 December, 1980.

144. Perspectives on Environmental and Development, Presidential Address, Indian Council of Geographers, Indian Science Congress Session, Mysore, 1982.

145. Comparative Issue of Migration and Urbanization in relation to Regional Development (with R. B. Singh), National Symposium on the Problems of Labour Migration and Regional Development in India, Visva Bharati, Santiniketan, February 1-3, 1982.

146. Forests in India and the Impact of Economic Developments, VIth IGU Symposium, Commission on Environmental Problems, Sao Paulo, Brazil, August, 1982, Paper sent, Jan. 1982.

147. Urban Environmental Quality in the Metropolises of Kanpur and Delhi (with V. K. Kumra), VIth I.G.U. Symposium, Commission on Environmental Problems, Sao Paulo, Brazil, August, 1982.

148. Role of Energy Resource in Rural Environment (with R.B. Singh), International Symposium I.G.U.-W.G. on Transformation of Rural Habitat in Developing Countries, Symposium on Energy Resources, Environment and Habitat Transformation in Developing Countries, Meerut, March 16-20, 1982 ; Invited to Chair the Session.

149. The Conditions of Domestic Water Supply Sources in Areas of Varanasi District, Uttar Pradesh (India) (with Ram Bilas), International Symposium on Energy Resources......Developing Countries, Meerut, March 16-20, 1982.

150. Tourism and Its Role in Accelerating the Economic Growth of Uttar Pradesh (with S. N. Singh), National

Workshop on Accelerating the Economic Growth of Uttar Pradesh, G. B. Pant Social Science Institute, Allahabad, April 17-18, 1982, Paper submitted, February, 1982.

151. Impact of Economic Development Upon Environment in South Mirzapur (with R. S. Yadava), Symposium 6 on The Environmental Impact of Economic Growth in Third World Countries, IGU Regional Conference, Main Session, Reo de Janeiro, August, 1982.

152. Forests Environment and Integrated Area Development in India, National Seminar on Forestry Integrated Area Development and Environment with special reference to Hill Tribal Areas, Himalaya Seva Sangh, New Delhi, March, 1982.

153. Industrial Potentials of Eastern Uttar Pradesh (with M.B. Singh), Prof. P. Dayal Felicitation Volume, Patna University, Patna.

154. Quality of Human Environment : A Case Study of Kanpur City (with V.K. Kumra), Professor K. Bagachi, Felicitation Volume. Calcutta, 1982.

155. Cumulative Index, *The National Geographical Journal of India*, Vols. XXI to XXV (1975—79), Tara Press, N.G.S.I., Varanasi, March, 1982.

156. Energy Resources and Technology Transfer in the Management of Rural Environment (with R. B. Singh), Symposium on Environmental Management.

157. A Methodological Approach to Identify the Functional Hierarchy of Rural Settlements : A Case Study of Kerakat Tahsil (Jaunpur) in the Middle Ganga Plain (with S. P. Mishra), *New Perspectives in Geography* (Editor L. R. Singh), Allahabad, 1981, pp. 123-133.

158. Floods in India : A Study of Their Occurrence, Causes, Forecasting and Control, Vth International Symposium of I.G.U. Commission on Environmental Problems, Mexico, October, 1981.

159. Supporting Services, Inputs and Food Production in relation to Rural Population : Case Study (with R. B. Singh), International Conference on Food Systems of the World, Aligarh Muslim University, Aligarh, 20-25, Feb. 1983.

160. River Ganga Pollution in Kanpur : A Case Study (with V.K. Kumra), National Conference on River Pollution and Human Health, Varanasi, February 20-21, 1983.
161. Public Awareness Towards River Pollution Along the Ghats of R. Ganga at Varanasi : A Case Study of Environmental Assessment and Perception (with Ram Bilas), National Conference on River Pollution and Human Health, Varanasi, February 20-21, 1983.
162. Flood Problems and Flood Control in India, National Seminar, ' Civil Engineering Institute of Technology, B.H.U., Varanasi, March 19-21, 1983.
163. Himachal Region of the Himalaya : A Study in Population Characteristics, Pressures and Problems, I.G.U. Commission on Population, Symposium in 'Mountain Population Pressure', Kathmandu, April, 1983.
164. Role of Biogas in Rural Development (with R.B. Singh), Indian Science Congress, Tirupati, 1983.
165. Power Plants : An Environmental Problem of Air Pollution (with Chandrasekhar), Indian Science Congress, Tirupati, 1983.
166. Environment and Development in India, with special reference to the Himalaya, VIIth I.G.U. Symposium on Environmental Problems, September, 198:, Palma de Mallorca (Spain),
167. Impact of Tourism in the Himalayan Region, VIIth I.G.U. Symposium, Commission on Environmental Problems, Sept. 1983, Palma de Mallorca (Spain).
168. Assessment and Utilization of Different Energy Programmes : Conservation and Development in Rural Perspectives (with R.B. Singh), International Conference on Energy Conservation and Environmental Management, Oct. 24-27, 1983, B.H.U., Varanasi, Abstract in Proceedings Abstracts.
169. Some Aspects of Development and Conservation in the Himalayan Region, Indian Science Congress, 1984, Ranchi (Submitted).
170. Quality of Human Environment : A Case Study of Kanpur (with V.K. Kumra), International Seminar on Maps and Atlases, National Atlas and Thematic Maps Organization, Calcutta, Dec. 1-3, 1983.

Western Uttar Pradesh (with M.B. Singh), U.G.C. Seminar, Patna University, Patna, March, 1984.

172. Spatio-Functional Analysis of Health Facilities in India (with M.B. Singh), Paper submitted for publication in the *National Geographer*, Allahabad.

173. Towards Searching a Viable Strategy for Reducing Regional Disparities in India (with R.B. Singh), Presented, Section IV, Theme 14, Int. Geog. Cong. Paris, 1984.

174. Forecasting the State of Environment in India at the Turn of the Century, Presented, VIII Symposium, I.G.U. Commission Symposium on Enviromental Problems, Caen (France), August, 1984.

175. Rural Environmental Management for Rural Development (with Chandrashekhar), National Seminar on Rural Area Development Strategies in India, B.H.U., Varanasi, February 15-19, 1984.

176. Service Centre Strategy in Integrated Rural Development: A Case Study (with S.P. Mishra), National Seminar on Rural Area Development Strategies in India, B.H.U., Varanasi, Feb. 15-19, 1984.

177. Environmental Problems : A Review Trend Report, *Trend Report on Research in Geography, 1976-82* for ICSSR, New Delhi.

178. Environmetal Studies in India (with Mehdi Raza) *Country Report*; *Trends in Geographical Research in India, 1980-83*, Presented to 25th IGC, Paris, Aug, 84, INSA, New Delhi, pp. 47-55.

179 New Dimensions in Tourism in U.P. Himalaya (with H.C. Rai), 6th NAGI Congress, 1984, Kharagpur.

180. Post-Independence Growth of Rural Industrial Units in Sewapuri, Varanasi Distt. U.P. (with L.S. Singh & A.K. Singh), 6th NAGI Congress, 1984, Kharagpur.

181. Environmental Policies in India, being published in *Environmental Policies and Global Strategies*, Croom Helm, U.K. (in press).

182. Tourism and Its Impact on Indian Economy (with V.K. Kumra, & H.C. Rai), Indian Science Congress, Lucknow, Jan. 85.

183. Impact of Tourism on Rural Development in Kumaon

Himalaya (with H.C. Rai), Indian Science Congress, Lucknow, Jan. 85.

184. Tribals in India : Perspectives on Problems and Policies, Seminar on Statistics, Population and Planning, Centre of Population Studies, B.H.U. Varanasi, July 29-30, 1984.

185. Environmental Quality in South Mirzapur (with Chadrashekhar), Indian Science Congress, Lucknow, Jan. 85.

# INDEX